REANALYSIS OF STRUCTURES

SOLID MECHANICS AND ITS APPLICATIONS
Volume 151

Aims and Scope of the Series

The fundamental questions arising in mechanics are: *Why?*, *How?*, and *How much?* The aim of this series is to provide lucid accounts written by authoritative researchers giving vision and insight in answering these questions on the subject of mechanics as it relates to solids.

The scope of the series covers the entire spectrum of solid mechanics. Thus it includes the foundation of mechanics; variational formulations; computational mechanics; statics, kinematics and dynamics of rigid and elastic bodies: vibrations of solids and structures; dynamical systems and chaos; the theories of elasticity, plasticity and viscoelasticity; composite materials; rods, beams, shells and membranes; structural control and stability; soils, rocks and geomechanics; fracture; tribology; experimental mechanics; biomechanics and machine design.

The median level of presentation is the first year graduate student. Some texts are monographs defining the current state of the field; others are accessible to final year undergraduates; but essentially the emphasis is on readability and clarity.

For a list of related mechanics titles, see final pages.

Reanalysis of Structures

A Unified Approach for Linear, Nonlinear, Static and Dynamic Systems

By

URI KIRSCH

Department of Civil and Environmental Engineering
Technion – Israel Institute of Technology
Haifa, Israel

ISBN 978-90-481-7803-2
ISBN 978-1-4020-8198-9 (e-book)

Published by Springer,
P.O. Box 17, 3300 AA Dordrecht, The Netherlands.

www.springer.com

Printed on acid-free paper

To Maya, Ofri, Lihi, Jonathan, Omer and Roy

Preface

This book deals with multiple repeated analyses (reanalysis) of structures. Reanalysis is needed in many problems of structural analysis, design and optimization. It is related to a wide range of applications in Aerospace Engineering, Civil Engineering, Mechanical Engineering and Naval Architecture. The book was developed while I was teaching graduate courses on analysis, design and optimization of structures, in the United States, Canada, Europe and Israel. It summarizes many years of research and developments in these areas. The purpose of the book is to collect together the main results of this work and to present them in a unified approach. It meets the need for a general text covering the basic concepts as well as recent developments on reanalysis of structures. This should prove useful to students, researchers, consultants and practicing engineers involved in structural analysis and design. Other books on structural analysis do not cover most of the topics presented in the book. Early developments on this subject are introduced in a previous book by the author (*Design-Oriented Analysis of Structures*, Kluwer Academic Publishers, Dordrecht 2002).

In general, the structural response cannot be expressed explicitly in terms of the structure properties, and structural analysis involves solution of a set of simultaneous equations. Reanalysis methods are intended to analyze efficiently structures that are modified due to changes in the structure properties. The object is to evaluate the structural response (e.g. displacements, stresses and forces) for such changes without solving the complete set of modified analysis equations. The solution procedures usually use the available response of the original structure.

Structural analysis is a main part of any design problem. The analysis often must be repeated many times during the design process due to changes in the design variables. The high computational cost involved in reanalysis is one of the main obstacles in the optimization of large-scale structures. In structural damage analysis, it is necessary to analyze the structure repeatedly for various changes due to damage. It is difficult to determine a priori what damage scenarios should be checked, and numerous analyses are required to evaluate various hypothetical scenarios. Reanalysis is involved also in the solution of various nonlinear and dynamic analysis problems. In nonlinear analysis, a set of updated linear equations must be solved repeatedly many times. Similarly, many of the vibration solution techniques are based on matrix iteration methods. To calculate the mode shapes it is often necessary to solve repeatedly a set of updated equations.

The significant increases in computer processing power, memory and storage space have not eliminated the computational cost and time constraints on the use of structural analysis. This is due to the constant increase in the required fidelity and complexity of analysis models and computational procedures. The model complexity is a function of various parameters such as the number of degrees of freedom and the topology of the structure. Complex analysis such as nonlinear and dynamic analysis use linearization algorithms that require linear analysis as a repeated step. History-dependent nonlinear analysis and nonlinear dynamic analysis are currently the extremes of analysis complexity. They typically require numerous linear analysis equivalents. Reanalysis methods are intended to overcome the difficulties involved in repeated analysis complexities.

The book introduces effective computational procedures for reanalysis. The necessary background material on structural analysis needed in the rest of the book is summarized in the first 3 chapters. However, the reader is expected to be familiar with the basic concepts of finite element analysis of structures. Various analysis models are considered in the book, including linear, nonlinear, static and dynamic analysis. In addition, design sensitivity analysis for all these models is introduced. Considerations related to the efficiency of the calculations, the accuracy of the results and the ease of implementation, are discussed in detail. To clarify the presentation, many illustrative examples and numerical results are demonstrated. No specific system of units is used in the examples. However, in some examples actual dimensions of the structure and specified magnitude of forces have been used.

Chapters 1–3 present introductory material on various analysis models. In Chap. 1 static analysis of framed structures and continuum structures is introduced. Considering the stiffness method, linear and nonlinear analysis formulations are discussed. Both geometric and material nonlinearities are considered, and common solution procedures are described.

Chap. 2 is devoted to vibration analysis. Free vibration is presented and properties of the eigenproblem are discussed. Various solution methods of the eigenproblem are introduced, including vector iteration methods, transformation methods, polynomial iterations, Rayleigh-Ritz analysis, the Lanczos method and subspace iteration.

In Chap. 3 dynamic analysis is developed. Solution methods of the equilibrium equations are presented, including direct iteration methods, mode superposition and special analysis procedures. The reduced basis approach for static and dynamic analysis is described and nonlinear dynamic analysis by implicit integration and mode superposition is discussed.

Chap. 4 deals with the statement and general solution approaches of reanalysis problems. Formulations of linear, nonlinear, static, vibration and dynamic reanalysis are presented. Various reanalysis methods are reviewed, including direct and approximate methods. Finally, developments in the unified approach considered in the rest of the text are described.

In Chaps. 5–7 the combined approximations approach for reanalysis is developed for the various analysis models presented in Chaps. 1–3. The basic concepts of combining various methods into a unified approach are introduced. The advantage is that the efficiency of local approximations and the improved accuracy of global approximations are combined to obtain effective solution procedures. The approach presented is suitable for a wide range of applications and different types of structures. Various types of changes in the structure can be considered, including changes in cross-sections, the material properties, the geometry and the topology of the structure. The solution procedures use finite-element stiffness analysis formulations. Calculation of derivatives is not required, and the approach is most attractive in cases where derivatives are not readily available or not easy to calculate. The accuracy of the results and the efficiency of the calculations can be controlled by the amount of information considered. Highly accurate results can be achieved at the expense of more computational effort by considering high-order approximations. On the other hand, very efficient procedures are obtained by simplified low-order approximations. It is shown that in certain cases exact solutions can be achieved with a small computational effort.

In Chap. 5 static reanalysis is discussed. The method for determining the basis vectors is developed, and solution procedures for linear and nonlinear reanalysis are described. The efficiency of the calculations and the accuracy of the results are demonstrated by numerical examples. Various cases of topological and geometrical changes are introduced.

In Chap. 6 vibration reanalysis is presented. It is shown how the problem can be formulated as a reduced small-scale eigenproblem and solved efficiently by combined approximations. Various solution procedures are developed and several techniques, intended to improve the accuracy of the results, are demonstrated.

In Chap. 7 dynamic reanalysis is introduced. Linear and nonlinear dynamic reanalysis problems are solved by the combined approximations approach, using procedures that are suitable for both direct integration and mode superposition.

Direct methods, giving exact closed-form solutions, are presented in Chap. 8. These methods are efficient in situations where a relatively small proportion of the structure is changed (e.g., changes in cross sections of a few elements). It is shown that in such cases the combined approximations approach provides also exact solutions. Exact solutions are demonstrated for various cases of topological and geometrical changes. Solution procedures are developed for the challenging problem where the number of degrees of freedom in the structure is changed.

In Chap. 9 effective procedures for repeated calculations of the response derivatives with respect to design variables are developed. It is shown that accurate derivatives can be obtained with a reduced computational effort for various analysis models. Procedures of sensitivity calculations are developed for linear and nonlinear static problems, vibration problems and dynamic problems.

In Chap. 10 efficiency and accuracy considerations are discussed. The efficiency is compared with various methods of analysis. Some cases of near exact solutions are demonstrated and procedures for error evaluation are presented.

The author wishes to express his appreciation to many graduate students and colleagues who helped in various ways. In particular, I am indebted to my former graduate students Michael Bogomolni and Oded Amir for the fruitful collaboration in research on dynamic and nonlinear reanalysis. Acknowledgement is due to various Foundations, Institutes and Organizations for the generous support of many research projects throughout the years that formed the basis of this text. Finally, the author gratefully acknowledges the assistance of Professors Donald Grierson and Raphael Haftka in reviewing the manuscript and offering critical comments

Uri Kirsch

Contents

1 Static Analysis

Structural analysis is a most exciting field of activity, but it is only a support activity in the field of structural design. Analysis is a main part of the formulation and the solution of any design problem, and it often must be repeated many times during the design process. The analysis process helps to identify improved designs with respect to performance and cost.

Referring to behavior under working loads, the objective of the analysis of a given structure is to determine the internal forces, stresses and displacements under application of the given loading conditions. In order to evaluate the response of the structure it is necessary to establish an analytical model, which represents the structural behavior under application of the loadings. An acceptable model must describe the physical behavior of the structure adequately, and yet be simple to analyze. That is, the basic assumptions of the analysis will ensure that the model represents the problem under consideration and that the idealizations and approximations used result in a simplified solution. This latter property is essential particularly in the design of complex or large systems.

Two categories of mathematical models are often considered:

- Lumped-parameter (discrete-system) models.
- Continuum-mechanics-based (continuous-system) models.

The solution of discrete analysis models involves the following steps:

- *Idealization of the system into a form that can be solved.* The actual structure is idealized as an assemblage of elements that are interconnected at the joints.
- *Formulation of the mathematical model.* The equilibrium requirements of each element are first established in terms of unknown displacements, and the element interconnection requirements are then used to establish the set of simultaneous analysis equations.
- *Solution of the model.* The response is calculated by solving the simultaneous equations for the unknown displacements; the internal forces and stresses of each element are calculated by using the element equilibrium requirements.

The overall effectiveness of an analysis depends to a large degree on the numerical procedures used for the solution of the equilibrium equations [1]. The accuracy of the analysis can, in general, be improved if a more refined model is used. In practice, there is a tendency to employ more and more refined models to approximate the actual structure. This means that the cost of an analysis and its practical feasibility depend to a considerable degree on the algorithms available for the solution of the resulting equations. The time required for solving the equilibrium equations can be a high percentage of the total solution time, particularly in nonlinear analysis or in dynamic analysis, when the solution must be repeated many times. An analysis may not be possible if the solution procedures are too costly. Because of the requirement to solve large systems, much research effort has been invested in equation solution algorithms.

In elastic analysis we refer to behavior under working loads. The forces must satisfy the conditions of equilibrium, and produce deformations compatible with the continuity of the structure and the support conditions. That is, any method must ensure that both conditions of equilibrium and compatibility are satisfied. In linear analysis we assume that displacements (translations or rotations) vary linearly with the applied forces. That is, any increment in a displacement is proportional to the force causing it. This assumption is based on the following two conditions:

- The material of the structure is elastic and obeys Hooke's law.
- All deformations are assumed to be small, so that the displacements do not significantly affect the geometry of the structure and hence do not alter the forces in the members. Thus, the changes in the geometry are small and can be neglected.

The majority of actual structures are designed to undergo only small and linear deformations. In such cases the principle of superposition can be applied. Linear elastic analysis involves the solution of a set of simultaneous linear equations. Framed structures are systems consisting of members that are long in comparison to the dimensions of their cross section. Typical framed structures are beams, grids, and plane and space trusses and frames. The displacement (stiffness) method [e.g. 2, 3], which is the most commonly used analysis method, is considered throughout this text. Linear analysis of framed structures is presented in Sect. 1.1

Elastic analysis of continuum structures, such as plates and shells, is usually performed by the numerical finite element method [e.g. 1, 4]. This method can be regarded as an extension of the displacement method to two- and three-dimensional continuum structures. The actual continuum is replaced by an equivalent idealized structure composed of discrete elements, referred to as finite elements, connected together at a number of nodes. By assuming displacement fields or stress patterns within an element, we can derive a stiffness matrix. A set of simultaneous algebraic equations is formed by applying conditions of equilibrium at every node of the idealized structure. The solution gives the nodal displacements, which in turn are used to determine the stresses. Linear analysis of continuum structures is described in Sect. 1.2.

In most structural analysis problems it is required to solve a set of linear equilibrium equations. Moreover, in various nonlinear and dynamic analysis problems it is necessary to repeat the solution many times for updated sets of linear equations. The solution process often involves factorization of the stiffness matrix, or iterative procedures. Solution of the linear equilibrium equations is discussed in Sect. 1.3.

A nonlinear relationship between the applied forces and the displacements exists under either geometric nonlinearity or material nonlinearity. The solution process can be carried out by various procedures, including an incremental solution scheme, the iterative Newton-Raphson method and combined incremental/iterative solutions. Solution procedures for nonlinear analysis of structures usually involve repeated solutions of sets of updated linear equations. Nonlinear analysis is introduced in Sect. 1.4.

1.1 Linear Analysis of Framed Structures

In the displacement method, joint displacements, chosen as the analysis unknowns, are determined from the conditions of equilibrium. The internal forces and stresses are then determined by superposition of the effects of the external loads and the separate joint displacements.

The equilibrium equations to be solved by the displacement method are

$$\mathbf{K}\,\mathbf{r} + \mathbf{R}_L = \mathbf{R}_E, \tag{1.1}$$

where $\mathbf{K}$ is the stiffness matrix, its elements K_{ij} represent the force in the ith coordinate due to unit displacement in the jth coordinate (K_{ij} are computed in the restrained structure, where i, j correspond to displacement degrees of freedom); $\mathbf{r}$ is the vector of unknown displacements; $\mathbf{R}_L$ is the vector of forces corresponding to the unknown displacements in the restrained structure; $\mathbf{R}_E$ is the vector of external loads corresponding to the unknown displacements ($\mathbf{R}_E = \mathbf{0}$ if no loads act directly in the directions of the unknown displacements). Defining the load vector $\mathbf{R} = \mathbf{R}_E - \mathbf{R}_L$, Eq. (1.1) becomes

$$\mathbf{K}\,\mathbf{r} = \mathbf{R}. \tag{1.2}$$

The vector of unknown displacements $\mathbf{r}$ is computed by solving the set of simultaneous equations (1.2). The vector of final member forces in the structure, $\mathbf{N}$, is given by the following superposition equations

$$\mathbf{N} = \mathbf{N}_L + \mathbf{N}_r\,\mathbf{r}, \tag{1.3}$$

where $\mathbf{N}_L$ is the vector of member forces due to loads and $\mathbf{N}_r$ is the matrix of member forces due to unit value of the elements of $\mathbf{r}$, both computed in the restrained structure.

If the loads act only in the directions of the unknown nodal displacements, $\mathbf{N}_L = \mathbf{0}$ and the stresses $\boldsymbol{\sigma}$ can be determined by

$$\boldsymbol{\sigma} = \mathbf{S}\,\mathbf{r}, \tag{1.4}$$

where $\mathbf{S}$ is the stress transformation matrix.

All equations are related to the action of a single loading. In the case of several loading conditions, all vectors will be transformed into matrices so that each of their columns will correspond to a certain loading condition.

The elements of $\mathbf{K}$, $\mathbf{R}$, $\mathbf{N}_L$, $\mathbf{N}_r$ and $\mathbf{S}$ are functions of the material properties, the geometry of the structure and members' cross-sections.

Example 1.1

To illustrate solution by the displacement method, consider the continuous beam shown in Fig. 1.1*a*. The beam has a constant flexural rigidity EI, the modulus of elasticity is E, the moment of inertia is I, the spans and the loads are represented by L and P, respectively. The object is to find the forces at the left-end support, N_1 and N_2. The two degrees of freedom are the two rotations at the joints r_1 and r_2, which are the unknown displacements. The coefficients computed in the restrained structure are (Figs. 1.1*b* and 1.1*c*)

$$\mathbf{K} = \begin{bmatrix} K_{11} & K_{12} \\ K_{21} & K_{22} \end{bmatrix} = \frac{4}{3}\frac{EI}{L}\begin{bmatrix} 5 & 1 \\ 1 & 5 \end{bmatrix}, \tag{a}$$

$$\mathbf{R}_L = \begin{Bmatrix} R_{L1} \\ R_{L2} \end{Bmatrix} = \frac{PL}{48}\begin{Bmatrix} 3 \\ -5 \end{Bmatrix} \qquad \mathbf{R}_E = \begin{Bmatrix} 0 \\ -PL \end{Bmatrix}, \tag{b}$$

$$\mathbf{R} = \mathbf{R}_E - \mathbf{R}_L = -\frac{PL}{48}\begin{Bmatrix} 3 \\ 43 \end{Bmatrix}, \tag{c}$$

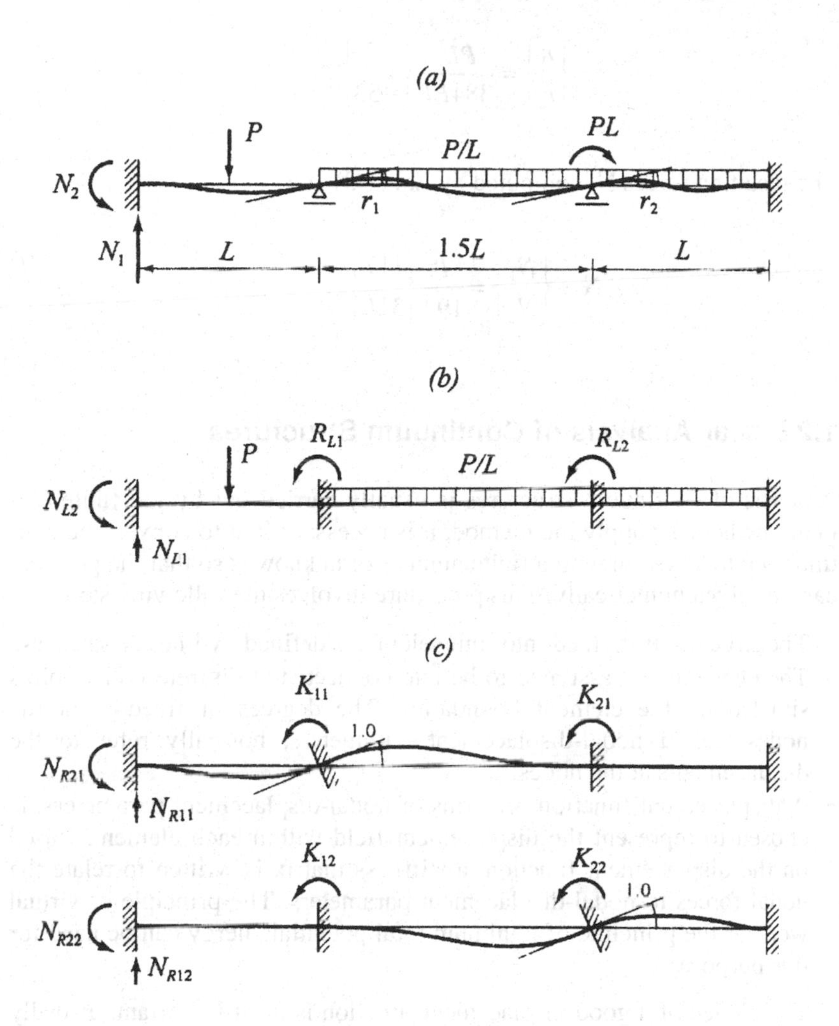

Fig. 1.1. ***a.*** Continuous beam example ***b.*** Loads on the restrained structure ***c.*** Unit displacements on the restrained structure

$$\mathbf{N}_L = \begin{Bmatrix} N_{L1} \\ N_{L2} \end{Bmatrix} = \begin{Bmatrix} P/2 \\ PL/8 \end{Bmatrix} \quad \mathbf{N}_r = \begin{bmatrix} N_{r11} & N_{r12} \\ N_{r21} & N_{r22} \end{bmatrix} = \frac{2EI}{L^2} \begin{bmatrix} 3 & 0 \\ L & 0 \end{bmatrix}. \qquad (d)$$

Substituting these coefficients into Eq. (1.2), we find the unknown displacements **r**

$$\mathbf{r} = \begin{Bmatrix} r_1 \\ r_2 \end{Bmatrix} = \frac{PL^2}{384EI} \begin{Bmatrix} 7 \\ -53 \end{Bmatrix}. \qquad (e)$$

The resulting forces **N**, computed by Eq. (1.3), are

$$\mathbf{N} = \begin{Bmatrix} N_1 \\ N_2 \end{Bmatrix} = \frac{P}{192} \begin{Bmatrix} 117 \\ 31L \end{Bmatrix}. \qquad (f)$$

1.2 Linear Analysis of Continuum Structures

Analysis of continuum structures is usually carried out by the finite element method. To apply the method, it is necessary first to convert the continuum into a system with a finite number of unknowns so that the problem can be solved numerically. This procedure involves the following steps:

- The structure is divided into finite elements defined by lines or surfaces.
- The elements are assumed to be interconnected at discrete nodal points situated on the element boundaries. The degrees of freedom at the nodes, called nodal-displacement parameters, normally refer to the displacements at the nodes.
- A displacement function, in terms of nodal-displacement parameters, is chosen to represent the displacement field within each element. Based on the displacement function, a stiffness matrix is written to relate the nodal forces to nodal-displacement parameters. The principle of virtual work or the principle of minimum total potential energy can be used for this purpose.

The choice of a good displacement function is most important, as badly chosen functions will lead to inaccurate analysis results. The displacement function must have the same number of unknown constants as the total number of degrees of freedom of the element. It must be balanced with respect to the coordinate axes and it must allow the element to undergo rigid-body movement without any internal strain. The displacement function also must be able to represent states of constant stress or strain. Otherwise, the stresses will not converge to a continuous function as progressively smaller elements are used in the idealization of the structure. It also must satisfy the compatibility of displacements along the boundaries with adjacent elements. That is, the function values, and sometimes their first derivatives, are required to be continuous along the boundaries. If all of the above conditions are satisfied, the idealization of the whole system will generally provide a lower bound to the strain energy, and convergence can be guaranteed as the mesh size is successively reduced. However, a number of elements that do not completely satisfy the conditions of compatibility along the boundaries have been successfully used, even though convergence is not assured.

Once the displacement function has been determined, it is possible to obtain all the strains and stresses within the element and to formulate the stiffness matrix and a consistent load matrix. The load matrix represents the equivalent nodal forces, which replace the external distributed loads.

Consider a linear elastic two-dimensional element (Fig. 1.2), for which the displacement function **f** can be written in the form

$$\mathbf{f} = \mathbf{P}\,\mathbf{H}, \tag{1.5}$$

where **f** may have three components (for a three-dimensional body), two translation components *u*, *v* (for plane stress, Fig. 1.2*a*), or simply be equal to the transverse deflection *w* (for a plate in bending, Fig. 1.2*b*); **P** is a function of the coordinates *x* and *y* only; and **H** is a vector of undetermined constants. Note that the element of Fig. 1.2*a* has six degrees of freedom representing the translations *u* and *v* for each node, while the element of Fig. 1.2*b* has nine degrees of freedom (vertical deflection *w*, and two rotations at each of the three nodes).

Applying Eq. (1.5) repeatedly to the nodes of the element one after the other, we obtain the following set of equations relating the nodal parameters $\mathbf{r}_e$ to the constants **H**

$$\mathbf{r}_e = \mathbf{C}\,\mathbf{H}. \tag{1.6}$$

The elements of matrix **C** are functions of the relevant nodal coordinates. From Eq. (1.6), the undetermined constants **H** can be expressed as

$$\mathbf{H} = \mathbf{C}^{-1}\,\mathbf{r}_e. \tag{1.7}$$

Substituting Eq (1.7) into Eq. (1.5), we have

$$\mathbf{f} = \mathbf{P}\,\mathbf{C}^{-1}\,\mathbf{r}_e. \tag{1.8}$$

In many cases, the displacement function is constructed directly in terms of the nodal parameters

$$\mathbf{f} = \mathbf{L}\,\mathbf{r}_e, \tag{1.9}$$

where **L** is a function of *x*, *y* and the coordinates of the nodes. Comparing Eq. (1.8) with Eq. (1.9), it is clear that

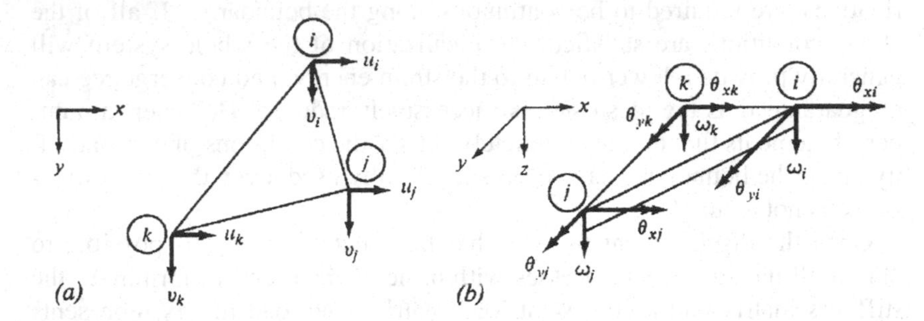

Fig. 1.2. ***a***. Plane stress ***b***. Bending element

$$\mathbf{L} = \mathbf{P}\,\mathbf{C}^{-1}. \tag{1.10}$$

The vector $\boldsymbol{\varepsilon}$ of generalized strains (normal strain, shear strain, bending or twisting curvature) can be expressed in the form

$$\boldsymbol{\varepsilon} = \mathbf{B}\,\mathbf{r}_e, \tag{1.11}$$

where the elements of $\mathbf{B}$ are derived by appropriate differentiation of $\mathbf{L}$ [Eq. (1.9)] with respect to x and y. The generalized stresses $\boldsymbol{\sigma}$ are given by

$$\boldsymbol{\sigma} = \mathbf{d}\boldsymbol{\varepsilon}, \tag{1.12}$$

where $\mathbf{d}$ is a symmetric elasticity matrix, representing the material properties of the element. Substituting Eq. (1.11) into Eq. (1.12) yields

$$\boldsymbol{\sigma} = \mathbf{d}\,\mathbf{B}\,\mathbf{r}_e. \tag{1.13}$$

The product $\mathbf{S} = \mathbf{dB}$ is the stress matrix, and Eq. (1.13) can be written as

$$\boldsymbol{\sigma} = \mathbf{S}\,\mathbf{r}_e. \tag{1.14}$$

To formulate the stiffness and the consistent load matrices, we consider an element subjected to concentrated forces $\mathbf{Q}_e$ at the nodes together with the uniformly distributed loads $\mathbf{q}$ per unit area. Writing the expression for the total potential energy of the element, differentiating with respect to the nodal parameters one after another and then (using the principle of minimum total potential energy) setting them equal to zero, we obtain the set of simultaneous equations

$$\left(\int_V \mathbf{B}^T \mathbf{d}\mathbf{B}\, dV\right)\mathbf{r}_e = \mathbf{Q}_e + \int_\Delta \mathbf{L}^T \mathbf{q}\, d\Delta, \tag{1.15}$$

where the two integrals are over the volume V and the area of the element Δ. Equation (1.15) can be written as

$$\mathbf{K}_e\,\mathbf{r}_e = \mathbf{R}_e, \tag{1.16}$$

where the element stiffness matrix $\mathbf{K}_e$ is defined by

$$\mathbf{K}_e = \int_V \mathbf{B}^T \mathbf{d}\,\mathbf{B}\,dV\,, \tag{1.17}$$

and the consistent load vector $\mathbf{R}_e$ is given by

$$\mathbf{R}_e = \mathbf{Q}_e + \mathbf{Q}_{qe}, \tag{1.18}$$

$\mathbf{Q}_{qe}$ being the consistent load vector for distributed loads $\mathbf{q}$

$$\mathbf{Q}_{qe} = \int_\Delta \mathbf{L}^T \mathbf{q}\,d\Delta\,. \tag{1.19}$$

The stiffness relationship can also be derived by the principle of virtual work, and Eq. (1.19) can be obtained by equating the virtual work done by the equivalent nodal forces and virtual work done by the distributed loads for the same set of permissible virtual displacements.

Example 1.2

Consider the simple element of Fig. 1.2*a* with the following displacement function [Eq. (1.5)]

$$\mathbf{f} = \begin{Bmatrix} u \\ v \end{Bmatrix} = \begin{bmatrix} 1 & x & y & 0 & 0 & 0 \\ 0 & 0 & 0 & 1 & x & y \end{bmatrix} \begin{Bmatrix} H_1 \\ H_2 \\ H_3 \\ H_4 \\ H_5 \\ H_6 \end{Bmatrix}. \tag{a}$$

Writing Eq. (*a*) for each of the nodes, we have

$$\begin{Bmatrix} u_i \\ u_j \\ u_k \end{Bmatrix} = \begin{bmatrix} 1 & x_i & y_i \\ 1 & x_j & y_j \\ 1 & x_k & y_k \end{bmatrix} \begin{Bmatrix} H_1 \\ H_2 \\ H_3 \end{Bmatrix} \qquad \begin{Bmatrix} v_i \\ v_j \\ v_k \end{Bmatrix} = \begin{bmatrix} 1 & x_i & y_i \\ 1 & x_j & y_j \\ 1 & x_k & y_k \end{bmatrix} \begin{Bmatrix} H_4 \\ H_5 \\ H_6 \end{Bmatrix}. \tag{b}$$

The constants $\mathbf{H}$ can be expressed in terms of nodal displacements [Eq. (1.7)] by inversion of Eq. (*b*)

$$\begin{Bmatrix} H_1 \\ H_2 \\ H_3 \end{Bmatrix} = \frac{1}{2\Delta} \begin{bmatrix} a_i & a_j & a_k \\ b_i & b_j & b_k \\ c_i & c_j & c_k \end{bmatrix} \begin{Bmatrix} u_i \\ u_j \\ u_k \end{Bmatrix} \quad \begin{Bmatrix} H_4 \\ H_5 \\ H_6 \end{Bmatrix} = \frac{1}{2\Delta} \begin{bmatrix} a_i & a_j & a_k \\ b_i & b_j & b_k \\ c_i & c_j & c_k \end{bmatrix} \begin{Bmatrix} v_i \\ v_j \\ v_k \end{Bmatrix}, \tag{c}$$

where a, b, c, Δ are constants expressed in terms of the nodal coordinates x_i, y_i, x_j, y_j, x_k, y_k (Δ is the area of the element). From Eq. (c) we may find the matrix $\mathbf{C}^{-1}$ corresponding to the vector of nodal-displacement parameters, $\mathbf{r}_e$. Thus $\mathbf{C}^{-1}$ and $\mathbf{r}_e$ are given by

$$\mathbf{C}^{-1} = \frac{1}{2\Delta} \begin{bmatrix} a_i & 0 & a_j & 0 & a_k & 0 \\ b_i & 0 & b_j & 0 & b_k & 0 \\ c_i & 0 & c_j & 0 & c_k & 0 \\ 0 & a_i & 0 & a_j & 0 & a_k \\ 0 & b_i & 0 & b_j & 0 & b_k \\ 0 & c_i & 0 & c_j & 0 & c_k \end{bmatrix} \qquad \mathbf{r}_e = \begin{Bmatrix} u_i \\ v_i \\ u_j \\ v_j \\ u_k \\ v_k \end{Bmatrix}. \tag{d}$$

From Eqs (a) and (d) we may find the matrix $\mathbf{L}$ [Eq. (1.10)]

$$\mathbf{L} = \frac{1}{2\Delta} \begin{bmatrix} (a_i + b_i x + c_i y) & 0 & (a_j + b_j x + c_j y) & 0 & (a_k + b_k x + c_k y) & 0 \\ 0 & (a_i + b_i x + c_i y) & 0 & (a_j + b_j x + c_j y) & 0 & (a_k + b_k x + c_k y) \end{bmatrix}. \tag{e}$$

The strains for a plane problem are given by

$$\varepsilon_x = \frac{\partial u}{\partial x} \qquad \varepsilon_y = \frac{\partial v}{\partial y} \qquad \gamma_{xy} = \frac{\partial u}{\partial y} + \frac{\partial v}{\partial x}. \tag{f}$$

Differentiation of the displacement function

$$\mathbf{f} = \begin{Bmatrix} u \\ v \end{Bmatrix} = \mathbf{L}\mathbf{r}_e, \tag{g}$$

gives the following expressions for $\boldsymbol{\varepsilon}$

$$\boldsymbol{\varepsilon} = \begin{Bmatrix} \varepsilon_x \\ \varepsilon_y \\ \gamma_{xy} \end{Bmatrix} = \begin{Bmatrix} \frac{1}{2\Delta}(b_i u_i + b_j u_j + b_k u_k) \\ \frac{1}{2\Delta}(c_i v_i + c_j v_j + c_k v_k) \\ \frac{1}{2\Delta}(c_i u_i + b_i v_i + c_j u_j + b_j v_j + c_k u_k + b_k v_k) \end{Bmatrix}. \quad (h)$$

The matrix **B** relating the strains $\boldsymbol{\varepsilon}$ to the nodal displacements $\mathbf{r}_e$ [Eq. (1.11)] is thus given by

$$\mathbf{B} = \frac{1}{2\Delta}\begin{bmatrix} b_i & 0 & b_j & 0 & b_k & 0 \\ 0 & c_i & 0 & c_j & 0 & c_k \\ c_i & b_i & c_j & b_j & c_k & b_k \end{bmatrix}. \quad (i)$$

Since the elements of **B** are constants, the strains inside the element must all be constant. This type of element is often called constant strain triangular element.

The elasticity matrix **d** [Eq. (1.12)] can be shown to be

$$\mathbf{d} = \begin{bmatrix} C_1 & C_1 C_2 & 0 \\ C_1 C_2 & C_1 & 0 \\ 0 & 0 & C_{12} \end{bmatrix}, \quad (j)$$

where

$$C_1 = E/(1-\upsilon^2) \quad C_2 = \upsilon \quad \text{for plane stress,} \quad (k)$$

$$C_1 = \frac{E(1-\upsilon)}{(1+\upsilon)(1-2\upsilon)} \quad C_2 = \frac{\upsilon}{1-\upsilon} \quad \text{for plane strain,}$$

$$C_{12} = C_1(1-C_2)/2 \quad \text{for both cases,}$$

υ = Poisson's ratio.

Since the elements of the stress matrix $\mathbf{S} = \mathbf{dB}$ [Eqs. (1.13), (1.14)] are constants, stresses within an element will also be constant and the result would be stress discontinuities from one element to the next. In practice, this can be overcome either by considering the stress values at the centroid of each element, or by using a method of averaging, in which the stresses of all the elements surrounding a node are summed and then averaged. These average stresses are then assigned to the node concerned.

The stiffness matrix, $\mathbf{K}_e$, can now be computed by Eq. (1.17). Since the elements of $\mathbf{B}$ and $\mathbf{d}$ are constants, the integration is equivalent to multiplying the integrand by the volume of the element V

$$\mathbf{K}_e = V\,\mathbf{B}^T\mathbf{d}\,\mathbf{B}. \tag{l}$$

For uniformly distributed body forces in the x and y directions of magnitude q_x and q_y per unit area, the consistent load vector, $\mathbf{Q}_{qe}$, is given by [see Eqs. (1.19) and (e)]

$$\mathbf{Q}_{qe} = \int_{\Delta} \mathbf{L}^T \begin{Bmatrix} q_x \\ q_y \end{Bmatrix} dx\,dy = \frac{\Delta}{3} \begin{Bmatrix} q_x \\ q_y \\ q_x \\ q_y \\ q_x \\ q_y \end{Bmatrix}. \tag{m}$$

That is, 1/3 of the load acting on the element is assigned to each node.

Consider the square isotropic plate of constant thickness $h = 1.0$ shown in Fig. 1.3a [2], which is free along three edges and fixed at the other edge. The displacement conditions are $u = v = 0$ at the fixed edge, and $u = 0$ along the line of symmetry. Poisson's ratio is $\upsilon = 0.25$, the modulus of elasticity is $E = 1.0$, and the object is to find the stress distribution in the plate. Using the finite-element idealization shown in Fig. 1.3b, the unknown displacements for the right-hand half of the plate are

$$\mathbf{r}^T = \{v_1, v_2, u_4, v_4, u_5, v_5, u_7, v_7, u_8, v_8\}, \tag{n}$$

and the corresponding external applied force vector is

$$\mathbf{R}^T = \{0.5, 0, 0, 1.0, 0, 0, 0, 0.5, 0, 0\}. \tag{o}$$

The stiffness matrices of the elements $\mathbf{K}_e$ are computed by Eq. (l), and the assembled matrix $\mathbf{K}$ is then computed by considering the contribution of the elements. Solving the equations

$$\mathbf{K}\,\mathbf{r} = \mathbf{R}, \tag{p}$$

we find the nodal displacements $\mathbf{r}$ shown in Fig. 1.3c. The element stresses, calculated by Eq. (1.13), are shown in Fig. 1.3d.

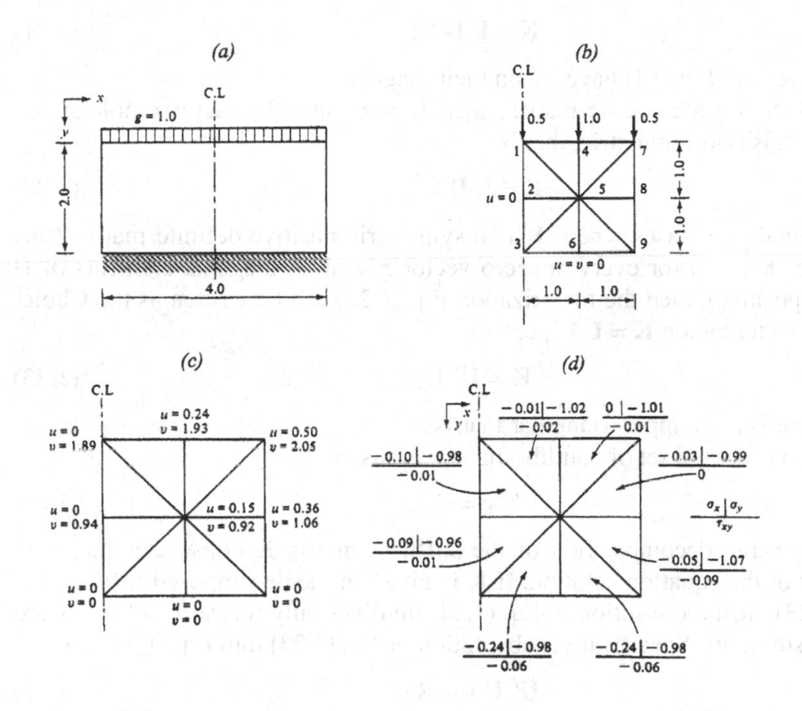

Fig. 1.3. ***a.*** Plate dimensions and loading ***b.*** Finite-element idealization ***c.*** Nodal displacements ***d.*** Element stresses

1.3 Solution of the Linear Equilibrium Equations

1.3.1 Matrix Factorization

In this section some common methods for solving the linear equilibrium equations are described. A square matrix **K** can be factorized into a product of two matrices **L**, **U**

$$\mathbf{K} = \mathbf{L}\,\mathbf{U}, \tag{1.20}$$

where **U** is an upper triangular matrix having pivots on its diagonal and **L** is a lower triangular matrix having 1's on its diagonal.

Divide **U** by a diagonal matrix **D** that contains the pivots (it is convenient to keep the same letter **U** for this new upper triangular matrix that has 1's on its diagonal). Then, the triangular factorization can be written as

$$\mathbf{K} = \mathbf{L\,D\,U}, \tag{1.21}$$

where both **L** and **U** have 1's on their diagonal.

If **A** is a symmetric matrix, then $\mathbf{U} = \mathbf{L}^T$ and the factorization of Eq. (1.21) is also symmetric, that is

$$\mathbf{K} = \mathbf{L\,D\,L}^T. \tag{1.22}$$

Finally, in cases where **K** is a symmetric positive-definite matrix (that is, $\mathbf{r}^T\,\mathbf{K}\,\mathbf{r} > 0$ for every nonzero vector **r**, and all diagonal elements of **D** are positive), then the factorization Eq. (1.22) can be written as the Cholesky factorization $\mathbf{K} = \mathbf{L\,L}^T$, or

$$\mathbf{K} = \mathbf{U}^T\,\mathbf{U}, \tag{1.23}$$

where **U** is an upper triangular matrix.

Consider the set of equilibrium equations (1.2)

$$\mathbf{K\,r} = \mathbf{R}. \tag{1.24}$$

In general, decomposition of the stiffness matrix **K** constitutes the main part of the equation solution. If **K** is given in the decomposed form of Eq. (1.23), then the solution of Eq. (1.24) involves only forward and backward substitutions. Specifically, substitution of Eq. (1.23) into Eq. (1.24) gives

$$\mathbf{U}^T\,\mathbf{U\,r} = \mathbf{R}. \tag{1.25}$$

We first solve for the vector of unknowns **t** by the forward substitution

$$\mathbf{U}^T\mathbf{t} = \mathbf{R}. \tag{1.26}$$

The unknown displacement vector **r** is then calculated by the backward substitution

$$\mathbf{U\,r} = \mathbf{t}. \tag{1.27}$$

Similarly, if **K** is given in the decomposed form of Eq. (1.22), substitution of Eq. (1.22) into Eq. (1.24) gives

$$\mathbf{L\,D\,L}^T\,\mathbf{r} = \mathbf{R}. \tag{1.28}$$

The vector **t** is first calculated by the forward substitution

$$\mathbf{L\,t} = \mathbf{R}. \tag{1.29}$$

The unknown displacement vector **r** is then calculated by the backward substitution

$$\mathbf{L}^T\,\mathbf{r} = \mathbf{D}^{-1}\,\mathbf{t}. \tag{1.30}$$

It should be noted that an efficient solution is often obtained because no operations are needed to be performed on zero elements. Thus, it is possible to avoid the storage of elements that remain zero and to skip the relevant operations.

1.3.2 Iterative Solution Procedures

A basic disadvantage of an iterative solution is that it is difficult to estimate the number of iterations required for convergence, which depends on the condition number of the matrix of coefficients (matrix **K** for **K r** = **R**). However, for very large systems a direct method of solution may require a large amount of storage and computer time. The required storage is proportional to $n\, m_k$, where n is the number of equations and m_k is the half-bandwidth. A measure of the number of operations is $1/2\; n\, m_k^2$. The half-bandwidth is roughly proportional to $n^{1/2}$. On the other hand, in an iterative solution the required storage is much less because we need to store only the actually nonzero matrix elements under the skyline of the matrix, a pointer array that indicates the location of each nonzero element, and some arrays also of small size. The number of nonzero elements under the skyline is only a small fraction of all the elements under the skyline.

The fact that effective procedures that accelerate convergence have become available for many applications has rendered iterative methods very attractive [1].

Conjugate Gradient Method with Preconditioning

The Preconditioned Conjugate Gradient (PCG) method presented in this section is one of the most effective and simple iterative methods. Moreover, the unified reanalysis approach developed in this text is closely related to this method, as will be shown later in Sect. 10.3.

The Conjugate Gradient (CG) method can be used as an iterative method for solving the linear set of Eq. (1.2)

$$\mathbf{K}\,\mathbf{r} = \mathbf{R}, \tag{1.31}$$

where **K** is an $n{\times}n$ symmetric and positive-definite matrix. This problem can be stated equivalently as minimization of the quadratic function

$$Q = 1/2\,\mathbf{r}^T\mathbf{K}\,\mathbf{r} - \mathbf{r}^T\mathbf{R}. \tag{1.32}$$

A set of n nonzero vectors $\mathbf{S}_1$, $\mathbf{S}_2$, ..., $\mathbf{S}_n$ is said to be conjugate with respect to **K**, if

$$\mathbf{S}_i^T \mathbf{K} \mathbf{S}_j = 0 \quad \text{for all } i \neq j. \tag{1.33}$$

It can be shown that any set of vectors satisfying this property is also linearly independent. A set of conjugate vectors possesses a powerful property [5]; namely, if the quadratic function Q is minimized sequentially, once along each of a set of n conjugate directions, the minimum of Q will be located at or before the nth step, regardless of the starting point. The order in which the directions are used is immaterial to this property.

There are many ways to choose the conjugate directions. The CG method is a conjugate directions method with a special property. In generating its set of conjugate vectors, it can compute a new vector by using only the previous vector and the current gradient. This property implies that the method requires little storage and computation. The method was developed in the 1950s as a method for finding exact solutions of symmetric positive definite systems [6]. Some years later the method came to be viewed as an iterative method that could give good approximate solutions to systems in many fewer than n steps [7, 8].

To solve the linear system of Eq. (1.31) by the CG method, we start with the initial estimate $\mathbf{r}_0$ at the initial point $\mathbf{K} = \mathbf{K}_0$

$$\mathbf{r}_0 = \mathbf{K}_0^{-1} \mathbf{R}. \tag{1.34}$$

The residual $\boldsymbol{\delta}_0$ of Eq. (1.31) is then calculated by

$$\boldsymbol{\delta}_0 = \mathbf{K}\, \mathbf{r}_0 - \mathbf{R}. \tag{1.35}$$

The first direction of minimization, $\mathbf{S}_0$, is the *steepest descent direction* of the quadratic function Q [Eq. (1.32)], at the initial point $\mathbf{r}_0$, given by

$$\mathbf{S}_0 = -\, \boldsymbol{\delta}_0. \tag{1.36}$$

A sequence of vectors $\mathbf{r}_k$ is then generated by the method. The vectors generated are given by the expression

$$\mathbf{r}_{k+1} = \mathbf{r}_k + \alpha_k\, \mathbf{S}_k, \tag{1.37}$$

where α_k is the one-dimensional minimizer of the quadratic function Q along $\mathbf{r}_k + \alpha_k\, \mathbf{S}_k$, given explicitly by

$$\alpha_k = \frac{\boldsymbol{\delta}_k^T \boldsymbol{\delta}_k}{\mathbf{S}_k^T \mathbf{K} \mathbf{S}_k}. \tag{1.38}$$

The residual $\boldsymbol{\delta}_{k+1}$ of the linear system of Eq. (1.31) is calculated by

$$\boldsymbol{\delta}_{k+1} = \mathbf{K}\, \mathbf{r}_{k+1} - \mathbf{R} = \boldsymbol{\delta}_k + \alpha_k\, \mathbf{K}\, \mathbf{S}_k. \tag{1.39}$$

The $(k+1)$th direction, $\mathbf{S}_{k+1}$, is selected by

$$\mathbf{S}_{k+1} = -\,\boldsymbol{\delta}_{k+1} + \beta_{k+1}\mathbf{S}_k. \tag{1.40}$$

The scalar β_{k+1} is determined by the requirement that $\mathbf{S}_k$ and $\mathbf{S}_{k+1}$ must be conjugate with respect to $\mathbf{K}$ [Eq. (1.33)]. By pre-multiplying Eq. (1.40) by $\mathbf{S}_k^T\mathbf{K}$ and imposing the condition $\mathbf{S}_k^T\mathbf{K}\mathbf{S}_{k+1} = 0$, we find the following expression for β_{k+1}

$$\beta_{k+1} = \frac{\boldsymbol{\delta}_{k+1}^T\mathbf{K}\,\mathbf{S}_k}{\mathbf{S}_k^T\mathbf{K}\,\mathbf{S}_k} = \frac{\boldsymbol{\delta}_{k+1}^T\boldsymbol{\delta}_{k+1}}{\boldsymbol{\delta}_k^T\boldsymbol{\delta}_k}. \tag{1.41}$$

In summary, we first calculate $\mathbf{r}_0$ [Eq. (1.34)], $\boldsymbol{\delta}_0$ [Eq. (1.35)] and $\mathbf{S}_0$ [Eq. (1.36)]. Then, the following calculations are carried out repeatedly:

- The scalar α_k is calculated by Eq. (1.38).
- The vector $\mathbf{r}_{k+1}$ is calculated by Eq. (1.37).
- The residual $\boldsymbol{\delta}_{k+1}$ of Eq. (1.31) is calculated by Eq. (1.39).
- The scalar β_{k+1} is calculated by Eq. (1.41).
- The direction $\mathbf{S}_{k+1}$ is calculated by Eq. (1.40).

Each search direction $\mathbf{S}_k$ and residual $\boldsymbol{\delta}_k$ are constrained by the Krylov subspace of degree k for $\boldsymbol{\delta}_0$, defined as

$$\mathcal{D}(\boldsymbol{\delta}_0, k) = \text{span}\,\{\boldsymbol{\delta}_0, \mathbf{K}\,\boldsymbol{\delta}_0, \ldots, \mathbf{K}^k\,\boldsymbol{\delta}_0\}. \tag{1.42}$$

The rate of convergence of the above procedure depends on the eigenvalue distribution of $\mathbf{K}$ and the initial approximation $\mathbf{r}_0$. The convergence is faster when the condition number of $\mathbf{K}$, defined by the ratio of the maximum and minimum eigenvalues, is smaller and/or when $\mathbf{K}$ has clustered eigenvalues.

The CG properties are valid only in exact arithmetic. For ill-conditioned problems the convergence of the method might be slow, mainly due to round-off errors, when working with inexact arithmetic. For such problems, the conjugate directions are no longer exactly conjugate after some iteration cycles. It is possible to accelerate the rate of convergence by transformation of the linear system of Eqs. (1.31) such that the eigenvalue distribution of $\mathbf{K}$ is improved [9]. The key to this process, which is known as *preconditioning*, is a change of variables from $\mathbf{r}$ to $\tilde{\mathbf{r}}$ via a nonsingular matrix $\mathbf{C}$ called the *pre-conditioner*

$$\mathbf{r} = \mathbf{C}\,\tilde{\mathbf{r}}. \tag{1.43}$$

Substituting Eq. (1.43) into Eq. (1.31) and pre-multiplying the resulting equation by $\mathbf{C}^T$, we obtain the following new system of equations

$$(\mathbf{C}^T\mathbf{K}\,\mathbf{C})\;\tilde{\mathbf{r}} = \mathbf{C}^T\mathbf{R}. \tag{1.44}$$

Defining

$$\tilde{\mathbf{K}} = \mathbf{C}^T\mathbf{K}\,\mathbf{C} \qquad \tilde{\mathbf{R}} = \mathbf{C}^T\,\mathbf{R}, \tag{1.45}$$

and substituting Eqs. (1.45) into Eq. (1.44), we obtain the new system

$$\tilde{\mathbf{K}}\,\tilde{\mathbf{r}} = \tilde{\mathbf{R}}. \tag{1.46}$$

Note that when $\mathbf{K}$ is symmetric and positive-definite and $\mathbf{C}$ has full rank, $\tilde{\mathbf{K}}$ is also symmetric and positive-definite. The convergence rate of the CG method applied to Eq. (1.46) will depend on the eigenvalues of the preconditioned matrix $\tilde{\mathbf{K}}$ rather than those of $\mathbf{K}$. The aim is to choose the pre-conditioner $\mathbf{C}$ such that the condition number of $\tilde{\mathbf{K}}$ is much smaller than the condition number of the original matrix $\mathbf{K}$. Alternatively, $\mathbf{C}$ could be chosen such that the eigenvalues of $\tilde{\mathbf{K}}$ are clustered.

It is possible to apply the CG method directly on the new system of Eq. (1.46). However, in this case it would be necessary to calculate and store the new (typically large) matrix $\tilde{\mathbf{K}}$ and to perform extensive computational effort at each iteration cycle. Alternatively, the tilde notation can be used during the solution procedure applied to Eq. (1.46), also for the auxiliary vectors $\tilde{\mathbf{r}}_k$ and $\tilde{\mathbf{S}}_k$, where

$$\mathbf{r}_k = \mathbf{C}\,\tilde{\mathbf{r}}_k \qquad \mathbf{S}_k = \mathbf{C}\tilde{\mathbf{S}}_k. \tag{1.47}$$

The relation between the residuals $\tilde{\boldsymbol{\delta}}_k$ and $\boldsymbol{\delta}_k$ is given by

$$\tilde{\boldsymbol{\delta}}_k = \tilde{\mathbf{K}}\,\tilde{\mathbf{r}} - \tilde{\mathbf{R}} = \mathbf{C}^T\left(\mathbf{K}\mathbf{C}\mathbf{C}^{-1}\mathbf{r} - \mathbf{R}\right) = \mathbf{C}^T\left(\mathbf{K}\mathbf{r} - \mathbf{R}\right) = \mathbf{C}^T\boldsymbol{\delta}_k, \tag{1.48}$$

$$\boldsymbol{\delta}_k = (\mathbf{C}^T)^{-1}\,\tilde{\boldsymbol{\delta}}_k.$$

Define an auxiliary vector

$$\mathbf{z}_k = \mathbf{C}\,\mathbf{C}^T\,\boldsymbol{\delta}_k. \tag{1.49}$$

If no preconditioning is used, then $\mathbf{CC}^T = \mathbf{I}$ and $\mathbf{z}_k = \boldsymbol{\delta}_k$. We can write the resulting Preconditioned Conjugate Gradient (PCG) method with the original variables $\mathbf{r}$, using modified formulas for α_k, β_{k+1} and $\mathbf{S}_{k+1}$

$$\alpha_k = \frac{\boldsymbol{\delta}_k^T\mathbf{z}_k}{\mathbf{S}_k^T\mathbf{K}\mathbf{S}_k}, \tag{1.50}$$

$$\beta_{k+1} = \frac{\boldsymbol{\delta}_{k+1}^T \mathbf{z}_{k+1}}{\boldsymbol{\delta}_k^T \mathbf{z}_k}, \tag{1.51}$$

$$\mathbf{S}_{k+1} = -\mathbf{z}_{k+1} + \beta_{k+1} \mathbf{S}_k. \tag{1.52}$$

For the calculated $\mathbf{r}_0$ [Eq. (1.34)], $\boldsymbol{\delta}_0$ [Eq. (1.35)], $\mathbf{S}_0$ [Eq. (1.36)] $\mathbf{z}_0$ [Eq. (1.49)], each iteration cycle involves the following steps:

- The scalar α_k is calculated by Eq. (1.50).
- The vector $\mathbf{r}_{k+1}$ is calculated by Eq. (1.37).
- The residual $\boldsymbol{\delta}_{k+1}$ of Eq. (1.31) is calculated by Eq. (1.39).
- The vector $\mathbf{z}_{k+1}$ is calculated by $\mathbf{z}_{k+1} = \mathbf{C}\,\mathbf{C}^T \boldsymbol{\delta}_{k+1}$ [Eq. (1.49)].
- The scalar β_{k+1} is calculated by Eq. (1.51).
- The direction $\mathbf{S}_{k+1}$ is calculated by Eq. (1.52).

Various pre-conditioners $\mathbf{C}$ have been proposed in the literature [9]. The best choice of $\mathbf{C}$ could be the inverse factor $\mathbf{C} = \mathbf{U}^{-1}$, where $\mathbf{U}$ is an upper triangular matrix given by the Cholesky factorization $\mathbf{K} = \mathbf{U}^T \mathbf{U}$. With this choice, the preconditioned matrix becomes equal to the identity matrix [Eq. (1.45)] $\tilde{\mathbf{K}} = (\mathbf{U}^{-1})^T \mathbf{K}\,\mathbf{U}^{-1} = (\mathbf{U}^T)^{-1} (\mathbf{U}^T \mathbf{U})\,\mathbf{U}^{-1} = \mathbf{I}$. However, in this case much computational effort is needed to calculate matrix $\mathbf{C} = \mathbf{U}^{-1}$. Alternatively, matrix $\mathbf{C}$ can be chosen as

$$\mathbf{C} = \mathbf{U}_0^{-1}, \tag{1.53}$$

where $\mathbf{U}_0$ is an upper triangular matrix, which is already given by the Cholesky factorization $\mathbf{K}_0 = \mathbf{U}_0^T \mathbf{U}_0$. In this case the only additional work, compared with the CG algorithm, is the computation of the auxiliary vector $\mathbf{z}_k$, i.e., the solution of the system of equations [see Eq. (1.49)]

$$\mathbf{K}_0\, \mathbf{z}_k = \boldsymbol{\delta}_k, \tag{1.54}$$

for $\mathbf{z}_k$, which guarantees the preconditioning effect. Solution of Eq. (1.54) is easy, as we already have the Cholesky factor $\mathbf{U}_0$ of $\mathbf{K}_0$. The preconditioned matrix $\tilde{\mathbf{K}}$ for the chosen $\mathbf{C}$ is

$$\tilde{\mathbf{K}} = \mathbf{C}^T \mathbf{K}\,\mathbf{C} = (\mathbf{U}_0^{-1})^T \mathbf{K}\,\mathbf{U}_0^{-1}. \tag{1.55}$$

It can be noted that for $\mathbf{K} = \mathbf{K}_0$ the preconditioned matrix is equal to the identity matrix $\tilde{\mathbf{K}} = (\mathbf{U}_0^{-1})^T \mathbf{K}_0 \mathbf{U}_0^{-1} = (\mathbf{U}_0^T)^{-1} (\mathbf{U}_0^T \mathbf{U}_0)\, \mathbf{U}_0^{-1} = \mathbf{I}$. If the eigenvalues of $\mathbf{K}$ are close to those of $\mathbf{K}_0$, the preconditioned matrix is close to the identity matrix, and the convergence of the PCG is extremely fast.

1.4 Nonlinear Analysis

In linear analysis the displacements $\mathbf{r}$, determined by solving the set of linear equations $\mathbf{K}\,\mathbf{r} = \mathbf{R}$, are linear functions of the loads $\mathbf{R}$. In various cases of large displacements, nonlinear stress-strain relations, or changes in the displacement boundary conditions during loading (e.g. contact problems), the displacements are some nonlinear functions of the loads. Considering separately material nonlinear effects and geometric nonlinear effects, we can categorize different nonlinear analysis problems. A nonlinear relationship between the applied forces and the resulting displacements exists under either of two conditions:

- The stress-strain relation is within the linear-elastic range, but the geometry of the structure changes significantly during an application of the loads (geometric nonlinearity).
- The stress-strain relation is nonlinear (material nonlinearity). Material non-linearity is considered in plastic analysis.

The most general analysis case is the one in which the structure is subjected to large displacements and nonlinear stress-strain relation. Nonlinear analysis is usually carried out in an iterative process. The study of nonlinear behavior includes plastic analysis and buckling of structures. The basic problem in nonlinear analysis is the solution of a set of nonlinear equations. Depending on the history of the loading, the stiffness of the structure may be softening or stiffening, the equilibrium path may be stable or unstable, and the structure itself may be at a stage of loading or unloading. All such phenomena are typified by the occurrence of critical points such as the limit points and snap-back points in the load-deflection curves.

The solution process can be carried out by different methods [1, 10, 11]. A requirement for the solution method is its ability to overcome the numerical problems associated with various types of behavior. In practical nonlinear analysis, the external forces are introduced in stages. Solution of the nonlinear set of equations is usually carried out by an incremental/iterative technique, such as a predictor-corrector method. This is accomplished by solving the equations for successive values of a load or displacement parameter, such that the solution corresponding to a particular value of the parameter is used to calculate a suitable approximation (predictor) for the displacements $\mathbf{r}$ at a different value of the parameter.

This approximation is then chosen as an initial estimate of $\mathbf{r}$ in a corrective-iterative procedure such as the Newton-Raphson Method.

1.4.1 Geometric Nonlinearity

For illustrative purposes consider the *Newton-Raphson method,* which is the most frequently used iteration scheme for the solution of the nonlinear analysis equations. Starting with linear analysis, we first calculate the initial displacements $\mathbf{r}_0$ by the linear analysis equations

$$\mathbf{K}_0\,\mathbf{r}_0 = \mathbf{R}_0, \tag{1.56}$$

where $\mathbf{K}_0$ is the given elastic stiffness matrix and $\mathbf{R}_0$ is the given external force vector. The matrix $\mathbf{K}_0$ is often factorized into the form of Eq. (1.23). The member forces $\mathbf{N}$ are calculated for the deformed geometry. Considering both the compatibility equations and the constitutive law, the member forces are some nonlinear functions $\mathbf{N}(\mathbf{r})$ of the displacements $\mathbf{r}$

$$\mathbf{N} = \mathbf{N}(\mathbf{r}). \tag{1.57}$$

The forces computed by Eq. (1.57) are not in equilibrium with the external forces $\mathbf{R}_0$. The internal forces $\mathbf{R}_I$, corresponding to the member forces $\mathbf{N}$, are determined by the equilibrium equations of the deformed geometry

$$\mathbf{R}_I = \mathbf{C}(\mathbf{r})\,\mathbf{N}, \tag{1.58}$$

where the elements of matrix $\mathbf{C}(\mathbf{r})$ depend on the deformed geometry.

The equilibrium equations, which are some nonlinear functions of $\mathbf{r}$, can be expressed as

$$\mathbf{f}(\mathbf{r}) = \mathbf{R}_0 - \mathbf{R}_I = \mathbf{0}. \tag{1.59}$$

Assume first-order (linear) Taylor series approximations about the initial displacements $\mathbf{r}_0$

$$\mathbf{f}(\mathbf{r}) = \mathbf{f}(\mathbf{r}_0) + \left.\frac{\partial \mathbf{f}}{\partial \mathbf{r}}\right|_{\mathbf{r}_0} (\mathbf{r} - \mathbf{r}_0)\,. \tag{1.60}$$

Substituting from Eq. (1.59) into Eq. (1.60) and assuming that that the external loads are deformation-independent, we obtain

$$\left.\frac{\partial \mathbf{R}_I}{\partial \mathbf{r}}\right|_{\mathbf{r}_0} (\mathbf{r} - \mathbf{r}_0) = \mathbf{R}_0 - \mathbf{R}_I\,. \tag{1.61}$$

Define the out-of-balance (residual) force vector $\delta\mathbf{R}$

$$\delta\mathbf{R} = \mathbf{R}_0 - \mathbf{R}_I. \tag{1.62}$$

The out-of-balance forces correspond to a load vector that is not yet balanced by element stresses, and hence an increment in the nodal-point displacements is required. The vector of incremental displacements due to the out-of-balance forces, $\delta\mathbf{r}$, is given by

$$\delta\mathbf{r} = \mathbf{r} - \mathbf{r}_0, \tag{1.63}$$

and the current tangent stiffness matrix $\mathbf{K}_T$ is defined as

$$\mathbf{K}_T = \left.\frac{\partial \mathbf{R}_I}{\partial \mathbf{r}}\right|_{\mathbf{r}_0}. \tag{1.64}$$

The tangent stiffness matrix can be expressed as

$$\mathbf{K}_T = \mathbf{K}_0 + \mathbf{K}_G, \tag{1.65}$$

where $\mathbf{K}_G = \mathbf{K}_G(\mathbf{r}, \mathbf{N})$ is the geometric stiffness matrix, whose elements are some functions of the deformed geometry and the member forces. Substituting Eqs. (1.62), (1.63), (1.64) into Eq. (1.61) gives

$$\mathbf{K}_T\, \delta\mathbf{r} = \delta\mathbf{R}. \tag{1.66}$$

The vector $\delta\mathbf{r}$ is calculated by the modified equilibrium equations (1.66), written for the deformed geometry. Starting with the initial displacements $\mathbf{r}_0$, we update the displacements iteratively by

$$\mathbf{r} = \mathbf{r}_0 + \delta\mathbf{r}, \tag{1.67}$$

where $\mathbf{r}_0$ is defined as the updated displacement at the previous cycle.

In summary, the following quantities are calculated repeatedly during the solution process:

- the member forces $\mathbf{N}$ [Eq. (1.57)];
- the corresponding internal force vector $\mathbf{R}_I$ [Eq. (1.58)];
- the out-of-balance force vector $\delta\mathbf{R}$ [Eq. (1.62)];
- the tangent stiffness matrix $\mathbf{K}_T$ [Eq. (1.64)];
- the incremental displacements $\delta\mathbf{r}$ [Eq. (1.66)]; and
- the updated displacements $\mathbf{r}$ [Eq. (1.67)]

These calculations are repeated until convergence occurs. Typical convergence conditions related to the norms of $\delta\mathbf{R}$ and $\delta\mathbf{r}$ are $\|\delta\mathbf{R}\| \le \varepsilon_R$ and $\|\delta\mathbf{r}\| \le \varepsilon_r$, where ε_R, ε_r are some small predetermined parameters.

This procedure is a simplified version of the Newton-Raphson method. Different variations and improvements might be considered [10], including an incremental solution, retaining the original decomposed tangential matrix during several iterations, or combined incremental-iterative methods. The correct evaluation of the tangent stiffness matrix is important. However, because of the expense involved in evaluating and factoring a new tangent stiffness matrix, it can be more efficient to evaluate the matrix only at certain times. In the *modified Newton-Raphson method* a new tangent stiffness matrix is established only at the beginning of each load step. Solution of the nonlinear equations is discussed later in Sect. 1.4.5.

Example 1.3

To illustrate the solution steps, consider the simple two-bar truss shown in Fig. 1.4. Assuming arbitrary units, the modulus of elasticity is $E = 10000$ and the cross sectional area of both members is $A_1 = A_2 = 0.01$. The elements of $\mathbf{K}_0$ and $\mathbf{R}_0$ are given by

$$\mathbf{K}_0 = \begin{bmatrix} EA_1/L_1 & \\ & EA_2/L_2 \end{bmatrix} = \begin{bmatrix} 5.0 & \\ & 10.0 \end{bmatrix}, \tag{a}$$

$$\mathbf{R}_0 = \begin{Bmatrix} R_{01} \\ R_{02} \end{Bmatrix} = \begin{Bmatrix} 10.0 \\ 5.0 \end{Bmatrix}.$$

Substituting Eqs. (*a*) into Eq. (1.56) and solving for $\mathbf{r}_0$, we find

$$\mathbf{r}_0 = \begin{Bmatrix} r_{01} \\ r_{02} \end{Bmatrix} = \begin{Bmatrix} 2.0 \\ 0.5 \end{Bmatrix}. \tag{b}$$

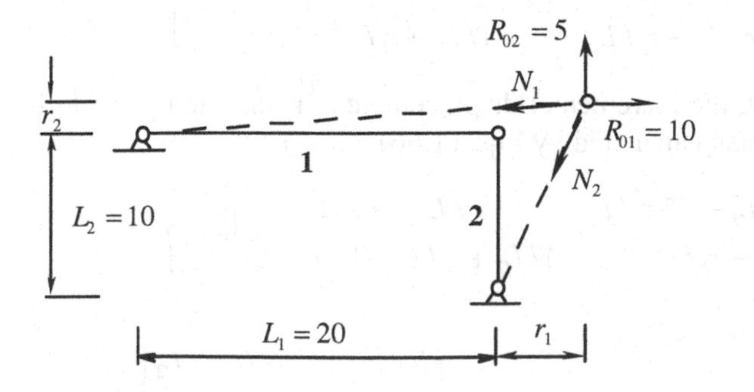

Fig. 1.4. Two-bar truss

Assume small strains such that $r_1^2/L_1^2 << r_1/L_1$, where L_1, L_2 are the members lengths and r_1, r_2 are the nodal displacements (Fig. 1.4). The force N_1 in member 1 can be expressed in terms of the displacements as

$$N_1 = EA_1\left(\left((L_1+r_1)^2+r_2^2\right)^{1/2}-L_1\right)/L_1 \approx EA_1\left(r_1/L_1+0.5r_2^2/L_1^2\right). \tag{c}$$

Using a similar expression for the force N_2 in member 2 gives the following vector of member forces $\mathbf{N}$ [Eq. (1.57)]

$$\mathbf{N} = \begin{Bmatrix} N_1 \\ N_2 \end{Bmatrix} = \begin{Bmatrix} EA_1\left(r_1/L_1+0.5r_2^2/L_1^2\right) \\ EA_2\left(r_2/L_2+0.5r_1^2/L_2^2\right) \end{Bmatrix}. \tag{d}$$

The internal force vector $\mathbf{R}_I$ [Eq. (1.58)] consists of the horizontal and vertical components of the member forces, i.e.,

$$\mathbf{R}_I = \begin{Bmatrix} R_{I1} \\ R_{I2} \end{Bmatrix} = \begin{bmatrix} 1 & r_1/L_2 \\ r_2/L_1 & 1 \end{bmatrix} \begin{Bmatrix} N_1 \\ N_2 \end{Bmatrix} = \begin{Bmatrix} N_1+N_2r_1/L_2 \\ N_2+N_1r_2/L_1 \end{Bmatrix}, \tag{e}$$

and the out-of-balance forces $\delta\mathbf{R}$ [Eq. (1.62)] are given by

$$\delta\mathbf{R} = \begin{Bmatrix} \delta R_1 \\ \delta R_2 \end{Bmatrix} = \begin{Bmatrix} R_{01} \\ R_{02} \end{Bmatrix} - \begin{Bmatrix} R_{I1} \\ R_{I2} \end{Bmatrix} = \begin{Bmatrix} R_{01} \\ R_{02} \end{Bmatrix} - \begin{Bmatrix} N_1+N_2r_1/L_2 \\ N_2+N_1r_2/L_1 \end{Bmatrix}. \tag{f}$$

For A_1=A_2=A, the tangent stiffness matrix $\mathbf{K}_T$ is [Eq. (1.64)]

$$\mathbf{K}_T = \begin{bmatrix} \partial R_1/\partial r_1 & \partial R_1/\partial r_2 \\ \partial R_2/\partial r_1 & \partial R_2/\partial r_2 \end{bmatrix} = \tag{g}$$

$$= EA\begin{bmatrix} 1/L_1+r_2/L_2^2+1.5r_1^2/L_2^3 & r_1/L_2^2+r_2/L_1^2 \\ r_1/L_2^2+r_2/L_1^2 & 1/L_2+r_1/L_1^2+1.5r_2^2/L_1^3 \end{bmatrix}.$$

Using Eqs. (*f*), (*g*), the incremental displacements $\delta\mathbf{r}$ and the updated displacements $\mathbf{r}$ are then calculated by Eqs. (1.66), (1.67)

$$EA\begin{bmatrix} 1/L_1+r_2/L_2^2+1.5r_1^2/L_2^3 & r_1/L_2^2+r_2/L_1^2 \\ r_1/L_2^2+r_2/L_1^2 & 1/L_2+r_1/L_1^2+1.5r_2^2/L_1^3 \end{bmatrix} \begin{Bmatrix} \delta r_1 \\ \delta r_2 \end{Bmatrix} = \tag{h}$$

$$= \begin{Bmatrix} R_{01} \\ R_{02} \end{Bmatrix} - \begin{Bmatrix} N_1+N_2r_1/L_2 \\ N_2+N_1r_2/L_1 \end{Bmatrix},$$

$$\mathbf{r} = \begin{Bmatrix} r_1 \\ r_2 \end{Bmatrix} = \begin{Bmatrix} r_{01} \\ r_{02} \end{Bmatrix} + \begin{Bmatrix} \delta r_1 \\ \delta r_2 \end{Bmatrix}. \qquad (i)$$

Starting with $\mathbf{r}_0$ [Eq. (*b*)] the iterative process converges very fast in this simple example. The results for 3 iteration cycles are shown in Table 1.1.

An alternative procedure is to improve the initial displacements by calculating first the tangential predictor [10]. Calculating the tangent stiffness matrix $\mathbf{K}_T$ for the given $\mathbf{r}_0$, we first calculate the displacements $\mathbf{r}$ by solving the set of equations

$$\mathbf{K}_T\,\mathbf{r} = \mathbf{R}_0. \qquad (j)$$

The result is $\mathbf{r}^T = \{1.5852, 0.1553\}$. The Newton-Raphson iteration then proceeds as shown in Table 1.2.

Table 1.1. Results, Newton-Raphson iteration

Iteration	$\mathbf{N}$ [Eq. (*d*)]	$\mathbf{R}_I$ [Eq. (*e*)]	$\delta\mathbf{R}$ [Eq. (*f*)]	$\delta\mathbf{r}$ [Eq. (*h*)]	$\mathbf{r}$ [Eq. (*i*)]
1	$\begin{Bmatrix} 10.0313 \\ 7.0 \end{Bmatrix}$	$\begin{Bmatrix} 11.4312 \\ 7.2508 \end{Bmatrix}$	$\begin{Bmatrix} -1.4312 \\ -2.2508 \end{Bmatrix}$	$\begin{Bmatrix} -0.1721 \\ -0.1794 \end{Bmatrix}$	$\begin{Bmatrix} 1.8279 \\ 0.3206 \end{Bmatrix}$
2	$\begin{Bmatrix} 9.1523 \\ 4.8766 \end{Bmatrix}$	$\begin{Bmatrix} 10.0437 \\ 5.0233 \end{Bmatrix}$	$\begin{Bmatrix} -0.0437 \\ -0.0233 \end{Bmatrix}$	$\begin{Bmatrix} -0.0072 \\ -0.0009 \end{Bmatrix}$	$\begin{Bmatrix} 1.8207 \\ 0.3197 \end{Bmatrix}$
3	$\begin{Bmatrix} 9.1163 \\ 4.8545 \end{Bmatrix}$	$\begin{Bmatrix} 10.0001 \\ 5.0002 \end{Bmatrix}$	$\begin{Bmatrix} -0.0001 \\ -0.0002 \end{Bmatrix}$	$\begin{Bmatrix} -0.0000 \\ -0.0000 \end{Bmatrix}$	$\begin{Bmatrix} 1.8207 \\ 0.3197 \end{Bmatrix}$

Table 1.2. Results, tangential predictor and Newton-Raphson iteration

Iteration	$\mathbf{N}$ [Eq. (*d*)]	$\mathbf{R}_I$ [Eq. (*e*)]	$\delta\mathbf{R}$ [Eq. (*f*)]	$\delta\mathbf{r}$ [Eq. (*h*)]	$\mathbf{r}$ [Eq. (*i*)]
1	$\begin{Bmatrix} 7.9290 \\ 2.8094 \end{Bmatrix}$	$\begin{Bmatrix} 8.3744 \\ 2.8710 \end{Bmatrix}$	$\begin{Bmatrix} 1.6256 \\ 2.1290 \end{Bmatrix}$	$\begin{Bmatrix} 0.2450 \\ 0.1665 \end{Bmatrix}$	$\begin{Bmatrix} 1.8302 \\ 0.3218 \end{Bmatrix}$
2	$\begin{Bmatrix} 9.1639 \\ 4.8928 \end{Bmatrix}$	$\begin{Bmatrix} 10.0594 \\ 5.0403 \end{Bmatrix}$	$\begin{Bmatrix} -0.0594 \\ -0.0403 \end{Bmatrix}$	$\begin{Bmatrix} -0.0095 \\ -0.0021 \end{Bmatrix}$	$\begin{Bmatrix} 1.8207 \\ 0.3197 \end{Bmatrix}$
3	$\begin{Bmatrix} 9.1163 \\ 4.8545 \end{Bmatrix}$	$\begin{Bmatrix} 10.0001 \\ 5.0002 \end{Bmatrix}$	$\begin{Bmatrix} -0.0001 \\ -0.0002 \end{Bmatrix}$	$\begin{Bmatrix} -0.0000 \\ -0.0000 \end{Bmatrix}$	$\begin{Bmatrix} 1.8207 \\ 0.3197 \end{Bmatrix}$

1.4.2 Material Nonlinearity

Material nonlinearity can arise when the stress-strain relationship of the material is nonlinear in the elastic and/or in the plastic range. In the simplest incremental method the applied forces **R** are divided into increments. The load increments are applied one at a time and an elastic analysis is carried out. For the *i*th load increment $\mathbf{R}_i$, the equilibrium equations $\mathbf{K}_i\ \mathbf{r}_i = \mathbf{R}_i$ are solved. The stiffness matrix $\mathbf{K}_i$ depends upon the stress level reached in the preceding increment. Thus, for the *i*th increment, the modulus of elasticity is the slope of the stress-strain diagram at the stress level reached in the increment i-1. The displacements obtained by the solution of the equilibrium equations for each load increment are summed to give the final displacements. The advantage of the incremental method is its simplicity. It can also be used for geometric nonlinear analysis. For this purpose, the stiffness matrix $\mathbf{K}_i$ for the increment i is based on the geometry of the structure and the internal forces determined in the preceding increment.

The Newton-Raphson or the modified Newton-Raphson methods can also be used to analyze structures with material non-linearity. In the Newton-Raphson method the full load is introduced, and an approximate solution is obtained and corrected by a series of iterations. A new tangent stiffness matrix is used in the solution of the linear equations at each iteration cycle. In the modified Newton-Raphson method the load is introduced in stages. To avoid generating a new stiffness matrix in each iteration cycle, the tangent stiffness matrix determined in the first cycle for each load stage is employed in all subsequent cycles, before proceeding to the next load stage. That is, a new tangent stiffness matrix is introduced only in the first cycle of each new load increment.

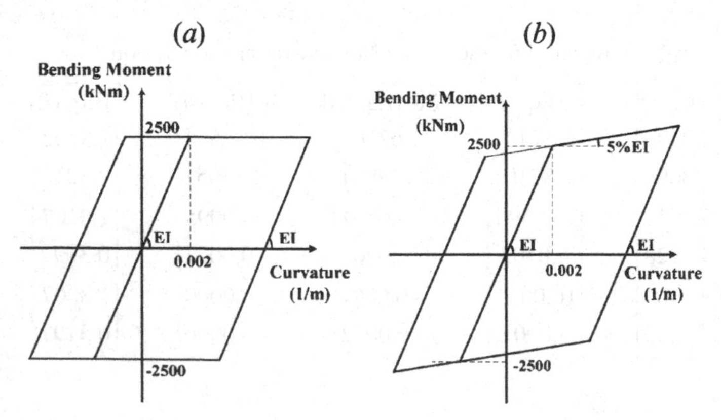

Fig. 1.5. Bi-linear moment-curvature relations

To simplify the analysis, approximate bi-linear relations are often assumed. Two such relations are shown in Fig. 1.5. Elasto-plastic moment-curvature relation with no hardening is shown in Fig. 1.5*a* (the modulus of elasticity is $E = 3\ 10^7 kN/m^2$, the elastic limit stress is $\sigma_Y = 20000 kN/m^2$, and the plastic hinge moment is $M_P = \sigma_Y\, bh^2/4 = 2500\ kNm$). Bi-linear moment-curvature relation with hardening of 5% is shown in Fig. 1.5*b*.

Plastic Analysis

Design of structures based on the plastic approach is increasingly used and has become accepted by various codes of practice. Considering elastic-perfectly-plastic model, the material is assumed to deform in an idealized manner such that the stress and strain are proportional to one another up to the yield stress. At this point the strain increases indefinitely without any increase in stress. For a bilinear moment-curvature relationship in plastic analysis of frames, for example, the structure behaves linearly until the first plastic hinge has developed. Under increasing load, the structure continues to behave linearly, with a reduced stiffness, until a second hinge is formed. The same behavior continues under increasing load until sufficient hinges have developed to form a failure mechanism. An elastic analysis is performed for each load increment, and the structure stiffness matrix is changed accordingly. Collapse is reached when:

- the stiffness matrix becomes singular;
- very large displacements are obtained.

For each load stage, the analysis is carried out for a proportionate increment in all loads. The corresponding internal forces are used to determine a load multiplier, which causes the yield stress to be reached at any one section. The sum of the multipliers in all stages is used to determine the value of the loads at collapse. For each loading stage, we generate the corresponding modified stiffness matrix.

Specifically, for the initial stiffness matrix $\mathbf{K}_1 = \mathbf{K}_0$ and load vector $\mathbf{R}_0$, we first calculate the displacements $\mathbf{r}_0$ by solving $\mathbf{K}_0\, \mathbf{r}_0 = \mathbf{R}_0$. The load vector $\mathbf{R}_0$ is then increased to obtain $\mathbf{R}_1 = \lambda_1\, \mathbf{R}_0$, which is the load that causes the yield stress to be reached at the first section. The displacements at this point are given by $\mathbf{r}_1 = \lambda_1\, \mathbf{r}_0$. The modified stiffness matrix $\mathbf{K}_2 = \mathbf{K}_1 + \Delta\mathbf{K}_1$ is then determined, considering the reduction $\Delta\mathbf{K}_1$ in the stiffness due to yield of the first section. The additional load vector $\mathbf{R}_2 = \lambda_2\, \mathbf{R}_0$ that causes the yield stress to be reached at the second section is determined from the modified equations $\mathbf{K}_2\, \mathbf{r}_2 = \mathbf{R}_2$, where $\mathbf{r}_2$ is the vector of additional displacements. We proceed with the modified stiffness matrix $\mathbf{K}_3 = \mathbf{K}_2 + \Delta\mathbf{K}_2$

considering the reduced stiffness due to yield of the first and the second sections. The additional load vector $\mathbf{R}_3 = \lambda_3\, \mathbf{R}_0$ that causes the yield stress to be reached at the third section is determined in a similar way from the modified equations $\mathbf{K}_3\, \mathbf{r}_3 = \mathbf{R}_3$, where $\mathbf{r}_3$ is the vector of additional displacements. These steps are repeated until collapse of the structure. If collapse occurs after the stress reached the yield stress at m sections, the value of the loads at collapse is given by

$$\mathbf{R}_{\text{collapse}} = \lambda \mathbf{R}_0 = \mathbf{R}_0 \sum_{i=1}^{m} \lambda_i \,, \tag{1.68}$$

where λ is the corresponding load factor.

Example 1.4

To illustrate plastic analysis consider the simple continuous beam, having two equal spans, shown in Fig. 1.6 and subjected to two concentrated loads at the middle of the spans. Denote the uniform plastic moment by M_P, the length of each span by L and the load factor by λ. The bending moments at the three critical sections, where plastic hinges can be formed, are as follows

$$M_1 = \frac{1.75}{16} PL \qquad M_2 = \frac{4.50}{16} PL \qquad M_3 = \frac{5.75}{16} PL \,. \tag{a}$$

In Eq. (a) positive values are assumed when tension is in the bottom fiber for sections 1, 3, and in the top fiber for section 2. Multiplying the loads by λ_1 the first plastic hinge is formed in section 3 when

$$M_3 = \frac{5.75}{16} \lambda_1 PL = M_P \,. \tag{b}$$

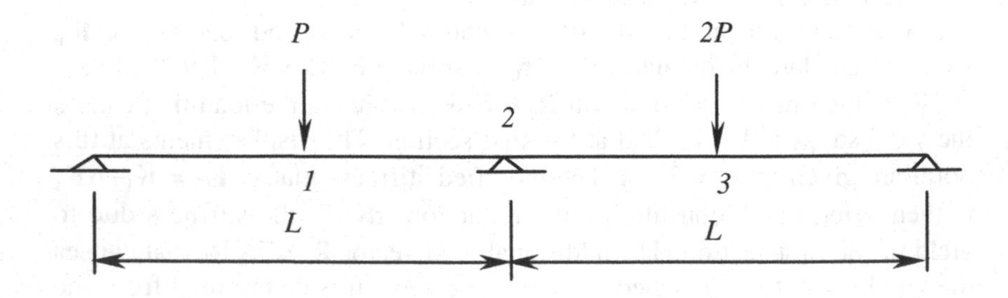

Fig. 1.6. Two-span continuous beam

Rearranging Eq. (*b*), we obtain the load factor λ_1

$$\lambda_1 = \frac{16}{5.75}\frac{M_P}{PL}. \tag{c}$$

Multiplying the terms in Eq. (*a*) by λ_1, we obtain the moments

$$M_1 = \frac{1.75}{5.75}M_P \qquad M_2 = \frac{4.50}{5.75}M_P \qquad M_3 = M_P. \tag{d}$$

The additional loads $\lambda_2 P$, $\lambda_2 2P$ are applied on a statically determinate beam obtained by forming a hinge in section 3. The corresponding additional moments are given by

$$M_1 = -\frac{1}{4}\lambda_2 PL \qquad M_2 = \lambda_2 PL \qquad M_3 = 0. \tag{e}$$

The second plastic hinge is formed in section 2 when [Eqs. (*d*), (*e*)]

$$M_2 = \frac{4.50}{5.75}M_P + \lambda_2 PL = M_P. \tag{f}$$

Solving Eq. (*f*) for λ_2 we find

$$\lambda_2 = \frac{1.25}{5.75}\frac{M_P}{PL}. \tag{g}$$

In this example collapse occurs after the second plastic hinge is formed. Therefore, the load factor λ corresponding to collapse is given by

$$\lambda = \lambda_1 + \lambda_2 = 3\frac{M_P}{PL}. \tag{h}$$

1.4.3 Time Varying Loads

Assume that the externally applied loads are described as a function of time. The time variable t may take on any value from zero to the maximum time of interest. In a static analysis without time effects other than the load level, time is only a convenient variable to describe the load. However, in a dynamic analysis and in static analysis with material time effects, the time variable describes the actual physical situation. In the incremental step-by-step solution we assume that the solution for the discrete time t is known and the solution for the time $t+\Delta t$ is required. The equilibrium conditions at time $t+\Delta t$, considering all nonlinearities, can be expressed as

$$^{t+\Delta t}\mathbf{R}_0 - {}^{t+\Delta t}\mathbf{R}_I = \mathbf{0}, \tag{1.69}$$

where $^{t+\Delta t}\mathbf{R}_0$ is the vector of external loads at time $t+\Delta t$, and $^{t+\Delta t}\mathbf{R}_I$ is the vector of nodal forces that correspond to the element stresses. Since the solution is known at time t, we can write

$$^{t+\Delta t}\mathbf{R}_I = {}^{t}\mathbf{R}_I + \delta\mathbf{R}_I, \tag{1.70}$$

where $\delta\mathbf{R}_I$ is the increment in the forces from time t to time $t+\Delta t$. This vector can be approximated by

$$\delta\mathbf{R}_I = {}^{t}\mathbf{K}\,\delta\mathbf{r}, \tag{1.71}$$

where $\delta\mathbf{r}$ is a vector of incremental nodal displacements and $^{t}\mathbf{K}$ is the tangent stiffness matrix, which corresponds to the geometric and material conditions at time t, given by

$$^{t}\mathbf{K} = \partial\,{}^{t}\mathbf{R}_I / \partial\,\mathbf{r}. \tag{1.72}$$

From Eqs. (1.70), (1.71) we obtain

$$^{t}\mathbf{K}\,\delta\mathbf{r} = {}^{t+\Delta t}\mathbf{R}_0 - {}^{t}\mathbf{R}_I. \tag{1.73}$$

Solving for $\delta\mathbf{r}$, we can calculate the approximate displacements at time $t+\Delta t$ by

$$^{t+\Delta t}\mathbf{r} = {}^{t}\mathbf{r} + \delta\mathbf{r}. \tag{1.74}$$

The exact displacements at time $t+\Delta t$ are those that correspond to the applied loads $^{t+\Delta t}\mathbf{R}_0$. We calculate in Eq. (1.74) only an approximation to these displacements because Eq. (1.71) was used. It is therefore necessary to iterate until the solution of Eq. (1.69) is sufficiently accurate.

1.4.4 Buckling Analysis

In many cases the objective of a nonlinear analysis is to estimate the maximum load that a structure can support prior to structural instability (or collapse). Assuming that the load distribution on the structure is known, the object is to find the load magnitude that the structure can sustain. The linearized buckling analysis presented in this section gives a reasonable estimate of the collapse load only if the collapse displacements are relatively small and the changes in the material properties are close to being linear. Denote the stiffness matrices at times $t - \Delta t$ and t as $^{t-\Delta t}\mathbf{K}$ and $^{t}\mathbf{K}$, respectively, and the corresponding vectors of external loads as $^{t-\Delta t}\mathbf{R}$ and $^{t}\mathbf{R}$. In the linearized buckling analysis we assume that at any time τ

$$^{\tau}\mathbf{K} = {}^{t-\Delta t}\mathbf{K} + \lambda({}^{t}\mathbf{K} - {}^{t-\Delta t}\mathbf{K}), \tag{1.75}$$

$$^{\tau}\mathbf{R} = {}^{t-\Delta t}\mathbf{R} + \lambda({}^{t}\mathbf{R} - {}^{t-\Delta t}\mathbf{R}), \tag{1.76}$$

where λ is a scaling factor, and we are interested in those values of λ that are greater than 1.

At collapse or buckling the tangent stiffness matrix becomes singular, and the condition for calculating the corresponding λ is

$$det\ {}^{\tau}\mathbf{K} = 0, \tag{1.77}$$

or, equivalently

$$^{\tau}\mathbf{K}\ \mathbf{\Phi} = 0, \tag{1.78}$$

where $\mathbf{\Phi}$ is a nonzero vector. Substituting Eq. (1.75) into Eq. (1.78), we obtain the eigenproblem

$$^{t-\Delta t}\mathbf{K}\ \mathbf{\Phi} = \lambda\,({}^{t-\Delta t}\mathbf{K} - {}^{t}\mathbf{K})\ \mathbf{\Phi}. \tag{1.79}$$

The eigenproblem properties and various solution procedures are discussed later in Chap. 2. The eigenvalues $\lambda_1, \lambda_2, \ldots, \lambda_n$ give the buckling loads [see Eq. (1.76)] and the eigenvectors $\mathbf{\Phi}_1, \mathbf{\Phi}_2, \ldots, \mathbf{\Phi}_n$ represent the corresponding buckling modes. The matrices $^{t-\Delta t}\mathbf{K}$ and $^{t}\mathbf{K}$ are both positive definite, but the matrix $^{t-\Delta t}\mathbf{K} - {}^{t}\mathbf{K}$ is indefinite. Therefore, some eigenvalues might be negative, but we are interested in only the smallest positive eigenvalues.

Rewrite Eq. (1.79) as

$$^{\tau}\mathbf{K}\mathbf{\Phi} = \gamma\ {}^{t-\Delta t}\mathbf{K}\mathbf{\Phi}, \tag{1.80}$$

where

$$\gamma = \frac{\lambda - 1}{\lambda}. \tag{1.81}$$

The eigenvalues $\gamma_1, \gamma_2, \ldots$ are all positive and usually only the smallest values are of interest. That is, γ_1 corresponds to the smallest positive value of λ in the problem of Eq. (1.79).

Having evaluated γ_1 and then λ_1 from Eq. (1.81), the buckling (or collapse) load is given by [Eq. (1.76)]

$$\mathbf{R}_{buckling} = {}^{t-\Delta t}\mathbf{R} + \lambda_1({}^{t}\mathbf{R} - {}^{t-\Delta t}\mathbf{R}). \tag{1.82}$$

Similarly, we can evaluate the linearized buckling loads corresponding to γ_2, γ_3, etc. The procedure presented can be used when geometric or material nonlinearities are considered.

1.4.5 Solution of the Nonlinear Equations

The equations used in the Newton-Raphson iteration to solve Eq. (1.69) for the kth iteration are obtained by linearizing the response about the conditions at time $t+\Delta t$ and iteration k–1. The resulting equations are

$$^{t+\Delta t}\mathbf{K}^{(k-1)}\,\delta\mathbf{r}^{(k)} = {}^{t+\Delta t}\mathbf{R}_0 - {}^{t+\Delta t}\mathbf{R}_I^{(k-1)} = \delta\mathbf{R}^{(k-1)}, \tag{1.83}$$

$$\delta\mathbf{r}^{(k)} = {}^{t+\Delta t}\mathbf{r}^{(k)} - {}^{t+\Delta t}\mathbf{r}^{(k-1)}, \tag{1.84}$$

with the initial conditions

$$^{t+\Delta t}\mathbf{r}^{(0)} = {}^{t}\mathbf{r} \qquad {}^{t+\Delta t}\mathbf{K}^{(0)} = {}^{t}\mathbf{K} \qquad {}^{t+\Delta t}\mathbf{R}_I^{(0)} = {}^{t}\mathbf{R}_I. \tag{1.85}$$

In the first iteration Eqs. (1.83), (1.84) reduce to Eqs. (1.73), (1.74). The out-of-balance load vector $\delta\mathbf{R}^{(k-1)}$ corresponds to a load vector that is not yet balanced, and hence an increment in the displacements is required.

The most powerful procedure for reaching convergence of Eqs. (1.83), (1.84) is the *full Newton-Raphson iteration*. Using this procedure, the current tangent stiffness matrix $^{t+\Delta t}\mathbf{K}^{(k-1)}$ is calculated as follows. Denoting the complete solution at time $t+\Delta t$ as $\mathbf{r}$, then the corresponding equilibrium equations can be expressed as

$$\mathbf{f}(\mathbf{r}) = {}^{t+\Delta t}\mathbf{R}_0(\mathbf{r}) - {}^{t+\Delta t}\mathbf{R}_I(\mathbf{r}) = \mathbf{0}. \tag{1.86}$$

Assuming that in the iterative solution we have evaluated $^{t+\Delta t}\mathbf{r}^{(k-1)}$, then a first-order Taylor series approximation gives

$$\mathbf{f}(\mathbf{r}) = \mathbf{f}({}^{t+\Delta t}\mathbf{r}^{(k-1)}) + \left.\frac{\partial \mathbf{f}}{\partial \mathbf{r}}\right|_{^{t+\Delta t}\mathbf{r}^{(k-1)}} (\mathbf{r} - {}^{t+\Delta t}\mathbf{r}^{(k-1)}) = \mathbf{0}. \tag{1.87}$$

Substituting from Eq. (1.86) into Eq. (1.87) and assuming that that the external loads are deformation-independent, we obtain

$$\left.\frac{\partial \mathbf{R}_I}{\partial \mathbf{r}}\right|_{^{t+\Delta t}\mathbf{r}^{(k-1)}} (\mathbf{r} - {}^{t+\Delta t}\mathbf{r}^{(k-1)}) = {}^{t+\Delta t}\mathbf{R}_0 - {}^{t+\Delta t}\mathbf{R}_I^{(k-1)}. \tag{1.88}$$

The current tangent stiffness matrix, $^{t+\Delta t}\mathbf{K}^{(k-1)}$, is given by

$$^{t+\Delta t}\mathbf{K}^{(k-1)} = \left.\frac{\partial \mathbf{R}_I}{\partial \mathbf{r}}\right|_{^{t+\Delta t}\mathbf{r}^{(k-1)}}. \tag{1.89}$$

Using Eqs. (1.84), (1.89) and substituting into Eq. (1.88), we obtain the full Newton-Raphson equations for calculating $\delta\mathbf{r}^{(k)}$ [Eq. (1.83)]

$$ {}^{t+\Delta t}\mathbf{K}^{(k-1)}\,\delta\mathbf{r}^{(k)} = {}^{t+\Delta t}\mathbf{R}_0 - {}^{t+\Delta t}\mathbf{R}_I^{(k-1)}\,. \tag{1.90} $$

The improved displacement solution, ${}^{t+\Delta t}\mathbf{r}^{(k)}$, is given by [Eq. (1.84)]

$$ {}^{t+\Delta t}\mathbf{r}^{(k)} = {}^{t+\Delta t}\mathbf{r}^{(k-1)} + \delta\mathbf{r}^{(k)}. \tag{1.91} $$

The relations in Eqs. (1.89), (1.91) constitute the Newton-Raphson solution of Eq. (1.69). In this iteration an incremental analysis is performed with time (or load) steps of size Δt and the initial conditions of Eq. (1.85). In the full Newton-Raphson iteration, a new tangent stiffness matrix is calculated in each iteration cycle. In general, the major computational cost per iteration lies in the calculation and factorization of the tangent stiffness matrix. Since these calculations can be expensive when large-order systems are considered, the use of modified algorithms can be effective.

In the *initial stress method* we use the initial stiffness matrix ${}^{0}\mathbf{K}$ in Eq. (1.83)

$$ {}^{0}\mathbf{K}\,\delta\mathbf{r}^{(k)} = {}^{t+\Delta t}\mathbf{R}_0 - {}^{t+\Delta t}\mathbf{R}_I^{(k-1)}\,. \tag{1.92} $$

In this case only the matrix ${}^{0}\mathbf{K}$ is factorized, thus avoiding multiple calculations and factorizations of the coefficient matrix in Eq. (1.83). This method corresponds to a linearization of the response about the initial configuration of the system and may converge very slowly and even diverge.

In the *modified Newton-Raphson iteration*, an approach somewhat in between the full Newton-Raphson iteration and the initial stress method is employed. In this method we use a stiffness matrix ${}^{\tau}\mathbf{K}$ that corresponds to one of the accepted equilibrium configurations at times 0, Δt, $2\Delta t$, ... or t. Thus, instead of solving Eq. (1.83), we solve

$$ {}^{\tau}\mathbf{K}\,\delta\mathbf{r}^{(k)} = {}^{t+\Delta t}\mathbf{R}_0 - {}^{t+\Delta t}\mathbf{R}_I^{(k-1)}\,. \tag{1.93} $$

The modified Newton-Raphson iteration involves fewer stiffness reformations than the full Newton-Raphson iteration and bases the stiffness matrix update on an accepted equilibrium configuration.

Quasi-Newton methods, or matrix update methods [12], have been developed as an alternative to the Newton-Raphson method for iteration on nonlinear systems of equations. These methods involve updating the inverse of the coefficient matrix to provide a secant approximation to the matrix. The quasi-Newton methods provide a compromise between the full reformulation of the stiffness matrix performed in the full Newton-Raphson method and the use of a previous stiffness matrix as is done in the modified Newton-Raphson method.

In the BFGS (Broyden-Fletcher-Goldfarb-Shanno) method [13], a search direction can be computed without explicitly calculating the updated matrices or performing any additional costly matrix factorizations as required in the full Newton-Raphson method. Line searches can also be used in the Newton-Raphson method. In this case the expense of the iterations increases, but less iteration cycles may be needed for convergence.

Load-Displacement-Constraint Methods

In nonlinear analysis it is frequently required to calculate the collapse load of a structure. Consider, for simplicity of presentation, a single displacement. For very small loads the load-displacement response is usually linear. As the load increases, the response becomes increasingly nonlinear until the collapse load is reached. The response beyond this point is referred to as post collapse or post buckling response. In some structures, the load first decreases in this regime and then increases again as the displacement increases. To calculate the response, we can employ initially relatively large load increments. As the collapse of the model is approached, the load increments must become smaller. At the collapse point the stiffness matrix is singular, and beyond that point special solution procedures, such as a load-displacement-constraint method [14], must be used. In this method we assume that the load vector varies proportionally during the response calculation. We introduce a load multiplier that increases or decreases the intensity of the applied loads, to obtain fast convergence in each load step and to be able to transverse the collapse point and evaluate the post collapse response.

The equilibrium conditions to be solved at time $t + \Delta t$ are

$$ {}^{t+\Delta t}\lambda \, \mathbf{R}_0 - {}^{t+\Delta t}\mathbf{R}_I = \mathbf{0}, \tag{1.94}$$

where ${}^{t+\Delta t}\lambda$ is a scalar load multiplier to be determined, $\mathbf{R}_0$ is the reference load vector, and ${}^{t+\Delta t}\mathbf{R}_I$ is the vector of nodal forces that correspond to the element stresses. Since Eq. (1.94) represents n equations in n+1 unknowns, we need an additional equation to determine the load multiplier.

Consider the following equations to be solved [see Eq. (1.93)]

$$ {}^{\tau}\mathbf{K} \, \delta\mathbf{r}^{(k)} = \left({}^{t+\Delta t}\lambda^{(k\text{-}1)} + \Delta\lambda^{(k)} \right) \mathbf{R}_0 - {}^{t+\Delta t}\mathbf{R}_I^{(k-1)} \, . \tag{1.95}$$

The additional equation is a constraint equation between $\Delta\lambda^{(k)}$ and $\delta\mathbf{r}^{(k)}$, of the form

$$ f(\Delta\lambda^{(k)}, \delta\mathbf{r}^{(k)}) = 0. \tag{1.96}$$

For a given load step we define

$$\mathbf{r}^{(k)} = {}^{t+\Delta t}\mathbf{r}^{(k)} - {}^{t}\mathbf{r}, \tag{1.97}$$

$$\lambda^{(k)} = {}^{t+\Delta t}\lambda^{(k)} - {}^{t}\lambda. \tag{1.98}$$

That is, $\mathbf{r}^{(k)}$, $\lambda^{(k)}$ are total increments in displacements and load multiplier, up to iteration k, within the load step. Effective constraint equations are given by the spherical constant arc length criterion [15, 16] or the scheme of constant increment of external work [17].

References

1. Bathe KJ (1996) Finite element procedures. Prentice Hall, NJ
2. Ghali A, Neville AM (1997). Structural analysis. E & FN SPON, London
3. Weaver WJr, Gere JM (1980) Analysis of framed structures, 2nd edn. Van Nostrand Reinhold, New York
4. Hughes TJR (1987) The finite element method - linear static and dynamic finite element analysis. Prentice Hall, NJ
5. Ortega JM, Rheinboldt WC (1970) Iterative solutions of nonlinear equations in several variables. Academic Press, New York
6. Hestenes MR, Stiefel E (1952) Methods of conjugate gradients for solving linear systems. J of Research of the National Bureau of Standards 49:409-436
7. Fletcher R, Reeves CM (1964) Function minimization by conjugate gradients. Comput J 7: 149.
8. Leunberger DG (1984) Introduction to linear and nonlinear programming. Addison-Wesley, Reading
9. Golub GH, Van Loan CF (1996) Matrix computations, 3rd edn. The Johns Hopkins University Press, Baltimore
10. Crisfield MA (1997) Nonlinear finite element analysis of solids and structures, Vol. 1: essentials. John Wiley &Sons, Chichester
11. Levy R, Spillers RS (2003) Analysis of geometrically nonlinear structures, 2nd edn. Kluwer Academic Publishers, Dordrecht
12. Dennis JEJr (1976) A brief survey of convergence results for quasi-Newton methods. SIAM-AMS Proceedings 10:185-200
13. Mathies H, Strang G (1979) The solution of finite element nonlinear equations, Int J Num Meth Engrg 14:1613-1626
14. Riks E (1979) An incremental approach to the solution of snapping and buckling problems. Int J Sol & Str 15:529-551
15. Crisfield MA (1981) A fast incremental/iterative solution procedure that handles snap-through. Computers & Structures 13:55-62
16. Ramm E (1981) Strategies for tracing nonlinear response near limit points, in nonlinear finite element analysis in structural mechanics. In Wunderlich W et al (eds) Springer Verlag, New York

17. Bathe KJ, Dvorkin EN (1983) On the automatic solution of nonlinear finite element equations. Computers & Structures 17:871-879

2 Vibration Analysis

2.1 Free Vibration

The purpose of dynamic analysis is to determine internal forces, stresses and displacements under application of dynamic (time varying) loads. In general, the structural response to any dynamic loading is expressed in terms of the displacements of the structure. Dynamic analysis is discussed in detail in various texts [e.g. 1–3]. Some basics of this topic are presented later in Chap. 3. The eigenproblem is to find the free-vibration frequencies and the mode shapes of the vibrating system. In this chapter eigenproblem analysis is introduced.

A dynamic response calculation is substantially more costly than a static analysis. Whereas in a static analysis the solution is obtained in one step, in dynamic analysis the solution is required at a number of discrete points over the time interval considered. The equations of motion for a multiple degrees of freedom system subjected to external dynamic forces are

$$\mathbf{M}\ddot{\mathbf{r}}(t) + \mathbf{C}\dot{\mathbf{r}}(t) + \mathbf{K}\mathbf{r}(t) = \mathbf{R}(t), \tag{2.1}$$

where $\mathbf{M}$ is the mass matrix, $\mathbf{C}$ is the damping matrix and $\mathbf{K}$ is the stiffness matrix. The unknown displacement vector $\mathbf{r}(t)$, the velocity vector $\dot{\mathbf{r}}(t)$, the acceleration vector $\ddot{\mathbf{r}}(t)$, and the load vector $\mathbf{R}(t)$ are functions of the time variable t. Eq. (2.1) may be written as

$$\mathbf{F}_I(t) + \mathbf{F}_D(t) + \mathbf{F}_R(t) = \mathbf{R}(t), \tag{2.2}$$

where the terms $\mathbf{F}_I(t)$, $\mathbf{F}_D(t)$, $\mathbf{F}_R(t)$, in the left-hand side of Eq. (2.2), represent the inertia forces, the damping forces and the resisting (elastic or inelastic) forces, respectively.

An elastic structure disturbed from its equilibrium condition by the application and removal of forces will oscillate about its position of static equilibrium. Thus, the displacements will vary periodically between specific limits in either direction. The distance of either of these limits from the position of equilibrium is the *amplitude of the vibration*. We may distinguish between the following two types of motion:

- *Free-vibration motion*, where no external forces or support motion act on the structure, and the motion may continue with the same amplitude for an indefinitely long time.
- *Damped free-vibration*, where forces tending to oppose the motion act on the structure. In practice there are always such forces, which cause the amplitude to diminish gradually until the motion ceases.

Considering a structure with no damping and ignoring the notation of the time variable (t), we obtain the equations of motion for a freely vibrating undamped system by omitting the load vector from Eq. (2.1)

$$\mathbf{M}\ddot{\mathbf{r}} + \mathbf{K}\mathbf{r} = \mathbf{0}. \tag{2.3}$$

The problem of *vibration analysis* consists of determining the conditions under which the equilibrium conditions (2.3) are satisfied.

Assuming that the free-vibration motion is simple harmonic, we find

$$\ddot{\mathbf{r}} = -\omega^2 \mathbf{r}, \tag{2.4}$$

where ω is the circular frequency. Substituting Eq. (2.4) into Eq. (2.3) and rearranging gives the eigenproblem

$$\mathbf{K}\mathbf{r} = \omega^2 \mathbf{M}\,\mathbf{r} = \lambda \mathbf{M}\,\mathbf{r}, \tag{2.5}$$

where the quantities $\lambda = \omega^2$ are the *eigenvalues*, indicating the square of the free-vibration frequencies, while the corresponding displacement vectors $\mathbf{r}$ express the *eigenvectors*, or *natural mode shapes* of the vibrating system. For a system having n degrees of freedom, the frequency vector $\boldsymbol{\omega}^T = \{\omega_1, \omega_2, \omega_3, ..., \omega_n\}$ represents the frequencies of the n modes of vibration possible in the system. The mode having the lowest frequency is called the first mode; the next higher frequency is the second mode, etc. If the solution is considered in order to obtain eigenvalues and eigenvectors, the problem is referred to as an *eigenproblem*, whereas if only eigenvalues are to be calculated, the problem is called an *eigenvalue problem*. The solution of the eigenproblem for large structures is often the most costly phase of a dynamic response analysis, and calculation of the eigenvalues and eigenvectors requires much computational effort. Solution of the eigenproblem is discussed in Sects. 2.2–2.8.

The shape of the vibrating system can be determined by solving for all the displacements in terms of any one coordinate. For convenience the displacement vector associated with the ith mode of vibration is often expressed in dimensionless form by dividing all the components by a reference component (e.g. the largest). The resulting ith mode shape $\boldsymbol{\Phi}_i$ is

$$\mathbf{\Phi}_i = \begin{Bmatrix} \Phi_{1i} \\ \Phi_{2i} \\ \vdots \\ \Phi_{ni} \end{Bmatrix} = \begin{Bmatrix} 1 \\ \Phi_{2i} \\ \vdots \\ \Phi_{ni} \end{Bmatrix}, \tag{2.6}$$

where Φ_{1i} is taken as the reference component. The square matrix made up of the n mode shapes, called the *modal matrix*, is represented by $\mathbf{\Phi}$

$$\mathbf{\Phi} = [\mathbf{\Phi}_1, \mathbf{\Phi}_2, ..., \mathbf{\Phi}_n]. \tag{2.7}$$

With the above definitions the problem of Eq. (2.5), called the *generalized eigenproblem*, is expressed as

$$\mathbf{K\Phi} = \mathbf{\Lambda M\Phi}. \tag{2.8}$$

The diagonal matrix of the eigenvalues, $\mathbf{\Lambda}$, is known as the *spectral matrix.*

2.1.1 Properties of the Eigenproblem

Some properties of the eigenproblem are discussed in the following. The natural modes can be shown to satisfy the following *orthogonality conditions* for any different mode shapes $\mathbf{\Phi}_i, \mathbf{\Phi}_j$

$$\mathbf{\Phi}_i^T \mathbf{M\Phi}_j = 0 \qquad \mathbf{\Phi}_i^T \mathbf{K\Phi}_j = 0. \tag{2.9}$$

The orthogonality of natural modes implies that the following square matrices are diagonal

$$\tilde{\mathbf{M}} = \mathbf{\Phi}^T \mathbf{M\Phi} \qquad \tilde{\mathbf{K}} = \mathbf{\Phi}^T \mathbf{K\Phi}, \tag{2.10}$$

where $\mathbf{\Phi}$ is the modal matrix [Eq. (2.7)]. The diagonal elements of $\tilde{\mathbf{M}}$ and $\tilde{\mathbf{K}}$ are given by

$$M_i = \mathbf{\Phi}_i^T \mathbf{M\Phi}_i \qquad K_i = \mathbf{\Phi}_i^T \mathbf{K\Phi}_i. \tag{2.11}$$

These diagonal elements are related by [see Eq. (2.8)]

$$K_i = \lambda_i M_i. \tag{2.12}$$

The matrices of Eq. (2.10) are positive definite, that is, the diagonal elements of both matrices are positive.

Consider the eigenpair λ_i, $\mathbf{\Phi}_i$ satisfying

$$\mathbf{K\Phi}_i = \lambda_i \mathbf{M\Phi}_i. \tag{2.13}$$

It is common to normalize modes so that the M_i have unit values

$$M_i = \mathbf{\Phi}_i^T \mathbf{M\Phi}_i = 1. \tag{2.14}$$

This fixes the lengths of the eigenvectors. The natural modes normalized by Eq. (2.14) are a *mass orthogonal set*. When the modes are normalized by this equation we obtain from Eq. (2.12)

$$K_i = \lambda_i. \tag{2.15}$$

Thus, the eigenvectors satisfy the orthogonality conditions

$$\mathbf{\Phi}_i^T \mathbf{M\Phi}_j = \delta_{ij} \qquad \mathbf{\Phi}_i^T \mathbf{K\Phi}_j = \lambda_i \delta_{ij}, \tag{2.16}$$

where δ_{ij} is the Kronecker delta, for which $\delta_{ij} = 0$ ($i \neq j$) and $\delta_{ii} = 1$. Using these relations, we may write the following conditions that the eigenvectors must satisfy

$$\mathbf{\Phi}^T \mathbf{M\Phi} = \mathbf{I}, \tag{2.17}$$

$$\mathbf{\Phi}^T \mathbf{K\Phi} = \mathbf{\Lambda}, \tag{2.18}$$

where **I** is the identity matrix and **Λ** is the spectral matrix.

2.1.2 The Standard Eigenproblem and the Rayleigh Quotient

Consider a *standard eigenproblem,* defined as

$$\mathbf{AV} = \lambda\, \mathbf{V}, \tag{2.19}$$

where matrix **A** is symmetric and **V** represents the eigenvectors. A change of basis is performed by using the *similarity transformation*

$$\mathbf{V} = \mathbf{P}\,\mathbf{v}, \tag{2.20}$$

where **P** is an orthogonal matrix, for which

$$\mathbf{P}^T = \mathbf{P}^{-1}, \tag{2.21}$$

and **v** represents the solution vector in the new basis. Substituting Eq. (2.20) into Eq. (2.19), pre-multiplying both sides of the resulting equation by $\mathbf{P}^T$ and using Eq. (2.21), we obtain the *generalized eigenproblem*

$$\mathbf{a}\,\mathbf{v} = \lambda\,\mathbf{b}\,\mathbf{v}, \tag{2.22}$$

where

$$\mathbf{a} = \mathbf{P}^T\mathbf{A}\,\mathbf{P} \qquad \mathbf{b} = \mathbf{P}^T\,\mathbf{P}. \tag{2.23}$$

Since matrix $\mathbf{P}$ is orthogonal $\mathbf{b} = \mathbf{I}$ and the generalized eigenproblem of Eq. (2.22) is reduced to the standard eigenproblem

$$\mathbf{a}\,\mathbf{v} = \lambda\,\mathbf{v}. \tag{2.24}$$

It can be proved that the problem of Eq. (2.24) has the same eigenvalues as the problem of Eq. (2.19), whereas the eigenvectors are related by Eq. (2.20).

The *Rayleigh quotient* $\rho(\mathbf{v})$ is defined as

$$\rho(\mathbf{v}) = \frac{\mathbf{v}^T\mathbf{a}\,\mathbf{v}}{\mathbf{v}^T\,\mathbf{v}}, \tag{2.25}$$

where

$$\lambda_1 \leq \rho(\mathbf{v}) \leq \lambda_n. \tag{2.26}$$

If $\mathbf{a}$ is positive definite then

$$0 < \rho(\mathbf{v}). \tag{2.27}$$

For any vector $\mathbf{v}$ the minimum of $\rho(\mathbf{v})$ will be reached when

$$\rho(\mathbf{v}) = \rho(\mathbf{v}_1) = \lambda_1. \tag{2.28}$$

Consider the generalized eigenproblem

$$\mathbf{K\Phi} = \lambda\mathbf{M\Phi}, \tag{2.29}$$

where $\mathbf{K}$, $\mathbf{M}$ are positive definite, which ensures that the eigenvalues are all positive, that is $\lambda_1 > 0$. The Rayleigh minimum principle states that

$$\lambda_1 = min\,\rho(\mathbf{\Phi}), \tag{2.30}$$

where the minimum is taken over all possible vectors $\mathbf{\Phi}$, and $\rho(\mathbf{\Phi})$ is the Rayleigh quotient

$$\rho(\mathbf{\Phi}) = \frac{\mathbf{\Phi}^T\mathbf{K\Phi}}{\mathbf{\Phi}^T\mathbf{M\Phi}}. \tag{2.31}$$

The bounds on the Rayleigh quotient are $0 < \lambda_1 \leq \rho(\mathbf{\Phi}) \leq \lambda_n$.

2.2 Solution of the Eigenproblem

2.2.1 Static Condensation

In dynamic analysis, those degrees of freedom that are not required to appear in the global finite element model can be eliminated by static condensation. In the calculation of frequencies and mode shapes, the basic assumption is that the mass of the structure can be lumped at only some specific degrees of freedom without much effect on the accuracy of the results. To illustrate reduction of the number of degrees of freedom, consider first the static equilibrium equations $\mathbf{K}\,\mathbf{r} = \mathbf{R}$. We assume that the stiffness matrix and corresponding displacement and load vectors can be partitioned into the form

$$\begin{bmatrix} \mathbf{K}_{aa} & \mathbf{K}_{ab} \\ \mathbf{K}_{ba} & \mathbf{K}_{bb} \end{bmatrix} \begin{Bmatrix} \mathbf{r}_a \\ \mathbf{r}_b \end{Bmatrix} = \begin{Bmatrix} \mathbf{R}_a \\ \mathbf{R}_b \end{Bmatrix} . \tag{2.32}$$

Using the second sub-matrix equation in Eq. (2.32), we obtain

$$\mathbf{r}_b = -\mathbf{K}_{bb}^{-1}(\mathbf{R}_b - \mathbf{K}_{ba}\mathbf{r}_a) . \tag{2.33}$$

Substituting Eq. (2.33) into the first sub-matrix equation in Eq. (2.32) we obtain the condensed equations

$$[\mathbf{K}_{aa} - \mathbf{K}_{ab}\mathbf{K}_{bb}^{-1}\mathbf{K}_{ba})]\mathbf{r}_a = \mathbf{R}_a - \mathbf{K}_{ab}\mathbf{K}_{bb}^{-1}\mathbf{R}_b . \tag{2.34}$$

The name static condensation refers to dynamic analysis, where it is often possible to reduce the size of the system before solving the eigenproblem by eliminating the mass-less degrees of freedom. Considering now the generalized eigenproblem and using partitioning, we rewrite the equation $\mathbf{K}\boldsymbol{\Phi} = \lambda\,\mathbf{M}\boldsymbol{\Phi}$ as

$$\begin{bmatrix} \mathbf{K}_{aa} & \mathbf{K}_{ab} \\ \mathbf{K}_{ba} & \mathbf{K}_{bb} \end{bmatrix} \begin{Bmatrix} \boldsymbol{\Phi}_a \\ \boldsymbol{\Phi}_b \end{Bmatrix} = \lambda \begin{bmatrix} 0 & 0 \\ 0 & \mathbf{M}_b \end{bmatrix} \begin{Bmatrix} \boldsymbol{\Phi}_a \\ \boldsymbol{\Phi}_b \end{Bmatrix} . \tag{2.35}$$

The first sub-matrix equation is

$$\mathbf{K}_{aa}\boldsymbol{\Phi}_a + \mathbf{K}_{ab}\boldsymbol{\Phi}_b = \mathbf{0} . \tag{2.36}$$

Therefore, the mass-less degrees of freedom $(\boldsymbol{\Phi}_a)$ are related to the degrees of freedom at the mass points $(\boldsymbol{\Phi}_b)$ by

$$\boldsymbol{\Phi}_a = -\mathbf{K}_{aa}^{-1}\mathbf{K}_{ab}\boldsymbol{\Phi}_b . \tag{2.37}$$

Substituting Eq. (2.37) into the second sub-matrix equation in Eq. (2.35), the resulting eigenproblem is reduced to the form

$$[\mathbf{K}_{bb} - \mathbf{K}_{ba}\mathbf{K}_{aa}^{-1}\mathbf{K}_{ab}]\mathbf{\Phi}_b = \lambda \mathbf{M}_b \mathbf{\Phi}_b . \tag{2.38}$$

One important disadvantage of the static condensation approach is that the reduced stiffness matrix $[\mathbf{K}_{bb} - \mathbf{K}_{ba}\mathbf{K}_{aa}^{-1}\mathbf{K}_{ab}]$ tends to fill as more massless degrees of freedom are eliminated. Therefore the reduction in size of the system may not be economical from a computational viewpoint.

2.2.2 Solution Methods

Solution of the eigenproblem requires considerably more effort than a static analysis. Since exact solution can be prohibitively expensive, approximate solution techniques have been developed for large scale systems. In such systems, the object is often to calculate the smallest eigenvalues and the corresponding eigenvectors.

In general, no explicit formulae are available, and all solution methods must be iterative in nature. To find an eigenpair λ_i, $\mathbf{\Phi}_i$, only one of them is calculated by iteration; the other can be obtained without further iteration. For example, if λ_i is obtained by iteration, then $\mathbf{\Phi}_i$ can be evaluated without iteration by solving the equations $(\mathbf{K} - \lambda_i \mathbf{M})\,\mathbf{\Phi}_i = \mathbf{0}$. On the other hand, if $\mathbf{\Phi}_i$ is determined by iteration, λ_i can be calculated by the Rayleigh quotient [Eqs. (2.30), (2.31)]

$$\lambda_i = \frac{\mathbf{\Phi}_i^T \mathbf{K} \mathbf{\Phi}_i}{\mathbf{\Phi}_i^T \mathbf{M} \mathbf{\Phi}_i} . \tag{2.39}$$

A basic question in considering an effective solution method is whether we should first solve for the eigenvalue and then calculate the eigenvector, or vice versa, or whether it is more economical to solve for both simultaneously. The answer to this question depends on various properties of the problem under consideration. The effectiveness of a solution method depends on the possibility of a reliable use of the procedure and the cost of solution, determined by the number of high-speed storage operations and an efficient use of backup storage devices.

The common solution methods can be subdivided into the following groups, corresponding to the properties used in the solution process [1, 4]:

- *Vector iteration methods* (e.g. inverse iteration), which work directly on the property of Eq. (2.13)

$$\mathbf{K}\mathbf{\Phi}_i = \lambda_i \mathbf{M}\mathbf{\Phi}_i \, . \tag{2.40}$$

- *Transformation methods* (e.g. Jacobi iteration), which use the basic orthogonality properties of the eigenvectors [Eqs. (2.17), (2.18)]

$$\mathbf{\Phi}^T \mathbf{M} \mathbf{\Phi} = \mathbf{I} \, , \tag{2.41}$$

$$\mathbf{\Phi}^T \mathbf{K} \mathbf{\Phi} = \mathbf{\Lambda} \, , \tag{2.42}$$

  where $\mathbf{\Phi}$ is the modal matrix [Eq. (2.7)], $\mathbf{\Lambda}$ is the spectral matrix (a diagonal matrix of the eigenvalues) and $\mathbf{I}$ is the identity matrix.
- *Polynomial iteration methods*, which work on the fact that

$$p(\lambda_i) = 0, \tag{2.43}$$

  where

$$p(\lambda) = det\,(\mathbf{K} - \lambda \mathbf{M}). \tag{2.44}$$

- Methods that employ the *Sturm sequence property* of the characteristic polynomials of Eq. (2.44)

$$p^{(r)}(\lambda^{(r)}) = \det\,(\mathbf{K}^{(r)} - \lambda^{(r)}\mathbf{M}^{(r)}) \qquad r = 1,\ldots, n\text{–}1, \tag{2.45}$$

  where $p^{(r)}(\lambda^{(r)})$ is the characteristic polynomial of the rth associated problem $\mathbf{K}^{(r)}\mathbf{\Phi}^{(r)} = \lambda^{(r)}\mathbf{M}^{(r)}\mathbf{\Phi}^{(r)}$. All matrices are of order $n-r$, and $\mathbf{K}^{(r)}$, $\mathbf{M}^{(r)}$ are obtained by deleting the last r rows and columns from **K**, **M**.
- There are many variants of these procedures. The *Lanczos method* and the *subspace iteration* method use a combination of the properties used in the above methods.

In this chapter only several methods are described. Vector iteration methods are presented in Sect. 2.3 and transformation methods are discussed in Sect. 2.4. Polynomial iterations are introduced in Sect. 2.5, Rayleigh-Ritz analysis is described in Sect. 6, the Lanczos method is presented in Sect. 2.7 and solution by subspace iteration is demonstrated in Sect. 2.8.

2.3 Vector Iteration Methods

2.3.1 Inverse Vector Iteration

In structural engineering we often analyze systems with narrowly banded stiffness matrix **K** and diagonal or narrowly banded mass matrix **M** subjected to excitations that excite primarily the lower natural modes of vibrations. Inverse vector iteration methods are usually effective for such cases.

Many of the eigenproblem solution techniques are based on the vector iteration approach. The use of iteration to evaluate the vibration mode of a structure is a very old concept that originally was called the Stodola method. To calculate the first-mode shape Eq. (2.5) can be rewritten in an iterative form. Since only thc shape is needed, the frequency is dropped from this equation to obtain

$$\mathbf{K}\overline{\mathbf{r}}^{(k)} = \mathbf{M}\mathbf{r}^{(k-1)}, \tag{2.46}$$

where k denotes the iteration number, $\mathbf{r}^{(k-1)}$ is the displacement vector in the previous iteration and $\overline{\mathbf{r}}^{(k)}$ is the resulting improved shape. To initiate the iteration procedure for evaluating the first mode shape, a trial displacement vector $\mathbf{r}^{(0)}$ is assumed that is a reasonable estimate of this shape. The improved iteration vector is then obtained by normalizing the shape $\overline{\mathbf{r}}^{(k)}$. There are various ways to obtain convenient normalized vectors. We can normalize the shape $\overline{\mathbf{r}}^{(k)}$ by

$$\mathbf{r}^{(k)} = \frac{\overline{\mathbf{r}}^{(k)}}{\left(\overline{\mathbf{r}}^{(k)T}\mathbf{M}\overline{\mathbf{r}}^{(k)}\right)^{1/2}}. \tag{2.47}$$

This equation assures that the new vector satisfies the mass orthogonality relation [see Eq. (2.17)]

$$\mathbf{r}^{(k)T}\mathbf{M}\mathbf{r}^{(k)} = 1. \tag{2.48}$$

Alternatively, normalizing the shape $\overline{\mathbf{r}}^{(k)}$ by dividing it by an arbitrary reference element of the vector, $ref(\overline{\mathbf{r}}^{(k)})$, gives

$$\mathbf{r}^{(k)} = \frac{\overline{\mathbf{r}}^{(k)}}{ref\left(\overline{\mathbf{r}}^{(k)}\right)}. \tag{2.49}$$

This operation has the effect of scaling the reference element to unity. In general, the vector is normalized with respect to its largest element.

It has been proven that the iteration process converges to the first-mode shape. The convergence is linear, thus if $\lambda_1 < \lambda_2$, the relative magnitude λ_1/λ_2 determines the rate of convergence; if $\lambda_1 = \lambda_2$ the rate of convergence is given by the ratio of λ_1 to the next distinct eigenvalue.

By repeating the process sufficiently, we can improve the mode-shape approximation to any desired level of accuracy. The eigenvalue can be estimated by the Rayleigh quotient [Eq. (2.39)]

$$\lambda^{(k)} = \frac{\overline{\mathbf{r}}^{(k)T}\mathbf{K}\overline{\mathbf{r}}^{(k)}}{\overline{\mathbf{r}}^{(k)T}\mathbf{M}\overline{\mathbf{r}}^{(k)}} = \frac{\overline{\mathbf{r}}^{(k)T}\mathbf{M}\mathbf{r}^{(k-1)}}{\overline{\mathbf{r}}^{(k)T}\mathbf{M}\overline{\mathbf{r}}^{(k)}}. \tag{2.50}$$

The approximation of this equation can be used to determine convergence of the iteration by comparing two successive values of λ.

Before starting the iteration procedure, the stiffness matrix $\mathbf{K}$ is often factorized, taking advantage of its narrow banded character. Equations (2.46), (2.47) state the basic iteration algorithm. However, it is more efficient to use the following iteration steps, starting with $\mathbf{r}^{(0)}$, $\mathbf{Y}^{(0)} = \mathbf{M}\,\mathbf{r}^{(0)}$,

$$\mathbf{K}\overline{\mathbf{r}}^{(k)} = \mathbf{Y}^{(k-1)}, \tag{2.51}$$

$$\overline{\mathbf{Y}}^{(k)} = \mathbf{M}\overline{\mathbf{r}}^{(k)}, \tag{2.52}$$

$$\rho(\overline{\mathbf{r}}^{(k)}) = \frac{\overline{\mathbf{r}}^{(k)T}\mathbf{Y}^{(k-1)}}{\overline{\mathbf{r}}^{(k)T}\overline{\mathbf{Y}}^{(k)}}, \tag{2.53}$$

$$\mathbf{Y}^{(k)} = \frac{\overline{\mathbf{Y}}^{(k)}}{(\overline{\mathbf{r}}^{(k)T}\overline{\mathbf{Y}}^{(k)})^{1/2}}. \tag{2.54}$$

Inverse iteration can be used to evaluate higher-order modes as well, by assuming shapes that contain no lower-mode components. Calculating the second mode shape $\mathbf{\Phi}_2$, we start with an arbitrary $\mathbf{r}$ and make it orthogonal to $\mathbf{\Phi}_1$ by the Gram-Schmidt orthogonalization presented later in Sect. 2.3.3. This process can also be used to orthogonalize a trial vector with respect to several eigenvectors that already have been determined. However, it should be noted that numerical problems may occur due to round-off errors, and convergence of the iteration process becomes slower for the higher mode shapes.

Example 2.1

To illustrate solution by the inverse vector iteration method, consider the three-bay eight-story frame shown in Fig. 2.1. The mass of the frame is lumped in the girders, with values M_1=1.0, M_2=1.5, M_3=2.0. The girders are assumed to be non-deformable and the lateral stiffness of each story is $EI/L^3 = 5.0$.

The initial displacement vector is assumed as

$$\mathbf{r}^{(0)T} = [1, 1, 1, 1, 1, 1, 1, 1]. \qquad (a)$$

The mode shapes obtained by the procedure of Eqs. (2.46), (2.49) for five iteration cycles and the exact solution $\mathbf{r}(\text{exact}) = \mathbf{\Phi}_1$ are shown in Table 2.1. The corresponding eigenvalue is $\lambda_1 = 1.5718$.

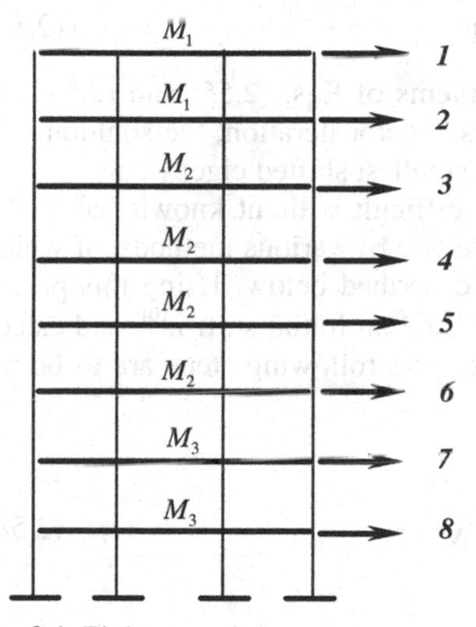

Fig. 2.1. Eight-story frame

Table 2.1. First mode shape, inverse vector iteration, five iteration cycles

$\mathbf{r}^{(1)}$	$\mathbf{r}^{(2)}$	$\mathbf{r}^{(3)}$	$\mathbf{r}^{(4)}$	$\mathbf{r}^{(5)}$	$\mathbf{r}$(exact)
1.0000	1.0000	1.0000	1.0000	1.0000	1.0000
0.9792	0.9747	0.9739	0.9738	0.9738	0.9738
0.9375	0.9245	0.9224	0.9221	0.9221	0.9221
0.8646	0.8388	0.8348	0.8342	0.8342	0.8342
0.7604	0.7202	0.7143	0.7136	0.7135	0.7134
0.6250	0.5727	0.5657	0.5648	0.5647	0.5647
0.4583	0.4015	0.3947	0.3939	0.3938	0.3937
0.2500	0.2071	0.2028	0.2022	0.2022	0.2022

2.3.2 Vector Iteration with Shifts

The inverse vector iteration procedure described above, combined with the concept of shifting the eigenvalue spectrum, provides an effective means to improve the convergence rate of the iteration process and to make it converge to an eigenvector other than the first mode. Considering the generalized eigenproblem

$$\mathbf{K\Phi} = \lambda \mathbf{M\Phi}\,, \tag{2.55}$$

introducing a shift μ, defining

$$\hat{\lambda} = \lambda - \mu \qquad \hat{\mathbf{K}} = \mathbf{K} - \mu\mathbf{M}\,, \tag{2.56}$$

and substituting Eqs. (2.56) into Eq. (2.55) yields

$$\hat{\mathbf{K}}\mathbf{\Phi} = \hat{\lambda}\,\mathbf{M\Phi}\,. \tag{2.57}$$

The eigenvectors of the two eigenproblems of Eqs. (2.55) and (2.57) are the same. Applying the standard inverse vector iteration, the solution will converge to the eigenvector having the smallest shifted eigenvalue.

Selection of an appropriate shift is difficult without knowledge of the eigenvalue. This difficulty can be overcome by various methods, of which the *Rayleigh quotient iteration* [5] is described below. Using this procedure, we start with an initial vector $\mathbf{r}^{(0)}$ and an initial shift $\lambda^{(0)}$, and calculate a new shift at each iteration cycle. The following steps are to be repeated until convergence:

- Determine $\bar{\mathbf{r}}^{(k)}$ by solving

$$(\mathbf{K} - \lambda^{(k-1)}\mathbf{M})\bar{\mathbf{r}}^{(k)} = \mathbf{M}\,\mathbf{r}^{(k-1)}\,. \tag{2.58}$$

- Estimate the eigenvalue and shift

$$\lambda^{(k)} = \frac{\bar{\mathbf{r}}^{(k)T}\mathbf{M}\mathbf{r}^{(k-1)}}{\bar{\mathbf{r}}^{(k)T}\mathbf{M}\bar{\mathbf{r}}^{(k)}} + \lambda^{k-1}\,. \tag{2.59}$$

- Normalize $\bar{\mathbf{r}}^{(k)}$ by Eq. (2.47)

$$\mathbf{r}^{(k)} = \frac{\bar{\mathbf{r}}^{(k)}}{\left(\bar{\mathbf{r}}^{(k)T}\mathbf{M}\bar{\mathbf{r}}^{(k)}\right)^{1/2}}\,. \tag{2.60}$$

The convergence of this iteration process depends on the initial vector $\mathbf{r}^{(0)}$ and an initial shift $\lambda^{(0)}$. The rate of convergence is faster than the standard vector iteration with shift described earlier, but at the expense of additional computation, because a new matrix $(\mathbf{K} - \lambda^{(k-1)}\mathbf{M})$ has to be factorized at each iteration cycle.

Example 2.2

To illustrate solution by inverse vector iteration with shifts, consider again the eight-story frame of example 2.1 shown in Fig. 2.1. The following two cases of initial values for the shift have been assumed:

- Case *a*. Initial shift $\lambda^{(0)} = 10$.
- Case *b*. Initial shift $\lambda^{(0)} = 30$.

The initial displacement vector in both cases is assumed as

$$\mathbf{r}^{(0)T} = [1, 1, 1, 1, 1, 1, 1, 1] \qquad (u)$$

Results obtained by the procedure described in this section are shown in Tables 2.2, 2.3. It is observed that the solution converges to the second mode shape in case *a*. In case *b* we obtain the third mode shape, and the corresponding eigenvalues are $\lambda_2 = 12.3880$ and $\lambda_3 = 31.1972$.

Table 2.2. Second mode shape, inverse iteration with an initial shift $\lambda^{(0)} = 10$

$\mathbf{r}^{(1)}$	$\mathbf{r}^{(2)}$	$\mathbf{r}^{(3)}$	$\mathbf{r}^{(4)}$	$\mathbf{r}^{(5)}$	**r**(exact)
1.0000	1.0000	1.0000	1.0000	1.0000	1.0000
0.8816	0.6375	0.8109	0.7933	0.7935	0.7935
0.6646	-0.0107	0.4713	0.4226	0.4232	0.4232
0.3538	-0.8797	0.0092	-0.0792	-0.0781	-0.0781
0.0270	-1.6846	-0.4346	-0.5568	-0.5553	-0.5553
-0.2342	-2.1482	-0.7258	-0.8622	-0.8605	-0.8605
-0.3644	-2.0865	-0.7778	-0.9008	-0.8992	-0.8992
-0.2765	-1.2830	-0.4948	-0.5675	-0.5666	-0.5666

Table 2.3. Third mode shape, inverse iteration with an initial shift $\lambda^{(0)} = 30$

$\mathbf{r}^{(1)}$	$\mathbf{r}^{(2)}$	$\mathbf{r}^{(3)}$	**r**(exact)
1.0000	1.0000	1.0000	1.0000
0.4155	0.4830	0.4800	0.4800
-0.4612	-0.2820	-0.2895	-0.2895
-1.1187	-0.8215	-0.8333	-0.8333
-1.0639	-0.7126	-0.7271	-0.7271
-0.3379	-0.0378	-0.0539	-0.0539
0.5148	0.6778	0.6614	0.6614
0.6838	0.7008	0.6889	0.6889

2.3.3 Matrix Deflation and Gram-Schmidt Orthogonalization

Assume that we have calculated a specific eigenpair $\lambda_i, \mathbf{\Phi}_i$. To ensure that the solution of another eigenpair does not converge again to $\lambda_i, \mathbf{\Phi}_i$, we need to deflate either the matrices or the iteration vectors. The basis of vector deflation is that in order for an iteration vector to converge to a required eigenvector, the iteration vector must not be orthogonal to it. Hence, if the iteration vector is orthogonal to the eigenvectors already calculated, convergence occurs to another eigenvector. The Gram-Schmidt orthogonalization can be used in the solution of the generalized eigenproblem $\mathbf{K\Phi} = \mathbf{\Lambda M \Phi}$ [Eq. (2.8)]. Assume that we have calculated the eigenvectors $\mathbf{\Phi}_1, \mathbf{\Phi}_2, ..., \mathbf{\Phi}_m$ by inverse iteration and that we want to $\mathbf{M}$-orthogonalize $\overline{\mathbf{r}}_1$ to these eigenvectors. A vector $\mathbf{r}_1$ which is $\mathbf{M}$-orthogonal to the eigenvectors $\mathbf{\Phi}_1, \mathbf{\Phi}_2, ..., \mathbf{\Phi}_m$ is calculated by

$$\mathbf{r}_1 = \overline{\mathbf{r}}_1 - \sum_{i=1}^{m} \alpha_i \mathbf{\Phi}_i \,. \tag{2.61}$$

The coefficients α_i are obtained using the orthogonality conditions

$$\mathbf{\Phi}_i^T \mathbf{M} \mathbf{r}_1 = 0 \qquad i = 1, ..., m, \tag{2.62}$$

$$\mathbf{\Phi}_i^T \mathbf{M} \mathbf{\Phi}_j = \delta_{ij} \qquad i, j = 1, ..., m, \tag{2.63}$$

where δ_{ij} is the kronecker delta, for which

$$\delta_{ij} = 0 \; (i \neq j) \qquad \delta_{ii} = 1 \,. \tag{2.64}$$

Pre-multiplying both sides of Eq. (2.61) by $\mathbf{\Phi}_i^T \mathbf{M}$ and using the orthogonality conditions of Eqs. (2.62), (2.63) we obtain

$$\alpha_i = \mathbf{\Phi}_i^T \mathbf{M} \overline{\mathbf{r}}_1 \qquad i = 1, ..., m. \tag{2.65}$$

The Gram-Schmidt orthogonalization is sensitive to round-off errors and must be used with care. If the technique is employed in inverse iteration without shifting, it is necessary to calculate the eigenvectors to high precision. In addition, the iteration vector should be orthogonalized at each iteration cycle to the eigenvectors already calculated.

Example 2.3

To illustrate solution by inverse vector iteration with the Gram-Schmidt orthogonalization, consider again the eight-story frame of example 2.1 shown in Fig. 2.1. The first mode shape $\mathbf{\Phi}_1$ is given by (see Table 2.1)

$$\mathbf{\Phi}_1^T = [1, 0.9738, 0.9221, 0.8342, 0.7134, 0.5647, 0.3937, 0.2022], \qquad (a)$$

and the initial displacement vector is assumed as

$$\mathbf{r}^{(0)T} = [1, 1, 1, 1, 1, 1, 1, 1]. \qquad (b)$$

We use the inverse vector iteration procedure described in Section 2.3.1 and employ the Gram-Schmidt orthogonalization such that the result at each iteration cycle is **M**-orthogonal to the eigenvector $\mathbf{\Phi}_1$. The results in Table 2.4 show slow convergence to the second mode shape, with small errors after seven iteration cycles.

Table 2.4. Second mode shape, inverse iteration with orthogonalization

$\mathbf{r}^{(1)}$	$\mathbf{r}^{(2)}$	$\mathbf{r}^{(3)}$	$\mathbf{r}^{(4)}$	$\mathbf{r}^{(5)}$	$\mathbf{r}^{(6)}$	$\mathbf{r}^{(7)}$	**r**(exact)
1.0000	1.0000	1.0000	1.0000	1.0000	1.0000	1.0000	1.0000
0.8398	0.8147	0.8028	0.7975	0.7952	0.7942	0.7938	0.7935
0.5372	0.4739	0.4450	0.4323	0.4269	0.4247	0.4238	0.4232
0.0739	-0.0162	-0.0531	-0.0680	-0.0741	-0.0765	-0.0775	-0.0781
-0.4604	-0.5269	-0.5463	-0.5523	-0.5542	-0.5549	0.5552	-0.5553
-0.9424	-0.9096	-0.8836	-0.8705	-0.8647	-0.8622	-0.8612	-0.8605
-1.2202	-1.0305	-0.9519	-0.9203	-0.9076	-0.9026	-0.9005	-0.8992
-0.9930	-0.6992	-0.6138	-0.5845	-0.5735	-0.5693	-0.5677	-0.5666

2.4 Transformation Methods

The transformation methods employ the orthogonality properties of the eigenvectors [Eqs. (2.41), (2.42)]

$$\mathbf{\Phi}^T \mathbf{M} \mathbf{\Phi} = \mathbf{I}, \qquad (2.66)$$

$$\mathbf{\Phi}^T \mathbf{K} \mathbf{\Phi} = \mathbf{\Lambda}. \qquad (2.67)$$

Since the matrix $\mathbf{\Phi}$ diagonalizes **K** and **M** we may construct it by iteration. Specifically, we define $\mathbf{K}_1 = \mathbf{K}$ and $\mathbf{M}_1 = \mathbf{M}$, and form the matrices

$$\mathbf{K}_{k+1} = \mathbf{P}_k^T \mathbf{K}_k \mathbf{P}_k, \qquad (2.68)$$

$$\mathbf{M}_{k+1} = \mathbf{P}_k^T \mathbf{M}_k \mathbf{P}_k \,, \tag{2.69}$$

The matrices $\mathbf{P}_k$ are selected to bring $\mathbf{K}_k$ and $\mathbf{M}_k$ closer to diagonal form

$$\mathbf{K}_{k+1} \rightarrow \Lambda, \tag{2.70}$$

$$\mathbf{M}_{k+1} \rightarrow \mathbf{I}, \tag{2.71}$$

as $k \rightarrow \infty$. Denoting the last iteration by l, then

$$\mathbf{\Phi} = \mathbf{P}_1 \mathbf{P}_2 ... \mathbf{P}_l \,. \tag{2.72}$$

In practice it is only necessary that matrices $\mathbf{K}_{k+1}$ and $\mathbf{M}_{k+1}$ converge to diagonal form. Two transformation methods discussed in detail elsewhere [1] are the Jacobi and the Householder-QR-inverse iteration methods.

In the Jacobi solution [6] the kth iteration step reduces to Eq. (2.68), where $\mathbf{P}_k$ is selected in such a way that an off-diagonal element in $\mathbf{K}_k$ is zeroed. Matrix $\mathbf{P}_k$ is an orthogonal rotation matrix satisfying the condition

$$\mathbf{P}_k^T \mathbf{P}_k = \mathbf{I} \,. \tag{2.73}$$

The Householder-QR-inverse iteration procedure is restricted to the solution of the standard eigenproblem. Householder transformations are employed to reduce the matrix $\mathbf{K}$ to tri-diagonal form. QR iteration is then used to evaluate the eigenvalues, and inverse iteration is used to calculate the eigenvectors of the tri-diagonal matrix. These vectors are transformed to obtain the eigenvectors of $\mathbf{K}$. Once the eigenvalues have been evaluated, we calculate the eignvectors of the tri-diagonal matrix by simple inverse iteration with shifts equal to the corresponding eigenvalues. The eigenvectors of the tri-diagonal matrix then need to be transformed with the Householder transformations used to obtain the eigenvectors of $\mathbf{K}$.

2.5 Polynomial Iterations

The set of homogeneous equations (2.55) has a nontrivial solution if

$$det\,[\mathbf{K} - \lambda \mathbf{M}] = 0. \tag{2.74}$$

The equation $(\mathbf{K} - \lambda_i \mathbf{M})\mathbf{\Phi}_i = \mathbf{0}$ is satisfied for nontrivial $\mathbf{\Phi}_i$ only if the matrix $\mathbf{K} - \lambda_i \mathbf{M}$ is singular. When the determinant is expanded, a polynomial known as *the characteristic equation*, or *frequency equation*, is obtained. The eigenvalues are the roots of the characteristic polynomial

$$p(\lambda) = det\,[\mathbf{K} - \lambda\mathbf{M}]. \tag{2.75}$$

Evaluation of the coefficients of the polynomial for large systems requires much computational effort and the roots of the equation are sensitive to numerical round-off errors in the coefficients. Using polynomial iterations, only the eigenvalues are calculated. The corresponding eigenvectors can then be obtained by inverse iteration with shifts. The solution by polynomial iteration can be carried out directly using matrices **K** and **M**, without transforming the problem into a different form. The solution is most effective if only a few eigenvalues are to be calculated.

The first step in explicit polynomial iteration is to write $p(\lambda)$ in the form

$$p(\lambda) = a_0 + a_1\lambda + a_2\lambda^2 + ... + a_n\lambda^n, \tag{2.76}$$

and evaluate the polynomial coefficients. The second step is to calculate the roots of the polynomial using, for example, a Newton iteration or secant iteration. A basic difficulty in this approach is that small errors in the coefficients cause large errors in the roots of the polynomial.

In implicit polynomial iteration we evaluate the value of $p(\lambda)$ directly, without calculating first the coefficients, by decomposing the $\mathbf{K} - \mu\mathbf{M}$ matrix into a lower-unit triangular matrix **L** and an upper triangular matrix **S** to obtain

$$\mathbf{K} - \lambda\mathbf{M} = \mathbf{LS}, \tag{2.77}$$

$$\det\,[\mathbf{K} - \lambda\mathbf{M}] = det\,\mathbf{LS} = S_{11}S_{22}\ldots S_{nn}, \tag{2.78}$$

where S_{ii} are the diagonal elements of **S**. This decomposition may require interchanges when $\lambda > \lambda_1$. Each row or column interchange merely affects a change in sign of the determinant. When row and corresponding column interchanges are carried out, the coefficient matrix $\mathbf{K} - \lambda\mathbf{M}$ in Eq. (2.78) remains symmetric. If the coefficient matrix is symmetric, then we have $\mathbf{S} = \mathbf{DL}^T$ and hence

$$det\,[\mathbf{K} - \lambda\mathbf{M}] = D_{11}D_{22}\ldots D_{nn}. \tag{2.79}$$

In this case, one polynomial evaluation requires about $1/2\; n\, m_k^2$ operations, where n is the order of **K** and **M**, and m_k is the half-bandwidth of **K**. A number of iteration schemes, such as the Newton method, can be used to calculate a root of the polynomial.

2.6 Rayleigh-Ritz Analysis

2.6.1 Approximate Eigenproblem Solution

Rayleigh-Ritz analysis is a general approach for finding approximations to the lowest eigenvalues and corresponding eigenvectors of the eigenproblem. The method has been used widely to reduce the dimension of the equations of motion. It is based on the assumption that the displacement vector can be approximated by a linear combination of *discrete Ritz vectors*. Various methods can be understood to be Ritz analysis, and the methods differ only in the choice of the Ritz basis vectors assumed in the analysis. Discrete Ritz vectors can be taken as approximations of the true vibration mode shapes. The Ritz vectors are an attractive alternative to standard normal modes, since they can be computed with significantly less computational effort. A proper choice of Ritz vectors, employed as starting vectors in various iterative procedures, can significantly accelerate the solution process.

In the Ritz analysis we consider a set of vectors $\overline{\mathbf{\Phi}}$, which are linear combination of the Ritz basis vectors $\mathbf{r}_1, \mathbf{r}_2, \ldots, \mathbf{r}_s$. A typical vector is given by

$$\overline{\mathbf{\Phi}} = y_1\mathbf{r}_1 + y_2\mathbf{r}_2 + \ldots + y_s\mathbf{r}_s = \mathbf{r}_B\,\mathbf{y}, \tag{2.80}$$

where $\mathbf{y}$ is the vector of *Ritz coordinates* y_i, $\mathbf{r}_B$ is the $n\times s$ matrix of the Ritz basis vectors $\mathbf{r}_i$ and s is smaller than the number of degrees of freedom n, that is

$$\mathbf{r}_B = [\mathbf{r}_1, \mathbf{r}_2, \ldots, \mathbf{r}_s] \qquad \mathbf{y}^T = \{y_1, y_2, \ldots, y_s\}. \tag{2.81}$$

The vector $\overline{\mathbf{\Phi}}$ lies in the subspace spanned by the Ritz basis vectors. To invoke the Rayleigh minimum principle, we first evaluate the Rayleigh quotient

$$\rho(\overline{\mathbf{\Phi}}) = \frac{\overline{\mathbf{\Phi}}^T\mathbf{K}\,\overline{\mathbf{\Phi}}}{\overline{\mathbf{\Phi}}^T\mathbf{M}\overline{\mathbf{\Phi}}}. \tag{2.82}$$

Substituting Eq. (2.80) into Eq. (2.82), differentiating with respect to $\mathbf{y}$ and setting the result equal to zero, we obtain the eigenproblem

$$\mathbf{K}_R\,\mathbf{y} = \rho\,\mathbf{M}_R\,\mathbf{y}, \tag{2.83}$$

where

$$\mathbf{K}_R = \mathbf{r}_B^T\mathbf{K}\mathbf{r}_B \qquad \mathbf{M}_R = \mathbf{r}_B^T\mathbf{M}\,\mathbf{r}_B. \tag{2.84}$$

The solution to Eq. (2.83) yields s eigenvalues $\rho_1, \ldots, \rho_s$, which are approximations to the eigenvalues $\lambda_1, \ldots, \lambda_s$, and s eigenvectors $\mathbf{y}_1, \ldots, \mathbf{y}_s$, which are used to evaluate the vectors $\overline{\mathbf{\Phi}}_1, \ldots, \overline{\mathbf{\Phi}}_s$. The latter vectors are approximations to the eigenvectors $\mathbf{\Phi}_1, \ldots, \mathbf{\Phi}_s$. In general, the approximations of the higher eigenvalues are less accurate than the approximations of the lower eigenvalues. It is instructive to note that *an eigenvalue calculated from the Ritz analysis is an upper bound on the corresponding exact eigenvalue of the system.* The success of the Rayleigh-Ritz method depends on how well linear combinations of Ritz vectors can approximate the natural modes of vibration.

2.6.2 Load-Dependent Ritz Vectors

Load-dependent Ritz vectors are a particular class of Ritz vectors where information about the loading on the structure is used to generate the vectors. A procedure for generating such vectors, called *derived Ritz vectors* [3] is described in this section. The initial vector of the coordinate sequence is the deflected shape resulting from static application of the dynamic load distribution, and the subsequent vectors account for inertial effects on the dynamic response. The following notation is used in the derivation of the vectors:

- $\overline{\mathbf{r}}_i$ is the preliminary deflected shape calculated first in the derivation of each vector, where the subscript is the number of the derived vector.
- $\tilde{\mathbf{r}}_i$ is obtained after orthogonalization with the preceding vectors.
- $\mathbf{r}_i$ is the final form of the derived vector obtained after normalization.

To derive the first vector, the static equilibrium equations,

$$\mathbf{K}\,\overline{\mathbf{r}}_1 = \mathbf{R}, \tag{2.85}$$

are first solved for the preliminary deflected shape $\overline{\mathbf{r}}_1$. The vector $\overline{\mathbf{r}}_1$ is then normalized to obtain the first derived Ritz vector $\mathbf{r}_1$

$$\mathbf{r}_1 = \frac{\overline{\mathbf{r}}_1}{\left(\overline{\mathbf{r}}_1^T \mathbf{M}\,\overline{\mathbf{r}}_1\right)^{1/2}}, \tag{2.86}$$

so that it provides a unit generalized mass $\mathbf{r}_1^T \mathbf{M} \mathbf{r}_1 = 1$.

Calculation of the second vector starts with solution of

$$\mathbf{K}\,\overline{\mathbf{r}}_2 = \mathbf{M}\,\mathbf{r}_1, \tag{2.87}$$

to obtain the deflected shape $\bar{\mathbf{r}}_2$ resulting from the inertial load $\mathbf{M}\mathbf{r}_1$. Then, the shape $\bar{\mathbf{r}}_2$ is made mass-orthogonal to the first vector $\mathbf{r}_1$ by the Gram-Schmidt procedure [Eqs. (2.61)–(2.65)] to obtain

$$\tilde{\mathbf{r}}_2 = \bar{\mathbf{r}}_2 - \left(\mathbf{r}_1^T \mathbf{M}\, \bar{\mathbf{r}}_2\right) \mathbf{r}_1 . \tag{2.88}$$

Finally, this shape is normalized to obtain the second derived Ritz vector

$$\mathbf{r}_2 = \frac{\tilde{\mathbf{r}}_2}{\left(\tilde{\mathbf{r}}_2^T \mathbf{M}\, \tilde{\mathbf{r}}_2\right)^{1/2}} , \tag{2.89}$$

so that $\mathbf{r}_2^T \mathbf{M} \mathbf{r}_2 = 1$. Derivation of further vectors proceeds in a similar way. It has been shown [3] that when a vector is made orthogonal to the two preceding shapes, it automatically is orthogonal to all preceding shapes.

In summary, derivation of $\mathbf{r}_{i+1}$ involves the following steps:

- The deflected shape $\bar{\mathbf{r}}_{i+1}$ is obtained by solving

$$\mathbf{K}\, \bar{\mathbf{r}}_{i+1} = \mathbf{M}\, \mathbf{r}_i . \tag{2.90}$$

- The deflected shape $\bar{\mathbf{r}}_{i+1}$ is made mass-orthogonal with respect to the two preceding vectors by the Gram-Schmidt procedure

$$\tilde{\mathbf{r}}_{i+1} = \bar{\mathbf{r}}_{i+1} - \alpha_i \mathbf{r}_i - \beta_i \mathbf{r}_{i-1} , \tag{2.91}$$

  where

$$\alpha_i = \mathbf{r}_i^T \mathbf{M} \bar{\mathbf{r}}_{i+1} \qquad \beta_i = \left(\tilde{\mathbf{r}}_i^T \mathbf{M} \tilde{\mathbf{r}}_i\right)^{1/2} . \tag{2.92}$$

- The shape is normalized to obtain $\mathbf{r}_{i+1}$ by calculating

$$\mathbf{r}_{i+1} = \frac{\tilde{\mathbf{r}}_{i+1}}{\left(\tilde{\mathbf{r}}_{i+1}^T \mathbf{M}\, \tilde{\mathbf{r}}_{i+1}\right)^{1/2}} . \tag{2.93}$$

The unique orthogonality property of these vectors makes it possible to organize the equations of motion in a special tri-diagonal form.

Example 2.4

Given a one-bay five-story shear frame, the following stiffness matrix **K**, mass matrix **M**, and load vector **R** are considered

$$\mathbf{K} = 31.56\begin{bmatrix} 2 & -1 & & & \\ -1 & 2 & -1 & & \\ & -1 & 2 & -1 & \\ & & -1 & 2 & -1 \\ & & & -1 & 1 \end{bmatrix}, \tag{a}$$

$$\mathbf{M} = 0.2591\begin{bmatrix} 1 & & & & \\ & 1 & & & \\ & & 1 & & \\ & & & 1 & \\ & & & & 1 \end{bmatrix} \qquad \mathbf{R} = 0.2591\begin{Bmatrix} 1 \\ 1 \\ 1 \\ 1 \\ 1 \end{Bmatrix}.$$

The object is to determine the first two modes, using two load-dependent Ritz vectors. The first Ritz vector is calculated by Eqs. (2.85), (2.86)

$$\bar{\mathbf{r}}_1^T = [0.0410, 0.0739, 0.0985, 0.1149, 0.1231], \tag{b}$$

$$\mathbf{r}_1^T = [0.3792, 0.6826, 0.9102, 1.062, 1.138].$$

The second Ritz vector is determined by Eqs. (2.87), (2.88), (2.89)

$$\bar{\mathbf{r}}_2^T = [0.0342, 0.0654, 0.0909, 0.1090, 0.1183], \tag{c}$$

$$\tilde{\mathbf{r}}_2^T = 0.01[-0.4134, -0.3705, -0.1204, 0.1500, 0.3164],$$

$$\mathbf{r}_2^T = [-1.217, -1.091, -0.3546, 0.4418, 0.9316].$$

The reduced stiffness matrix and mass matrix are calculated by Eq. (2.84)

$$\mathbf{K}_R = \begin{bmatrix} 9.986 & -3.086 \\ -3.086 & 91.95 \end{bmatrix} \qquad \mathbf{M}_R = \begin{bmatrix} 1.0 & \\ & 1.0 \end{bmatrix}. \tag{d}$$

Solving the reduced Eigenproblem [Eq. (2.83)] we obtain

$$\mathbf{y}_1 = \begin{bmatrix} 0.9993 \\ 0.0376 \end{bmatrix} \qquad \mathbf{y}_2 = \begin{bmatrix} -0.0376 \\ 0.9993 \end{bmatrix}, \tag{e}$$

$$\omega_1 = \rho_1^{1/2} = 3.142 \qquad \omega_2 = \rho_2^{1/2} = 9.595.$$

The approximate mode shapes are obtained by Eq. (2.80)

$$\overline{\mathbf{\Phi}}_1^T = [0.3332, 0.6412, 0.8962, 1.078, 1.172], \tag{f}$$

$$\overline{\mathbf{\Phi}}_2^T = [-1.230, -1.116, -0.3886, 0.4016, 0.8882].$$

The exact solution is given by

$$\mathbf{\Phi}_1^T = [0.3338,\ 0.6405,\ 0.8954,\ 1.078,\ 1.173], \tag{g}$$

$$\mathbf{\Phi}_2^T = [-0.8954,\ -1.173,\ -0.6411,\ 0.3338,\ 1.078],$$

$$\omega_1 = \lambda_1^{1/2} = 3.142 \qquad \omega_2 = \lambda_2^{1/2} = 9.170. \tag{h}$$

It is observed that, using two load-dependent Ritz vectors, better accuracy is obtained for the first mode shape.

2.7 The Lanczos Method

If the objective is to calculate only a few eigenvalues and corresponding eigenvectors, an iteration analysis based on the *Lanczos transformation* can be very efficient [7]. The *Lanczos coordinates* are an effective set of Ritz vectors, where the vectors are derived by a procedure that is similar in many respects to vector iteration analysis of the fundamental vibration modes. The Lanczos method is equivalent to the discrete Rayleigh-Ritz reduction with vectors in the *Krylov sequence* selected as the global approximation vectors. The method generates the latter vectors, and the sequence converges to the eigenvector corresponding to the smallest eigenvalue. The basis vectors of the Krylov subspace consist of the vectors

$$\mathbf{r}_1,\ \mathbf{D}\,\mathbf{r}_1,\ \mathbf{D}^2\,\mathbf{r}_1,\ \ldots,\ \mathbf{D}^{s-1}\mathbf{r}_1, \tag{2.94}$$

where the starting vector $\mathbf{r}_1$ is the static response due to the loads, s is the number of vectors considered and matrix $\mathbf{D}$ is defined as $\mathbf{D} = \mathbf{K}^{-1}\mathbf{M}$.

The Lanczos algorithm involves supplementing the Krylov sequence with the Gram-Schmidt orthogonalization process at each step. The result is a set of **M**-orthonormal vectors that is used to reduce the dimension of the set of equations. These vectors do not have the full uncoupling property of the mode shapes, but they are much less expensive to generate.

Each Lanczos step, for derivation of the $(i+1)$th vector $\mathbf{r}_{i+1}$, $i = 1, \ldots, n$, involves the following basic operations [1, 8]:

- The deflected shape $\bar{\mathbf{r}}_{i+1}$ is obtained by solving

$$(\mathbf{K} - \mu\mathbf{M})\ \bar{\mathbf{r}}_{i+1} = \mathbf{M}\,\mathbf{r}_i, \tag{2.95}$$

  where μ is a shift (usually the initial μ equals zero). This calculation is very similar to the inverse vector iteration with shifts.
- The shape $\bar{\mathbf{r}}_{i+1}$ is made mass-orthogonal with respect to the two preceding vectors by the Gram-Schmidt procedure [Eqs. (2.91), (2.92)].
- The shape is normalized to obtain $\mathbf{r}_{i+1}$ [Eq. (2.93)]. The coefficients α_i, β_i are used to form matrix $\mathbf{T}_n$

$$\mathbf{T}_n = \begin{bmatrix} \alpha_1 & \beta_1 & & & \\ \beta_1 & \alpha_2 & \beta_2 & & \\ & & \cdots & & \\ & & & \cdots & \\ & & & \alpha_{n-1} & \beta_{n-1} \\ & & & \beta_{n-1} & \alpha_n \end{bmatrix}. \tag{2.96}$$

The Lanczos method transforms the generalized eigenproblem,

$$\mathbf{K}\,\boldsymbol{\Phi} = \lambda\,\mathbf{M}\boldsymbol{\Phi}, \tag{2.97}$$

into a standard form with a tri-diagonal coefficient matrix $\mathbf{T}_n$

$$\mathbf{T}_n\tilde{\boldsymbol{\Phi}} = (1/\lambda)\,\tilde{\boldsymbol{\Phi}}. \tag{2.98}$$

The eigenvalues of $\mathbf{T}_n$ are the reciprocals of the eigenvalues of Eq. (2.97). The eigenvectors of the two problems of Eqs. (2.97), (2.98) are related by

$$\boldsymbol{\Phi} = \mathbf{r}_B\tilde{\boldsymbol{\Phi}}, \tag{2.99}$$

where matrix $\mathbf{r}_B$ is defined as [see Eq. (2.95) and Eqs. (2.91)–(2.93)]

$$\mathbf{r}_B = [\mathbf{r}_1, \mathbf{r}_2, \ldots, \mathbf{r}_n]. \tag{2.100}$$

Theoretically, the vectors $\mathbf{r}_{i+1}$ generated by the procedure of Eq. (2.95) and Eqs. (2.91)–(2.93) are **M**-orthogonal, that is $\mathbf{r}_i^T\mathbf{M}\,\mathbf{r}_j = \delta_{ij}$. In practice, the tri-diagonalization does not produce the desired orthogonal vectors because of round-off errors. If we perform the transformation for $i = 1, \ldots, q$ ($q << n$), we calculate $\mathbf{T}_q$ corresponding to the eigenproblem

$$\mathbf{T}_q\,\mathbf{S} = \nu\,\mathbf{S}. \tag{2.101}$$

The solution of Eq. (2.101) may yield good approximations to the smallest eigenvalues and corresponding eigenvectors of the original eigenproblem.

The following definitions are used in describing the solution approach:

- A *Lanczos step* is the use of Eq. (2.95) and Eqs. (2.91)–(2.93). Thus, at each Lanczos step it is necessary to solve Eq. (2.95) for $\overline{\mathbf{r}}_{i+1}$, where the shift μ is changed at each Lanczos stage.
- A *Lanczos stage* consists of q Lanczos steps and the calculation of the eigenpairs of the problem of Eq. (2.101).

The solution algorithm involves the following operations.

- Start a new Lanczos stage. Choose a starting vector that is orthogonal to all previous eigenvector approximations and normalize it [Eq. (2.93)]. Choose a shift μ (usually $\mu = 0$ for the first Lanczos stage).
- Perform the Lanczos steps. Although a Gram Schmidt orthogonalization has been performed, the vector $\mathbf{r}_{i+1}$ is checked for loss of orthogonality. If the loss occurs, this Lanczos stage is terminated with $q = i$. Otherwise perform a maximum number of steps q_{max} and set $q = q_{max}$. Compute additional converged eigenpairs by solving Eq. (2.101). Reset the number of converged eigenvalues in the preceding stages to the new value. If the required eigenpairs have not yet been obtained, restart for an additional Lanczos stage.
- Continue until all required eigenpairs have been calculated or until the maximum number of assigned Lanczos steps has been reached.

2.8 Subspace Iteration

Similar to the Lanczos method, a combination of the properties used in various methods are considered in the subspace iteration method [1, 9]. Specifically, the method is based on simultaneous vector iteration, Sturm sequence information and Rayleigh-Ritz analysis. The method is particularly suitable for the calculation of a few eigenvectors and eigenvalues of large systems. The trial vectors are all subjected to inverse iteration combined with some technique that forces convergence to independent shapes. The convergence is to the lowest undamped vibration mode shapes. The method consists of the following steps.

- Establish the matrix $\mathbf{r}^{(1)}$, consisting of q starting vectors ($q > p$, where p is the number of vectors to be calculated).
- Use simultaneous inverse iteration and Ritz analysis to extract the best eigenvalue and eigenvector approximations from the q iteration vectors. The solution involves the following operations:

– Calculate the matrix of new vectors $\bar{\mathbf{r}}^{(k+1)}$ by simultaneous inverse iteration

$$\mathbf{K}\bar{\mathbf{r}}^{(k+1)} = \mathbf{M}\mathbf{r}^{(k)}. \tag{2.102}$$

– Find the projections of the matrices **K** and **M** by

$$\mathbf{K}_R^{(k+1)} = \bar{\mathbf{r}}^{T(k+1)}\mathbf{K}\,\bar{\mathbf{r}}^{(k+1)} \qquad \mathbf{M}_R^{(k+1)} = \bar{\mathbf{r}}^{T(k+1)}\mathbf{M}\,\bar{\mathbf{r}}^{(k+1)}. \tag{2.103}$$

– Solve the eigensystem of the projected matrices

$$\mathbf{K}_R^{(k+1)}\,\mathbf{y}^{(k+1)} = \mathbf{\Lambda}^{(k+1)}\mathbf{M}_R^{(k+1)}\mathbf{y}^{(k+1)}, \tag{2.104}$$

and find the eigenpairs $\mathbf{y}^{(k+1)}$ and $\mathbf{\Lambda}^{(k+1)}$.

– Calculate the improved approximation to the matrix of eigenvectors, $\mathbf{r}^{(k+1)}$

$$\mathbf{r}^{(k+1)} = \bar{\mathbf{r}}^{(k+1)}\mathbf{y}^{(k+1)}. \tag{2.105}$$

Then, provided that the vectors of matrix $\mathbf{r}^{(1)}$ are not orthogonal to one of the required eigenvectors, the solution process will converge to the set of eigenvalues and corresponding eigenvectors

$$\mathbf{\Lambda}^{(k+1)} \to \mathbf{\Lambda} \qquad \mathbf{r}^{(k+1)} \to \mathbf{\Phi}. \tag{2.106}$$

The iteration is equivalent to iteration with a q-dimensional subspace and should not be regarded as a simultaneous iteration with q individual iteration vectors

- After convergence of the iteration, use the Surm sequence to verify that the required eigenvalues and vectors have been calculated.

It has been noted [1] that the method presented is most effective for evaluating a relatively small number of eigenpairs. For a larger number of eigenpairs the cost of the solution rises rapidly. Various acceleration procedures for the basic subspace iteration method have been proposed.

Example 2.5

To illustrate solution by subspace iteration, consider again the eight-story frame of example 2.1, shown in Fig. 2.1. The object is to calculate 3 eigenvectors. The matrix $\mathbf{r}^{(1)}$, consisting of 4 starting vectors, is chosen as

$$\mathbf{r}^{(1)} = \begin{bmatrix} 1 & 0 & 0 & 0 \\ 1 & 0 & 0 & 0 \\ 0 & 1 & 0 & 0 \\ 0 & 1 & 0 & 0 \\ 0 & 0 & 1 & 0 \\ 0 & 0 & 1 & 0 \\ 0 & 0 & 0 & 1 \\ 0 & 0 & 0 & 1 \end{bmatrix}. \qquad (a)$$

Using the solution procedure described in this section, the results obtained for the first 3 mode shapes in 4 iteration cycles are given in Tables 2.5 through 2.7. The corresponding eigenvalues are summarized in Table 2.8. It is observed that accurate results are achieved for the first mode shape after 2 iteration cycles and for the second mode shape after 4 iteration cycles. Some errors are obtained for the third mode shape after 4 iteration cycles. To reduce these errors, 5 vectors (instead of 4 vectors) could be considered.

Table 2.5. First mode shape, subspace iteration

$\mathbf{r}^{(1)}$	$\mathbf{r}^{(2)}$	$\mathbf{r}^{(3)}$	$\mathbf{r}^{(4)}$	$\mathbf{r}$(exact)
1.0000	1.0000	1.0000	1.0000	1.0000
0.9739	0.9738	0.9738	0.9738	0.9738
0.9217	0.9221	0.9221	0.9221	0.9221
0.8346	0.8342	0.8342	0.8342	0.8342
0.7128	0.7134	0.7134	0.7134	0.7134
0.5653	0.5647	0.5647	0.5647	0.5647
0.3922	0.3937	0.3937	0.3937	0.3937
0.2037	0.2022	0.2022	0.2022	0.2022

Table 2.6. Second mode shape, subspace iteration

$\mathbf{r}^{(1)}$	$\mathbf{r}^{(2)}$	$\mathbf{r}^{(3)}$	$\mathbf{r}^{(4)}$	$\mathbf{r}$(exact)
1.0000	1.0000	1.0000	1.0000	1.0000
0.5398	0.7937	0.7935	0.7935	0.7935
-0.3806	0.4219	0.4231	0.4232	0.4232
-0.7795	-0.0762	-0.0781	-0.0782	-0.0781
-0.6569	-0.5579	-0.5556	-0.5554	-0.5553
-0.1728	-0.8595	-0.8603	-0.8605	-0.8605
0.6728	-0.8959	-0.8987	-0.8991	-0.8992
0.7304	-0.5702	-0.5671	-0.5667	-0.5666

Table 2.7. Third mode shape, subspace iteration

$\mathbf{r}^{(1)}$	$\mathbf{r}^{(2)}$	$\mathbf{r}^{(3)}$	$\mathbf{r}^{(4)}$	r(exact)
1.0000	1.0000	1.0000	1.0000	1.0000
0.2996	0.4817	0.0717	0.4786	0.4800
-1.1012	-0.3151	-0.9756	-0.2918	-0.2895
-0.8873	-0.8098	-0.6245	-0.8300	-0.8333
0.9413	-0.6967	0.6950	-0.7237	-0.7271
1.1097	-0.0841	0.9430	-0.0568	-0.0539
-0.3821	0.6528	-0.1826	0.6595	0.6614
-0.7520	0.6992	-0.8342	0.6914	0.6889

Table 2.8. Summary of eigenvalues, subspace iteration

Mode	1	2	3
λ(subspace)	1.5718	12.3000	31.1980
λ(exact)	1.5718	12.3880	31.1972

References

1. Bathe KJ (1996) Finite element procedures. Prentice Hall, NJ
2. Chopra AK (2001) Dynamics of structures. Prentice Hall, NJ
3. Clough RW, Penzien JP (1993) Dynamics of structures. McGraw-Hill, NY
4. Wilkinson W (1965) The algebraic eigenvalue problem. Oxford University Press, Oxford
5. Ostrowski AM (1958-59) On the convergence of the Rayleigh quotient iteration for the computation of the characteristic roots and vectors, parts I-VI. Archive for Rational Mechanics and Analysis 1-3
6. Jacobi CGJ (1846) Ube rein leichtes verfahren die in der theoie der sacularstorungen gleichungen numerisch aufzulosen. Crelle's J. 30:51-94
7. Lanczos C (1950) An iteration method for the solution of the eigenvalue problem of linear differential and integral operators. J Research of the National Bureau of Standards 45:255-281
8. Ericson T, Ruhe KJ (1980) The spectral transformation Lanczos method for the numerical solution of large sparse generalized symmetric eigenvalue problems. Mathematics of Computation 35:1251-1268
9. Bathe KJ (1971) Solution methods of large generalized eigenvalue problems in structural engineering. Report UC SESM 71-20, Civil Engineering Department, University of California Berkeley

3 Dynamic Analysis

3.1 Linear Dynamic Analysis

Linear dynamic analysis is discussed in detail in various texts [e.g. 1–3]. In this section some basics of this topic are presented. Consider the equations of motion for a system subjected to external dynamic forces

$$\mathbf{M}\ddot{\mathbf{r}}(t)+\mathbf{C}\dot{\mathbf{r}}(t)+\mathbf{K}\mathbf{r}(t)=\mathbf{R}(t)\,, \tag{3.1}$$

where **M** is the mass matrix, **C** is the damping matrix and **K** is the stiffness matrix. The displacement vector $\mathbf{r}(t)$, the velocity vector $\dot{\mathbf{r}}(t)$, the acceleration vector $\ddot{\mathbf{r}}(t)$ and the load vector $\mathbf{R}(t)$ are functions of the time t.

For linear analysis, Eq. (3.1) represents a system of linear differential equations of second order and, in principle, the solution can be obtained by standard procedures for the solution of differential equations with constant coefficients. In practical analysis, the common procedures can be divided into two methods of solution (the choice of one method or the other is determined by their relative numerical effectiveness):

- *Direct integration*, where Eqs. (3.1) are integrated using a numerical step-by-step procedure. The term "direct" means that prior to the numerical integration, no transformation of the equations into different form is carried out. Direct numerical integration is based on the idea that the equations are satisfied only at discrete time intervals. In addition, the method is based on the assumption that a variation of displacements, velocities and accelerations within each time interval has a certain form. The form of this assumption determines the accuracy, stability, and cost of the solution procedure.
- *Mode superposition*, where the equilibrium equations are transformed into a form in which the step-by-step solution is less costly. In a practical finite element analysis only the lowest modes are considered. In general the object is to obtain a good approximation to the actual exact response. The finite element analysis approximates the lowest exact frequencies best, and little or no accuracy can be expected in approximating the higher frequencies and mode shapes. In a mode

superposition analysis only a few modes may need to be considered. In such cases this procedure can be much more effective than direct integration. The mode superposition procedure may be more effective also when the integration must be carried out for many time steps.

3.1.1 Direct Integration

We assume that the displacement, velocity and acceleration vectors at time t=0 are known. The derivatives are considered for a constant time-step Δt. The cost of a direct integration analysis is directly proportional to the number of time steps required for solution. The commonly used effective direct integration methods are briefly described in the following. A detailed discussion is given elsewhere [4]. We distinguish between *explicit integration methods*, where the solution for ${}^{t+\Delta t}\mathbf{r}$ is based on using the equilibrium conditions at time t, and *implicit integration methods* such as the Houbolt [5], Wilson [6] and Newmark [7] methods, where the equilibrium conditions at time $t+\Delta t$ are used.

Explicit Integration Methods

The equilibrium conditions at time t are

$$\mathbf{M}\,{}^{t}\ddot{\mathbf{r}}+\mathbf{C}\,{}^{t}\dot{\mathbf{r}}+\mathbf{K}\,{}^{t}\mathbf{r}={}^{t}\mathbf{R}\,. \tag{3.2}$$

Considering the *central-difference method*, we assume that the accelerations are approximated by the expression for $t-\Delta t$, t, $t+\Delta t$

$${}^{t}\ddot{\mathbf{r}}=\frac{1}{\Delta t^{2}}\left({}^{t-\Delta t}\mathbf{r}-2\,{}^{t}\mathbf{r}+{}^{t+\Delta t}\mathbf{r}\right). \tag{3.3}$$

The assumed expression for the velocities is

$${}^{t}\dot{\mathbf{r}}=\frac{1}{2\Delta t}\left(-{}^{t-\Delta t}\,\mathbf{r}+{}^{t+\Delta t}\mathbf{r}\right). \tag{3.4}$$

Defining the integration constants a_0, …, a_3

$$a_0=\frac{1}{\Delta t^{2}}\qquad a_1=\frac{1}{2\Delta t}\qquad a_2=2a_0\qquad a_3=\frac{1}{a_2}\,, \tag{3.5}$$

then the effective mass matrix, $\hat{\mathbf{M}}$, the displacement vector, ${}^{-\Delta t}\mathbf{r}$, and the effective load vector at time t, ${}^{t}\hat{\mathbf{R}}$, are given by

$$\hat{\mathbf{M}} = a_0\mathbf{M} + a_1\mathbf{C}\ , \tag{3.6}$$

$$^{-\Delta t}\mathbf{r} = {}^{0}\mathbf{r} - \Delta t\ {}^{0}\dot{\mathbf{r}} + a_3\ {}^{0}\ddot{\mathbf{r}}\ , \tag{3.7}$$

$$^{t}\hat{\mathbf{R}} = {}^{t}\mathbf{R} - (\mathbf{K} - a_2\mathbf{M})\,{}^{t}\mathbf{r} - (a_0\mathbf{M} - a_1\mathbf{C})\,{}^{t-\Delta t}\mathbf{r}\ . \tag{3.8}$$

Substituting Eqs. (3.3) through (3.8) into Eq. 3.2, we find the displacements at time $t+\Delta t$, ${}^{t+\Delta t}\mathbf{r}$, from the equilibrium conditions at time t

$$\hat{\mathbf{M}}\ {}^{t+\Delta t}\mathbf{r} = {}^{t}\hat{\mathbf{R}}\ . \tag{3.9}$$

Integration methods that require a time step smaller than a critical time step, such as the central difference method, are *conditionally stable*. If a time step larger than the critical time step is used the integration is unstable. For example, any errors resulting from round-off in the computer grow and make the response calculations worthless in many cases.

The complete algorithm using the central difference method is as follows. For the given $\mathbf{K}$, $\mathbf{M}$, $\mathbf{C}$, ${}^{0}\mathbf{r}$, ${}^{0}\dot{\mathbf{r}}$, ${}^{0}\ddot{\mathbf{r}}$, we start with selecting the time step Δt smaller than the critical time step

$$\Delta t_{cr} = T_n/\pi\ , \tag{3.10}$$

where T_n is the smallest period of the system. Then we calculate the integration constants a_0, …, a_3 [Eq. (3.5)], the effective mass matrix ${}^{t}\hat{\mathbf{M}}$ [Eq. (3.6)], the displacements ${}^{-\Delta t}\mathbf{r}$ [Eq. (3.7)], and triangularize matrix $\hat{\mathbf{M}}$ by

$$\hat{\mathbf{M}} = \mathbf{LDL}^T\ . \tag{3.11}$$

The solution process involves the following stages for each time step:

- Calculation of the effective load vector at time t, ${}^{t}\hat{\mathbf{R}}$ [Eq. (3.8)].
- Solving for the displacements at time $t+\Delta t$, ${}^{t+\Delta t}\mathbf{r}$ [Eq. (3.9)].
- Evaluation of the accelerations at time t [Eq. (3.3)].
- Evaluation of the velocities at time t [Eq. (3.4)].

It is observed that the integration scheme does not require a factorization of the effective stiffness matrix. The method is effective when each time step solution can be performed very efficiently, e.g., when a lumped diagonal mass matrix can be assumed and damping can be neglected. In such cases the system of equations can be solved without factorizing a matrix. Another advantage of the method is that no stiffness matrix of the complete assemblage needs to be calculated. The solution can be carried out on the element level and relatively little high-speed storage is required. Thus, systems of very large order can be solved effectively.

Example 3.1

Consider a system with the following equilibrium equations

$$\begin{bmatrix} 2 & 0 \\ 0 & 1 \end{bmatrix} \ddot{\mathbf{r}} + \begin{bmatrix} 6 & -2 \\ -2 & 4 \end{bmatrix} \mathbf{r} = \begin{Bmatrix} 0 \\ 10 \end{Bmatrix}. \qquad (a)$$

The free-vibration periods are T_1=4.45, T_2=2.8. Assuming Δt=T_2/10=0.28, the object is to evaluate the displacements for 12 time steps and the initial values ${}^0\mathbf{r} = {}^0\dot{\mathbf{r}} = \mathbf{0}$ by the central-difference method. Using Eq. (*a*) we find

$$\ddot{\mathbf{r}} = \begin{Bmatrix} 0 \\ 10 \end{Bmatrix}. \qquad (b)$$

The integration constants are [Eq. (3.5)]

$$a_0 = 12.8 \quad a_1 = 1.79 \quad a_2 = 25.5 \quad a_3 = 0.0392. \qquad (c)$$

The effective mass matrix $\hat{\mathbf{M}}$ and the effective load vector ${}^t\hat{\mathbf{R}}$ at time t are given by [Eqs. (3.6), (3.8)]

$$\hat{\mathbf{M}} = \begin{bmatrix} 25.5 & 0 \\ 0 & 12.8 \end{bmatrix}, \qquad (d)$$

$$\hat{\mathbf{R}} = \begin{Bmatrix} 10 \\ 0 \end{Bmatrix} + \begin{bmatrix} 45.0 & 2 \\ 2 & 21.5 \end{bmatrix} {}^t\mathbf{r} - \begin{bmatrix} 25.5 & 0 \\ 0 & 12.8 \end{bmatrix} {}^{t-\Delta t}\mathbf{r}. \qquad (e)$$

Table 3.1. Displacements ${}^t\mathbf{r}$ as functions of time t, central-difference method

t	Δt	$2\Delta t$	$3\Delta t$	$4\Delta t$	$5\Delta t$	$6\Delta t$
${}^t\mathrm{r}$	$\begin{Bmatrix} 0 \\ 0.392 \end{Bmatrix}$	$\begin{Bmatrix} 0.0307 \\ 1.45 \end{Bmatrix}$	$\begin{Bmatrix} 0.168 \\ 2.83 \end{Bmatrix}$	$\begin{Bmatrix} 0.487 \\ 4.14 \end{Bmatrix}$	$\begin{Bmatrix} 1.02 \\ 5.02 \end{Bmatrix}$	$\begin{Bmatrix} 1.70 \\ 5.26 \end{Bmatrix}$

t	$7\Delta t$	$8\Delta t$	$9\Delta t$	$10\Delta t$	$11\Delta t$	$12\Delta t$
${}^t\mathrm{r}$	$\begin{Bmatrix} 2.40 \\ 4.90 \end{Bmatrix}$	$\begin{Bmatrix} 2.91 \\ 4.17 \end{Bmatrix}$	$\begin{Bmatrix} 3.07 \\ 3.37 \end{Bmatrix}$	$\begin{Bmatrix} 2.77 \\ 2.78 \end{Bmatrix}$	$\begin{Bmatrix} 2.04 \\ 2.54 \end{Bmatrix}$	$\begin{Bmatrix} 1.02 \\ 2.60 \end{Bmatrix}$

Solving the resulting equations [Eq. (3.9)],

$$\begin{bmatrix} 25.5 & 0 \\ 0 & 12.8 \end{bmatrix} {}^{t+\Delta t}\mathbf{r} = {}^{t}\hat{\mathbf{R}}, \tag{f}$$

for each time step, we obtain the displacements shown in Table 3.1.

Implicit Integration Methods

In these methods we satisfy the equilibrium conditions at time $t+\Delta t$. The methods are unconditionally stable, and their effectiveness derives from the fact that to obtain accuracy in the integration the time step Δt can be very large. However, the integration methods are implicit and a factorization of the effective stiffness matrix is required for the solution.

In the *Houbolt method* [5] backward-difference formulas are used. If mass and damping effects are neglected ($\mathbf{M} = \mathbf{0}$ and $\mathbf{C} = \mathbf{0}$), the method reduces directly to a static analysis. That is, we obtain the static solution for time-dependent loads. On the other hand, the central difference method could not be used in this case. The *Wilson method* [6] is an extension of the linear acceleration method, in which a linear variation of acceleration from time t to time $t+\Delta t$ is assumed. In this method no special starting procedures are needed since displacements, velocities, and accelerations at time $t+\Delta t$ are expressed in terms of the same quantities at time t only. The *Newmark integration scheme* [7] can also be understood to be an extension of the linear acceleration method. There is a close relationship between the implementation of this method and the Wilson method.

Considering the Newmark method, we use the following assumptions.

$$ {}^{t+\Delta t}\dot{\mathbf{r}} = {}^{t}\dot{\mathbf{r}} + [(1-\delta)\,{}^{t}\ddot{\mathbf{r}} + \delta\,{}^{t+\Delta t}\ddot{\mathbf{r}}]\Delta t, \tag{3.12}$$

$$ {}^{t+\Delta t}\mathbf{r} = {}^{t}\mathbf{r} + {}^{t}\dot{\mathbf{r}}\,\Delta t + [(1/2-\alpha)\,{}^{t}\ddot{\mathbf{r}} + \alpha\,{}^{t+\Delta t}\ddot{\mathbf{r}}]\Delta t^2, \tag{3.13}$$

where α and δ are parameters determined to obtain integration accuracy and stability. For solution at time $t+\Delta t$, we consider the equations

$$\mathbf{M}\,{}^{t+\Delta t}\ddot{\mathbf{r}} + \mathbf{C}\,{}^{t+\Delta t}\dot{\mathbf{r}} + \mathbf{K}\,{}^{t+\Delta t}\mathbf{r} = {}^{t+\Delta t}\mathbf{R}. \tag{3.14}$$

Solving Eq. (3.13) for ${}^{t+\Delta t}\ddot{\mathbf{r}}$ in terms of ${}^{t+\Delta t}\mathbf{r}$ and substituting for ${}^{t+\Delta t}\ddot{\mathbf{r}}$ into Eq. (3.12), we can express ${}^{t+\Delta t}\ddot{\mathbf{r}}$ and ${}^{t+\Delta t}\dot{\mathbf{r}}$ in terms of ${}^{t+\Delta t}\mathbf{r}$ only. These two relations for ${}^{t+\Delta t}\ddot{\mathbf{r}}$ and ${}^{t+\Delta t}\dot{\mathbf{r}}$ are substituted into Eq. (3.14) to solve for ${}^{t+\Delta t}\mathbf{r}$. We then calculate ${}^{t+\Delta t}\dot{\mathbf{r}}$, ${}^{t+\Delta t}\ddot{\mathbf{r}}$ by Eqs. (3.12), (3.13).

The complete algorithm using the Newmark integration scheme is as follows. For the given **K**, **M**, **C**, ${}^0\mathbf{r}$, ${}^0\dot{\mathbf{r}}$, ${}^0\ddot{\mathbf{r}}$, we start with selecting the time step Δt, and the parameters $\delta \geq 0.5$ and $\alpha \geq 0.25(0.5+\delta)^2$. We calculate the integration constants $a_0, \ldots, a_7$, which are functions of Δt, α, δ, by

$$a_0 = \frac{1}{\alpha \Delta t^2} \qquad a_1 = \frac{\delta}{\alpha \Delta t} \qquad a_2 = \frac{1}{\alpha \Delta t} \qquad a_3 = \frac{1}{2\alpha} - 1, \tag{3.15}$$

$$a_4 = \frac{\delta}{\alpha} - 1 \qquad a_5 = \frac{\Delta t}{2}\left(\frac{\delta}{\alpha} - 2\right) \qquad a_6 = \Delta t(1-\delta) \qquad a_7 = \delta \Delta t.$$

Then we form and triangularize the effective stiffness matrix $\hat{\mathbf{K}}$ by

$$\hat{\mathbf{K}} = \mathbf{K} + a_0\mathbf{M} + a_1\mathbf{C}, \tag{3.16}$$

$$\hat{\mathbf{K}} = \mathbf{LDL}^T. \tag{3.17}$$

For each time step we calculate the effective loads ${}^{t+\Delta t}\hat{\mathbf{R}}$, the displacements ${}^{t+\Delta t}\mathbf{r}$, the accelerations ${}^{t+\Delta t}\ddot{\mathbf{r}}$ and the velocities ${}^{t+\Delta t}\dot{\mathbf{r}}$ at time $t+\Delta t$ by the following expressions

$${}^{t+\Delta t}\hat{\mathbf{R}} = {}^{t+\Delta t}\mathbf{R} + \mathbf{M}(a_0{}^t\mathbf{r} + a_2{}^t\dot{\mathbf{r}} + a_3{}^t\ddot{\mathbf{r}}) + \mathbf{C}(a_1{}^t\mathbf{r} + a_4{}^t\dot{\mathbf{r}} + a_5{}^t\ddot{\mathbf{r}}), \tag{3.18}$$

$$\mathbf{LDL}^T\,{}^{t+\Delta t}\mathbf{r} = {}^{t+\Delta t}\hat{\mathbf{R}}, \tag{3.19}$$

$${}^{t+\Delta t}\ddot{\mathbf{r}} = a_0({}^{t+\Delta t}\mathbf{r} - {}^t\mathbf{r}) - a_2{}^t\dot{\mathbf{r}} - a_3{}^t\ddot{\mathbf{r}}, \tag{3.20}$$

$${}^{t+\Delta t}\dot{\mathbf{r}} = {}^t\dot{\mathbf{r}} + a_6{}^t\ddot{\mathbf{r}} + a_7{}^{t+\Delta t}\ddot{\mathbf{r}}. \tag{3.21}$$

Example 3.2

Consider again the system of example 3.1 with the equilibrium equations

$$\begin{bmatrix} 2 & 0 \\ 0 & 1 \end{bmatrix}\ddot{\mathbf{r}} + \begin{bmatrix} 6 & -2 \\ -2 & 4 \end{bmatrix}\mathbf{r} = \begin{Bmatrix} 0 \\ 10 \end{Bmatrix}. \tag{a}$$

The object is to calculate the displacement response of the system by the Newmark method for 12 time steps, $\Delta t = 0.28$, and the initial values

$$ {}^0\mathbf{r} = \begin{Bmatrix} 0 \\ 0 \end{Bmatrix} \quad {}^0\dot{\mathbf{r}} = \begin{Bmatrix} 0 \\ 0 \end{Bmatrix} \quad {}^0\ddot{\mathbf{r}} = \begin{Bmatrix} 0 \\ 10 \end{Bmatrix}. \tag{b} $$

Assuming $\delta = 0.5$, $\alpha = 0.25$, the integration constants are [Eq. (3.15)]

$$ a_0 = 51.0 \quad a_1 = 7.14 \quad a_2 = 14.3 \quad a_3 = 1.00, \tag{c} $$

$$ a_4 = 1.00 \quad a_5 = 0.00 \quad a_6 = 0.14 \quad a_7 = 0.14, $$

and the effective stiffness matrix is [Eq. (3.16)]

$$ \hat{\mathbf{K}} = \begin{bmatrix} 6 & -2 \\ -2 & 4 \end{bmatrix} + 51.0 \begin{bmatrix} 2 & 0 \\ 0 & 1 \end{bmatrix} = \begin{bmatrix} 108 & -2 \\ -2 & 55 \end{bmatrix}. \tag{d} $$

For each time step we evaluate the effective loads ${}^{t+\Delta t}\hat{\mathbf{R}}$ [Eq. (3.18)], the displacements ${}^{t+\Delta t}\mathbf{r}$ [Eq. (3.19)], the accelerations ${}^{t+\Delta t}\ddot{\mathbf{r}}$ [Eq. (3.20)] and the velocities ${}^{t+\Delta t}\dot{\mathbf{r}}$ [Eq. (3.21)] by the following expressions

$$ {}^{t+\Delta t}\hat{\mathbf{R}} = \begin{Bmatrix} 0 \\ 10 \end{Bmatrix} + \begin{bmatrix} 2 & 0 \\ 0 & 1 \end{bmatrix} (51.0\,{}^t\mathbf{r} + 14.3\,{}^t\dot{\mathbf{r}} + 1.0\,{}^t\ddot{\mathbf{r}}), \tag{e} $$

$$ \mathbf{LDL}^T\,{}^{t+\Delta t}\mathbf{r} = {}^{t+\Delta t}\hat{\mathbf{R}}, \tag{f} $$

$$ {}^{t+\Delta t}\ddot{\mathbf{r}} = 51.0({}^{t+\Delta t}\mathbf{r} - {}^t\mathbf{r}) - 14.3\,{}^t\dot{\mathbf{r}} - 1.0\,{}^t\ddot{\mathbf{r}}, \tag{g} $$

$$ {}^{t+\Delta t}\dot{\mathbf{r}} = {}^t\dot{\mathbf{r}} + 0.14\,{}^t\ddot{\mathbf{r}} + 0.14\,{}^{t+\Delta t}\ddot{\mathbf{r}}. \tag{h} $$

The resulting displacements are summarized in Table 3.2.

Table 3.2. Displacements ${}^t\mathbf{r}$ as functions of time t, Newmark method

t	Δt	$2\Delta t$	$3\Delta t$	$4\Delta t$	$5\Delta t$	$6\Delta t$
${}^t\mathbf{r}$	$\begin{Bmatrix} 0.00673 \\ 0.364 \end{Bmatrix}$	$\begin{Bmatrix} 0.0505 \\ 1.35 \end{Bmatrix}$	$\begin{Bmatrix} 0.189 \\ 2.68 \end{Bmatrix}$	$\begin{Bmatrix} 0.485 \\ 4.00 \end{Bmatrix}$	$\begin{Bmatrix} 0.961 \\ 4.95 \end{Bmatrix}$	$\begin{Bmatrix} 1.58 \\ 5.34 \end{Bmatrix}$

t	$7\Delta t$	$8\Delta t$	$9\Delta t$	$10\Delta t$	$11\Delta t$	$12\Delta t$
${}^t\mathbf{r}$	$\begin{Bmatrix} 2.23 \\ 5.13 \end{Bmatrix}$	$\begin{Bmatrix} 2.76 \\ 4.48 \end{Bmatrix}$	$\begin{Bmatrix} 3.00 \\ 3.64 \end{Bmatrix}$	$\begin{Bmatrix} 2.85 \\ 2.90 \end{Bmatrix}$	$\begin{Bmatrix} 2.28 \\ 2.44 \end{Bmatrix}$	$\begin{Bmatrix} 1.40 \\ 2.31 \end{Bmatrix}$

3.1.2 Mode Superposition

In this approach a change of basis from the finite element nodal displacements to the eigenvectors of the generalized eigenproblem is performed prior to the time integration. We use the transformation

$$\mathbf{r}(t) = \mathbf{P}\,\mathbf{Z}(t)\,, \tag{3.22}$$

where $\mathbf{P}$ is an $n{\times}n$ transformation matrix and the components of $\mathbf{Z}$ are the *generalized displacements*, also called *modal coordinates* – or *normal coordinates*. The objective of the transformation is to obtain new system stiffness, mass and damping matrices, which have a smaller band width than the original system matrices. Substituting Eq. (3.22) into Eq. (3.1) and pre-multiplying by $\mathbf{P}^T$, we obtain

$$\tilde{\mathbf{M}}\ddot{\mathbf{Z}}(t) + \tilde{\mathbf{C}}\dot{\mathbf{Z}}(t) + \tilde{\mathbf{K}}\mathbf{Z}(t) = \tilde{\mathbf{R}}(t)\,, \tag{3.23}$$

where

$$\tilde{\mathbf{M}} = \mathbf{P}^T\mathbf{M}\,\mathbf{P} \qquad \tilde{\mathbf{C}} = \mathbf{P}^T\mathbf{C}\,\mathbf{P} \qquad \tilde{\mathbf{K}} = \mathbf{P}^T\mathbf{K}\,\mathbf{P} \qquad \tilde{\mathbf{R}} = \mathbf{P}^T\mathbf{R}\,. \tag{3.24}$$

An effective transformation matrix $\mathbf{P}$ is the displacement solutions of the free-vibration equations [Eq. (2.3)]

$$\mathbf{M}\ddot{\mathbf{r}} + \mathbf{K}\mathbf{r} = \mathbf{0}\,. \tag{3.25}$$

The result is the generalized eigenproblem [Eq. (2.8)]

$$\mathbf{K}\boldsymbol{\Phi} = \boldsymbol{\Lambda}\,\mathbf{M}\boldsymbol{\Phi}\,, \tag{3.26}$$

where $\boldsymbol{\Lambda}$ is the spectral matrix (diagonal matrix of the eigenvalues). Thus, for $\mathbf{P} = \boldsymbol{\Phi}$ the transformation of Eq. (3.22) becomes

$$\mathbf{r}(t) = \boldsymbol{\Phi}\,\mathbf{Z}(t)\,. \tag{3.27}$$

For the complete response, the displacements are obtained by superposition of the response in each mode

$$\mathbf{r}(t) = \sum_{i=1}^{n} \boldsymbol{\Phi}_i\, Z_i(t)\,. \tag{3.28}$$

Using the orthogonality properties [Eq. (2.9)], multiplying both sides of Eq. (3.28) by $\boldsymbol{\Phi}_i^T\mathbf{M}$ and rearranging, we obtain for Z_i

$$Z_i = \frac{\boldsymbol{\Phi}_i^T\mathbf{M}\,\mathbf{r}(t)}{\boldsymbol{\Phi}_i^T\mathbf{M}\boldsymbol{\Phi}_i}\,. \tag{3.29}$$

Substituting [Eqs. (2.17), (2.18)]

$$\mathbf{\Phi}^T \mathbf{M} \mathbf{\Phi} = \mathbf{I} \qquad \mathbf{\Phi}^T \mathbf{K} \mathbf{\Phi} = \mathbf{\Lambda}\,, \tag{3.30}$$

into Eqs. (3.24), and substituting the resulting expressions and

$$\mathbf{T}(t) = \mathbf{\Phi}^T \mathbf{R}(t)\,, \tag{3.31}$$

into Eq. (3.23), we obtain the equilibrium equations that correspond to the modal generalized displacements

$$\ddot{\mathbf{Z}}(t) + \mathbf{\Phi}^T \mathbf{C} \mathbf{\Phi}\, \dot{\mathbf{Z}}(t) + \mathbf{\Lambda} \mathbf{Z}(t) = \mathbf{T}(t)\,. \tag{3.32}$$

Solution without Damping

If damping effects are not considered, Eq. (3.32) becomes

$$\ddot{\mathbf{Z}}(t) + \mathbf{\Lambda} \mathbf{Z}(t) = \mathbf{T}(t)\,. \tag{3.33}$$

Since $\mathbf{\Lambda}$ is a diagonal matrix we obtain decoupled equilibrium equations. The individual equations are of the form

$$\ddot{Z}_i(t) + \lambda_i Z_i(t) = T_i(t)\,, \tag{3.34}$$

where

$$T_i(t) = \mathbf{\Phi}_i^T \mathbf{R}(t)\,. \tag{3.35}$$

The initial conditions at time 0 are obtained by [Eqs. (3.29), (3.30)]

$${}^0\mathbf{Z} = \mathbf{\Phi}^T \mathbf{M}\, {}^0\mathbf{r} \qquad {}^0\dot{\mathbf{Z}} = \mathbf{\Phi}^T \mathbf{M}\, {}^0\dot{\mathbf{r}}\,. \tag{3.36}$$

In summary, the response analysis by mode superposition involves the following steps:

- Solution of the eigenproblem [Eq. (3.26)].
- Solution of the decoupled equations of motion [Eq. (3.34)].
- Superposition of the response [Eq. (3.28)].

Example 3.3

Consider again the system of example 3.1 with the following given data

$$\mathbf{K} = \begin{bmatrix} 6 & -2 \\ -2 & 4 \end{bmatrix} \qquad \mathbf{M} = \begin{bmatrix} 2 & 0 \\ 0 & 1 \end{bmatrix} \qquad \mathbf{R} = \begin{Bmatrix} 0 \\ 10 \end{Bmatrix}. \tag{a}$$

The object is to calculate the displacement response of the system using mode superposition and the Newmark method with time step $\Delta t = 0.28$ for the time integration. The generalized eigenproblem is

$$\begin{bmatrix} 6 & -2 \\ -2 & 4 \end{bmatrix} \boldsymbol{\Phi} = \lambda \begin{bmatrix} 2 & 0 \\ 0 & 1 \end{bmatrix} \boldsymbol{\Phi}, \tag{b}$$

and the two solutions obtained are

$$\boldsymbol{\Lambda} = \begin{bmatrix} \lambda_1 & \\ & \lambda_2 \end{bmatrix} = \begin{bmatrix} 2 & \\ & 5 \end{bmatrix}, \tag{c}$$

$$\boldsymbol{\Phi} = [\boldsymbol{\Phi}_1, \boldsymbol{\Phi}_2] = \begin{bmatrix} 1/\sqrt{3} & 1/2\sqrt{2/3} \\ 1/\sqrt{3} & -\sqrt{2/3} \end{bmatrix}.$$

Using Eq. (3.31) we find

$$\mathbf{T} = \boldsymbol{\Phi}^T \mathbf{R} = \begin{bmatrix} 1/\sqrt{3} & 1/\sqrt{3} \\ 1/2\sqrt{2/3} & -\sqrt{2/3} \end{bmatrix} \begin{Bmatrix} 0 \\ 10 \end{Bmatrix} = \begin{Bmatrix} 10/\sqrt{3} \\ -10\sqrt{2/3} \end{Bmatrix}, \tag{d}$$

and the decoupled equilibrium equations are given by [Eq. (3.33)]

$$\ddot{\mathbf{Z}}(t) + \begin{bmatrix} 2 & \\ & 5 \end{bmatrix} \mathbf{Z}(t) = \begin{Bmatrix} 10/\sqrt{3} \\ -10\sqrt{2/3} \end{Bmatrix}. \tag{e}$$

Solving the decoupled equations by the Newmark method we obtain the results shown in Table 3.3. Substituting these results in [see Eq. (3.27)]

Table 3.3. Generalized displacements $\mathbf{Z}(t)$ as functions of time t

t	Δt	$2\Delta t$	$3\Delta t$	$4\Delta t$	$5\Delta t$	$6\Delta t$
$\mathbf{Z}(t)$	$\begin{Bmatrix} 0.2258 \\ -0.3046 \end{Bmatrix}$	$\begin{Bmatrix} 0.8199 \\ -0.7920 \end{Bmatrix}$	$\begin{Bmatrix} 1.8070 \\ -2.1239 \end{Bmatrix}$	$\begin{Bmatrix} 2.379 \\ -2.939 \end{Bmatrix}$	$\begin{Bmatrix} 4.123 \\ -3.258 \end{Bmatrix}$	$\begin{Bmatrix} 5.064 \\ -2.632 \end{Bmatrix}$

t	$7\Delta t$	$8\Delta t$	$9\Delta t$	$10\Delta t$	$11\Delta t$	$12\Delta t$
$\mathbf{Z}(t)$	$\begin{Bmatrix} 5.579 \\ -2.161 \end{Bmatrix}$	$\begin{Bmatrix} 5.774 \\ -1.156 \end{Bmatrix}$	$\begin{Bmatrix} 5.521 \\ -0.3307 \end{Bmatrix}$	$\begin{Bmatrix} 4.855 \\ -0.0041 \end{Bmatrix}$	$\begin{Bmatrix} 3.866 \\ -0.2482 \end{Bmatrix}$	$\begin{Bmatrix} 2.773 \\ -1.088 \end{Bmatrix}$

$$\mathbf{r}(t) = \begin{bmatrix} 1/\sqrt{3} & 1/2\sqrt{2/3} \\ 1/\sqrt{3} & -\sqrt{2/3} \end{bmatrix} \mathbf{Z}(t)\,, \tag{f}$$

we obtain the displacement response as shown in Table 3.2 (example 3.2). As expected, the response is the same as the response obtained in example 3.2 when the Newmark method is used in direct integration.

Consideration of Damping

When damping is considered, the equations of motion for a freely vibrating system are

$$\mathbf{M}\ddot{\mathbf{r}} + \mathbf{C}\dot{\mathbf{r}} + \mathbf{K}\mathbf{r} = \mathbf{0}\,. \tag{3.37}$$

Expressing the displacements in terms of the natural modes of the system without damping we obtain [Eqs. (3.27), (3.28)]

$$\mathbf{r} = \sum_{i=1}^{n} \mathbf{\Phi}_i Z_i = \mathbf{\Phi}\mathbf{Z}\,. \tag{3.38}$$

Substituting Eq. (3.38) into Eq. (3.37) and pre-multiplying by $\mathbf{\Phi}^T$ gives

$$\tilde{\mathbf{M}}\ddot{\mathbf{Z}} + \tilde{\mathbf{C}}\dot{\mathbf{Z}} + \tilde{\mathbf{K}}\mathbf{Z} = \mathbf{0}\,, \tag{3.39}$$

where

$$\tilde{\mathbf{M}} = \mathbf{\Phi}^T\mathbf{M}\mathbf{\Phi} \qquad \tilde{\mathbf{C}} = \mathbf{\Phi}^T\mathbf{C}\mathbf{\Phi} \qquad \tilde{\mathbf{K}} = \mathbf{\Phi}^T\mathbf{K}\mathbf{\Phi}\,. \tag{3.40}$$

We still would like to deal with a decoupled system in Eqs. (3.39). It has been noted that matrices $\tilde{\mathbf{M}}$ and $\tilde{\mathbf{K}}$ are diagonal [Eqs. (3.30)]. If matrix $\tilde{\mathbf{C}}$ is also diagonal, Eq. (3.39) represents uncoupled differential equations in the modal coordinates Z_i.

We may distinguish between *classical damping* and *non-classical damping*. The system is said to have classical damping when classical modal analysis is applicable. Such systems possess the same natural modes as those of the un-damped system. The dynamic response of linear systems with classical damping can be determined by classical modal analysis. Thus the response in each natural vibration mode can be computed independently of the others, and the modal responses can be combined to determine the total response. Systems with damping such that matrix $\tilde{\mathbf{C}}$ is non-diagonal are said to have non-classical damping. These systems are not amenable to classical modal analysis, and they do not possess the same natural modes as the un-damped system. Classical modal analysis is not applicable to a structure consisting of subsystems with very different levels of damping. For such systems the equations of motion cannot be uncoupled by transforming to modal coordinates of the system without damping. Such systems can be analyzed by transforming the equations of motion to the eigenvectors of the complex problem, or by solution of the coupled system of differential equations. Classical modal analysis is also not applicable to inelastic systems.

The mode superposition analysis is particularly effective in cases where damping is proportional, that is, the eigenvectors are also **C**-orthogonal

$$\mathbf{\Phi}_i^T \mathbf{C} \mathbf{\Phi}_j = 2M_i \omega_i \zeta_i \delta_{ij} , \tag{3.41}$$

where ω_i is the circular frequency, δ_{ij} is the Kronecker delta and ζ_i is a modal damping parameter, called the *damping ratio*. Denoting

$$C_i = \mathbf{\Phi}_i^T \mathbf{C} \mathbf{\Phi}_i , \tag{3.42}$$

substituting Eq. (3.42) into Eq. (3.41) and rearranging, we obtain for $i = j$

$$\zeta_i = \frac{C_i}{2M_i \omega_i} . \tag{3.43}$$

The damping ratio ζ_i is usually not computed by Eq. (3.43). It is estimated using experimental data for structures similar to the one being analyzed.

Each of the differential equations (3.39) in modal coordinates is

$$M_i \ddot{Z}_i + C_i \dot{Z}_i + K_i Z_i = 0 , \tag{3.44}$$

where M_i and K_i are defined by Eqs. (2.11). Substituting Eq. (3.41), $\omega^2 = \lambda$ and $K_i = \lambda_i M_i$ [Eq. (2.12)] into Eq. (3.44), and dividing the resulting equation by M_i, we obtain

$$\ddot{Z}_i + 2\omega_i \zeta_i \dot{Z}_i + \omega_i^2 Z_i = 0 . \tag{3.45}$$

Equation (3.45) can be solved for $Z_i(t)$ by methods suitable for free vibration problems of a single degree of freedom.

For the case of proportional damping under consideration, the equations of motion are reduced to [see Eqs. (3.32), (3.45)]

$$\ddot{Z}_i(t) + 2\omega_i \zeta_i \dot{Z}_i(t) + \omega_i^2 Z_i(t) = T_i(t) , \tag{3.46}$$

where $T_i(t) = \mathbf{\Phi}_i^T \mathbf{R}(t)$ [Eq. (3.35)]. The decoupled equation (3.46) can be expressed in matrix form as

$$\mathbf{I}\,\ddot{\mathbf{Z}} + \mathbf{C}_d\,\dot{\mathbf{Z}} + \boldsymbol{\Omega}^2\mathbf{Z} = \mathbf{T}\,, \tag{3.47}$$

where $\mathbf{T} = \boldsymbol{\Phi}^T\mathbf{R}(t)$ [Eq. (3.31)] and the identity matrix $\mathbf{I} = \boldsymbol{\Phi}^T\mathbf{M}\boldsymbol{\Phi}$ is the mass matrix in normalized coordinates [Eq. (3.30)]. The damping matrix $\mathbf{C}_d = \boldsymbol{\Phi}^T\mathbf{C}\boldsymbol{\Phi}$ and the stiffness matrix $\boldsymbol{\Omega}^2 = \boldsymbol{\Phi}^T\mathbf{K}\boldsymbol{\Phi}$ in these coordinates are diagonal low-order matrices, given by

$$\mathbf{C}_d = \boldsymbol{\Phi}^T\mathbf{C}\boldsymbol{\Phi} = \begin{bmatrix} 2\omega_1\zeta_1 & & \\ & \ddots & \\ & & 2\omega_p\zeta_p \end{bmatrix}, \tag{3.48}$$

$$\boldsymbol{\Omega}^2 = \boldsymbol{\Phi}^T\mathbf{K}\boldsymbol{\Phi} = \begin{bmatrix} \omega_1^2 & & \\ & \ddots & \\ & & \omega_p^2 \end{bmatrix}. \tag{3.49}$$

In general, the damping matrix cannot be constructed from element damping matrices, such as the mass and stiffness matrices. Introduction of the matrix is often based on approximation of the overall energy dissipation during the system response.

The solution of Eq. (3.46) can be accomplished by several methods. For certain loadings which can be expressed as an analytic function, exact mathematical solutions are possible. Otherwise, one of the following methods is often used:

- *Direct step-by-step* solution, using a numerical finite difference method.
- *Duhamel integral*, which is numerically integrated.
- *Transformation to the frequency domain.*
- *Piecewise exact method*, using a series of straight lines for representing the loading between unequal time intervals.
- *Response spectra analysis*, where the load is specified as a response spectra.

In summary, computation of the dynamic response by modal analysis involves the following steps:

- Determine the eigenpairs λ_i, $\boldsymbol{\Phi}_i$ by solving the eigenproblem [Eq. (3.26)]

$$\mathbf{K}\boldsymbol{\Phi} = \boldsymbol{\Lambda}\,\mathbf{M}\boldsymbol{\Phi}\,. \tag{3.50}$$

- Compute the modal coordinates $Z_i(t)$. If damping effects are not considered, solve the individual decoupled equations [Eq. (3.34)]

$$\ddot{Z}_i(t) + \lambda_i Z_i(t) = T_i(t) . \tag{3.51}$$

In case of proportional damping, solve the decoupled equations [Eq. (3.46)]

$$\ddot{Z}_i(t) + 2\omega_i \zeta_i \dot{Z}_i(t) + \omega_i^2 Z_i(t) = T_i(t) . \tag{3.52}$$

- Compute the nodal displacements by [Eq. (3.28)]

$$\mathbf{r}(t) = \sum_{i=1}^{n} \mathbf{\Phi}_i \, Z_i(t) . \tag{3.53}$$

3.1.3 Special Analysis Procedures

Modal Response Contributions

Consider a common case where the loadings are given in the form

$$\mathbf{R}(t) = \mathbf{S}\, P(t) , \tag{3.54}$$

and the spatial distribution **S** can be expressed as

$$\mathbf{S} = \sum_{i=1}^{n} \mathbf{S}_i = \sum_{i=1}^{n} \Gamma_i \mathbf{M} \mathbf{\Phi}_i , \tag{3.55}$$

where $\mathbf{S}_i$ is the contribution of the *i*th mode to **S**. The factor Γ_i is called a *modal participation factor*. Multiplying both sides of Eq (3.55) by $\mathbf{\Phi}_i^T$, assuming $\mathbf{\Phi}_i^T \mathbf{M} \mathbf{\Phi}_i = 1$ and rearranging, we obtain for Γ_i

$$\Gamma_i = \mathbf{\Phi}_i^T \mathbf{S} . \tag{3.56}$$

Substituting the contribution, $T_i(t)$, of the *i*th mode,

$$T_i(t) = \mathbf{\Phi}_i^T \mathbf{R}_i(t) = \mathbf{\Phi}_i^T \mathbf{S}_i P(t) = \Gamma_i P(t) , \tag{3.57}$$

into Eq. (3.46), we obtain

$$\ddot{Z}_i(t) + 2\omega_i \zeta_i \dot{Z}_i(t) + \omega_i^2 Z_i(t) = \Gamma_i P(t) . \tag{3.58}$$

The solution $Z_i(t)$ can expressed in terms of a single degree of freedom system with properties (ω_i, ζ_i) of the ith mode of the original system excited by the force $P(t)$. The response of this system is given by

$$\ddot{D}_i(t) + 2\omega_i\zeta_i\dot{D}_i(t) + \omega_i^2 D_i(t) = P(t), \tag{3.59}$$

where D_i denotes the solution of the single degree of freedom system. Comparing Eqs. (3.58) and (3.59), we obtain

$$Z_i(t) = \Gamma_i D_i(t). \tag{3.60}$$

Thus, once D_i is known, Z_i is readily available.

The contribution $\mathbf{r}_i(t)$ of the ith mode to the nodal displacements $\mathbf{r}(t)$ [Eqs. (3.53), (3.60)] is given by

$$\mathbf{r}_i(t) = \mathbf{\Phi}_i Z_i(t) = \Gamma_i \mathbf{\Phi}_i D_i(t). \tag{3.61}$$

Substituting $\mathbf{K\Phi}_i = \omega_i^2 \mathbf{M\Phi}_i$ and Eqs. (3.38), (3.55), (3.61) into the expression $\mathbf{R}_{sti}(t) = \mathbf{K}\,\mathbf{r}_i(t)$ of the *equivalent static forces* associated with the ith mode, we obtain

$$\mathbf{R}_{sti}(t) = \mathbf{K}\,\mathbf{r}_i(t) = \omega_i^2 \mathbf{M\Phi}_i Z_i(t) = \mathbf{S}_i \omega_i^2 D_i(t). \tag{3.62}$$

Thus, the total response $r(t)$ can be expressed now as the sum of the contributions $r_i(t)$ of the modes by

$$r(t) = \sum_{i=1}^{n} r_i(t) = \sum_{i=1}^{n} r_i^{st} \omega_i^2 D_i(t). \tag{3.63}$$

where r_i^{st} is the static value of the response due to the external forces $\mathbf{S}_i$.

In summary, the contribution $r_i(t)$ of the ith mode to the dynamic response is obtained by multiplying the results of two analyses:

- Static analysis of the structure subjected to external forces $\mathbf{S}_i$.
- Dynamic analysis of the ith mode single degree of freedom system excited by the force $P(t)$.

Thus, modal analysis requires static analysis for n sets of external forces $\mathbf{S}_i$, and dynamic analysis of n different single degree of freedom systems. The dynamic response is obtained by combining the effect of the modes.

Define the *dynamic response factor* for the ith mode as the ratio of dynamic amplitude to corresponding static amplitude. This factor can help to identify the modes that may contribute significantly to the response. Using this definition, the two methods of static correction and mode acceleration superposition are described in the following subsections.

Static Correction Method

The *static correction method* is effective when many modes must be included to represent the distribution **S** of the applied forces, but the exciting force $P(t)$ is such that the dynamic response factor for only a few lower modes is significantly larger than unity. That is, for some of the higher modes the dynamic response factor may be only slightly larger than unity. This is the case when the higher-mode period is much shorter than the period of the harmonic excitation. The response in such a higher mode could be determined by static analysis.

The modal contributions to the response can be expressed as

$$r(t) = \sum_{i=1}^{nd} r_i(t) + \sum_{i=nd+1}^{n} r_i(t), \tag{3.64}$$

where *nd* is the number of modes with natural periods such that the dynamic effects are significant. Defining $\bar{r}_i = r_i^{st} / r^{st}$ and substituting into Eq. (3.63), we obtain the following expression for the dynamic response $r_i(t)$

$$r_i(t) = r^{st} \bar{r}_i \omega_i^2 D_i(t). \tag{3.65}$$

A static solution of Eq. (3.59) gives $D_i(t)$ for modes nd+1 to n. Dropping the velocity $\dot{D}_i(t)$ and the acceleration , we obtain

$$\omega_i^2 D_i(t) = P(t). \tag{3.66}$$

Substituting Eqs. (3.65) and (3.66) into Eq. (3.64) gives

$$r(t) = r^{st} \sum_{i=1}^{nd} \bar{r}_i \omega_i^2 D_i(t) + r^{st} P(t) \sum_{i=nd+1}^{n} \bar{r}_i . \tag{3.67}$$

Thus, Eq. (3.59) needs to be solved by dynamic analysis procedures only for the first *nd* modes. The second term in Eq. (3.67) is the static response solution for the higher modes nd+1 to n, which may be considered as the static correction to the dynamic response solution given by the first term. Equation (3.67) can be expressed in the following final form

$$r(t) = r^{st} \left[P(t) + \sum_{i=1}^{nd} \bar{r}_i \left(\omega_i^2 D_i(t) - P(t) \right) \right]. \tag{3.68}$$

Mode Acceleration Superposition Method

The *mode acceleration superposition method* can provide a similar effect as the static correction method. Using Eq. (3.65), the total response is

$$r(t) = r^{st}\sum_{i=1}^{n}\bar{r}_i\omega_i^2 D_i(t) . \tag{3.69}$$

Rearranging Eq. (3.59) gives

$$\omega_i^2 D_i(t) = P(t) - \ddot{D}_i(t) - 2\omega_i\zeta_i\dot{D}_i(t) . \tag{3.70}$$

Substituting Eq. (3.70) into Eq. (3.69) and noting that the sum of all $\bar{r}_i$ equals unity, we obtain

$$r(t) = r^{st}\left\{P(t) - \sum_{i=1}^{n}\bar{r}_i[\ddot{D}_i(t) + 2\omega_i\zeta_i\dot{D}_i(t)]\right\}. \tag{3.71}$$

This expression can be interpreted as the static solution given by the first term on the right side, modified by the second term to obtain the dynamic response of the system. If the response in all modes higher than the first *nd* modes is essentially static, the summation can be truncated accordingly.

The mode acceleration superposition method is equivalent to the static correction method. Thus, the two methods should provide identical results. The static correction method is usually easier to implement, because it requires simple modification of classical *modal analysis* (or the classical *mode displacement* superposition method).

3.2 Reduced Basis

In many structural analysis problems, a large system of simultaneous equations must be solved repeatedly in order to evaluate the response of the structure. This process involves much computational effort, particularly for large-scale, nonlinear and time-dependent (dynamic) problems.

The basic idea of the reduced-basis approach is that of transforming a problem with a large number of Degrees Of Freedom (DOF) into one with a much smaller number of DOF. The response of the system, which was originally described by a large number of DOF, is approximated by a linear combination of a few pre-selected basis vectors. The problem is then stated in terms of a small number of unknown coefficients of the basis vectors. The resulting analysis model is more efficient, since only the corresponding small system of equations must be solved. This approach is most effective in cases where highly accurate approximations can be achieved by solving the reduced system of equations.

The solution process involves the following two main stages:

- Generation of basis vectors for approximating the response.
- Determination of the unknown coefficients of the basis vectors, using a variational technique.

Reduced-basis has been used in various problems and applications [8], including the unified approach presented in this text. Formulations of static and dynamic analysis are described in Sects. 3.2.1 and 3.2.2, respectively.

3.2.1 Static Analysis

We assume that the displacement vector **r** can be approximated by a linear combination of pre-selected s linearly independent *basis vectors*, also called *global approximation vectors*, $\mathbf{r}_1, \mathbf{r}_2, ..., \mathbf{r}_s$

$$\mathbf{r} = y_1\mathbf{r}_1 + y_2\mathbf{r}_2 + ... + y_s\mathbf{r}_s = \mathbf{r}_B\,\mathbf{y}, \tag{3.72}$$

where s is much smaller than the number of degrees of freedom n, $\mathbf{r}_B$ is the $n{\times}s$ matrix of the basis vectors and **y** is a vector of unknown coefficients

$$\mathbf{r}_B = [\mathbf{r}_1, \mathbf{r}_2, \ldots, \mathbf{r}_s] \qquad \mathbf{y}^T = \{y_1, y_2, \ldots, y_s\}. \tag{3.73}$$

The space spanned by the global approximation vectors (matrix $\mathbf{r}_B$), is usually referred to as the reduced basis subspace.

The justification for using this approach is that the large number of degrees of freedom describing the response of the system is often dictated by such considerations as complex topology or numerous changes in the system properties, rather than by the complexity of the response. The power of the reduced basis method derives from the fact that for many systems of practical interest, the transformation of Eq. (3.72) can provide highly accurate approximations of **r**, even when s is very much smaller than n.

The equilibrium equations are now approximated by a smaller system of equations in the new unknowns **y**. Substituting Eq. (3.72) into Eq. (1.2) and pre-multiplying the resultant equation by $\mathbf{r}_B^T$ gives the $s{\times}s$ system

$$\mathbf{r}_B^T\mathbf{K}\,\mathbf{r}_B\mathbf{y} = \mathbf{r}_B^T\mathbf{R}\,. \tag{3.74}$$

Introducing the notation

$$\mathbf{K}_R = \mathbf{r}_B^T\mathbf{K}\,\mathbf{r}_B \qquad \mathbf{R}_R = \mathbf{r}_B^T\mathbf{R}\,, \tag{3.75}$$

and substituting Eqs. (3.75) into Eq. (3.74), we obtain

$$\mathbf{K}_R\,\mathbf{y} = \mathbf{R}_R. \tag{3.76}$$

The $s{\times}s$ matrix $\mathbf{K}_R$ is full but is symmetric and much smaller in size than the $n{\times}n$ matrix $\mathbf{K}$ of the original system. That is, rather than computing the exact solution by solving the large $n{\times}n$ system in the original equations, we first solve the smaller $s{\times}s$ system in Eq. (3.76) for $\mathbf{y}$, and then evaluate the approximate displacements $\mathbf{r}$ for the computed $\mathbf{y}$ by Eq. (3.72).

The reduced set of Eqs. (3.76) can be obtained from the total potential energy expression

$$U = 1/2\,\mathbf{r}^T\,\mathbf{K}\,\mathbf{r} - \mathbf{r}^T\mathbf{R}. \tag{3.77}$$

Substituting Eq. (3.72) into Eq. (3.77) gives

$$U = \mathbf{y}^T\,\mathbf{r}_B^T\,(1/2\,\mathbf{K}\,\mathbf{r}_B\,\mathbf{y} - \mathbf{R}). \tag{3.78}$$

Differentiating Eq. (3.78) with respect to $\mathbf{y}$, setting the result equal to zero and using the symmetry of $\mathbf{K}$, we obtain the following conditions for minimum potential energy

$$\mathbf{r}_B^T\,\mathbf{K}\,\mathbf{r}_B\,\mathbf{y} - \mathbf{r}_B^T\,\mathbf{R} = \mathbf{0}. \tag{3.79}$$

These conditions are equivalent to Eq. (3.76) [see Eq. (3.75)].

The effectiveness of the reduced-basis approach depends, to a great extent, on the appropriate choice of the basis vectors $\mathbf{r}_1$, $\mathbf{r}_2$, ..., $\mathbf{r}_s$, which span the reduced basis subspace. Proper selection of the basis vectors is perhaps the most important factor affecting the successful application of the approach. Displacement vectors of previously analyzed designs can be used, but it should be emphasized that an ad hoc or intuitive choice of such vectors may not lead to satisfactory approximations. In addition, calculation of the basis vectors requires several exact analyses of the structure for the basis design points, which might involve extensive computational effort.

An ideal set of basis vectors will provide accurate results with a small computational effort. Specifically, the following criteria must be satisfied in arriving at the basis vectors to be used [9]:

- Linear independence.
- Low computational expense in their generation, and simplicity of automatic selection of their number.
- Good approximation properties, in the sense of high accuracy of the solution obtained by using these vectors.
- Simplicity of obtaining the system response characteristics using these vectors.

The first criterion is necessary for convergence of the approximation process and the latter three criteria govern the computational efficiency of the method and its effectiveness. The proper selection of the basis vectors depends on the system response characteristics being approximated, as well as the particular application.

3.2.2 Dynamic Analysis

From the viewpoint of computational effort, the reduction of degrees of freedom is more important in dynamic problems than in static problems, because the solution must be performed successively at many different times to generate the time history of the response. The discretized model of a complicated system may have numerous degrees of freedom. Therefore, it is customary in dynamic analysis to reduce the equations of motion to a much smaller number before the dynamic response is calculated.

The earliest applications of these methods have been to eigenvalue problems, and date back to the early 1960's. At that time, the calculation of the eigenvalues and eigenvectors of large systems by the available algorithms required much computational effort. The earliest reduction method applied to linear dynamic problems is the classical modal superposition technique, in which the global approximation vectors are selected to be linear vibration modes. It has been noted that in the analysis of linear structures the response is often expressed in terms of the un-damped free vibration mode shapes, using only the lower modes. The main analytical problem then becomes the evaluation of the mode shapes, and the problem of reducing the number of degrees of freedom is transferred to this phase of the analysis.

The different reduced-basis methods for eigenproblems can be classified into two general categories according to the selection of the basis vectors:

- *Single step methods*, where both the basis vectors and the reduced equations are generated at a single step and then used to evaluate the approximate eigenvectors and eigenvalues. Examples for this class of methods include *Rayleigh-Ritz reduction* (projection method), *static condensation*, and *dynamic substructuring* or component mode synthesis techniques.
- *Multi-step methods*, or generalized reduction methods, in which the basis vectors and the reduced equations are modified in successive iterations. In some of these methods the initial matrix consists of very simple vectors. Examples of these methods include the *Lanczos method* and *subspace iteration*.

Two of the most widely used modal methods for transient structural analysis are the *mode-displacement* method and the *mode-acceleration* method. In these two methods the dynamic response is approximated by a linear combination of modal displacements and modal accelerations, respectively. To improve the convergence rate of modal methods, a higher-order modal method can be used. In an effort to avoid the cost associated with calculating the modes, other choices have been proposed for the reduced-basis subspace. These include Taylor subspace, in which the global approximation vectors are the various-order time derivatives of the response, Lagrange subspace, in which the response vectors at various times are selected as the approximation vectors, Runge-Kutta subspace and Krylov subspace. A comprehensive discussion and references on these and other methods are given elsewhere [8].

Considering again the equations of motion [Eq. (3.1)] with no damping and ignoring the notation of the time variable (t) we have

$$\mathbf{M}\ddot{\mathbf{r}} + \mathbf{K}\,\mathbf{r} = \mathbf{R}\,. \tag{3.80}$$

A standard model reduction process for the system of Eqs. (3.80) can be described as the simple coordinate transformation of Eq. (3.72). Substituting the latter equation into Eq. (3.80) and pre-multiplying the resultant equation by $\mathbf{r}_B^T$, we obtain

$$\mathbf{M}_R\ddot{\mathbf{y}} + \mathbf{K}_R\mathbf{y} = \mathbf{R}_R\,, \tag{3.81}$$

where the reduced mass matrix, stiffness matrix and load vector are defined as

$$\mathbf{M}_R = \mathbf{r}_B^T\mathbf{M}\,\mathbf{r}_B \qquad \mathbf{K}_R = \mathbf{r}_B^T\mathbf{K}\,\mathbf{r}_B \qquad \mathbf{R}_R = \mathbf{r}_B^T\mathbf{R}\,. \tag{3.82}$$

The transformation matrix $\mathbf{r}_B$ could be a modal matrix containing normal modes of the system. The model reduction process is then just the standard mode-displacement method, and Eq. (3.81) is a set of uncoupled linear ordinary differential equations. More generally, $\mathbf{r}_B$ could contain general Ritz vectors, as described in Sect. 2.6. Both matrices $\mathbf{M}_R$ and $\mathbf{K}_R$ are full if the vectors are not orthogonal.

Rayleigh-Ritz Analysis

Consider again the equations of motion for a system subjected to forces $\mathbf{R}(t) = \mathbf{S}\,P(t)$ [Eq. 3.54)]

$$\mathbf{M}\ddot{\mathbf{r}} + \mathbf{C}\dot{\mathbf{r}} + \mathbf{K}\,\mathbf{r} = \mathbf{S}\,P(t)\,. \tag{3.83}$$

In the Rayleigh-Ritz method, the displacements are expressed as a linear combination of several Ritz vectors $\mathbf{r}_1, \mathbf{r}_2, \ldots, \mathbf{r}_s$

$$\mathbf{r}(t) = y_1(t)\,\mathbf{r}_1 + y_2(t)\,\mathbf{r}_2 + \ldots + y_s(t)\mathbf{r}_s = \mathbf{r}_B\,\mathbf{y}(t), \tag{3.84}$$

where $\mathbf{y}(t)$ is the vector of the *generalized Ritz coordinates* $y_i(t)$ and $\mathbf{r}_B$ is the $n{\times}s$ matrix of the Ritz basis vectors $\mathbf{r}_i$. Substituting the Ritz transformation of Eq. (3.84) into Eq. (3.83) and pre-multiplying by $\mathbf{r}_B^T$, we obtain

$$\tilde{\mathbf{M}}\ddot{\mathbf{y}} + \tilde{\mathbf{C}}\dot{\mathbf{y}} + \tilde{\mathbf{K}}\mathbf{y} = \tilde{\mathbf{T}}P(t)\,, \tag{3.85}$$

where

$$\tilde{\mathbf{M}} = \mathbf{r}_B^T\mathbf{M}\mathbf{r}_B \qquad \tilde{\mathbf{C}} = \mathbf{r}_B^T\mathbf{C}\mathbf{r}_B \qquad \tilde{\mathbf{K}} = \mathbf{r}_B^T\mathbf{K}\mathbf{r}_B \qquad \tilde{\mathbf{T}} = \mathbf{r}_B^T\mathbf{S}\,. \tag{3.86}$$

Equation (3.85) is a system of s differential equations in the s generalized coordinates. It is observed that Eq. (3.85) in generalized coordinates is similar to Eq. (3.47) in modal coordinates. However, both sets of equations differ in an important sense. The Ritz vectors are used for transformation in one case, whereas the natural vibration modes are used in the other case. In general, the Ritz vectors are different from the natural modes, therefore the matrices $\tilde{\mathbf{M}}, \tilde{\mathbf{C}}, \tilde{\mathbf{K}}$ are not diagonal (whereas matrices $\mathbf{I}$, $\mathbf{C}_d$, $\mathbf{\Omega}^2$ are diagonal).

In summary, the Ritz transformation of Eq. (3.84) has made it possible to reduce the original set of n equations in the nodal displacements $\mathbf{r}$ to a smaller set of s equations in the generalized coordinates $\mathbf{y}(t)$.

3.3 Nonlinear Dynamic Analysis

3.3.1 Implicit Integration

The implicit time integration schemes discussed in Sect. 3.1.1 for linear dynamic analysis can also be used in nonlinear dynamic analysis. In this section we consider for illustrative purpose the trapezoidal rule, which is Newmark's method with $\delta = 0.5$ and $\alpha = 0.25$, and the modified Newton-Raphson iteration. Neglecting the effect of damping, the equilibrium equations at time $t{+}\Delta t$ are expressed as

$$\mathbf{M}\,{}^{t+\Delta t}\ddot{\mathbf{r}}^{(k)} + {}^{t}\mathbf{K}\,\delta\mathbf{r}^{(k)} = {}^{t+\Delta t}\mathbf{R}_0 - {}^{t+\Delta t}\mathbf{R}_I^{(k-1)}\,, \tag{3.87}$$

where ${}^{t}\mathbf{K}$ is the stiffness matrix at some previous time, superscript k denotes the iteration cycle and $\mathbf{R}_0$, $\mathbf{R}_I$ are the internal and external force vectors, respectively. The vector of displacements due to the out-of-balance forces, $\delta \mathbf{r}^{(k)}$, is defined as

$$\delta \mathbf{r}^{(k)} = {}^{t+\Delta t}\mathbf{r}^{(k)} - {}^{t+\Delta t}\mathbf{r}^{(k-1)}. \tag{3.88}$$

Using the trapezoidal rule of time integration, we employ the assumptions

$$ {}^{t+\Delta t}\mathbf{r} = {}^{t}\mathbf{r} + \frac{\Delta t}{2}\left({}^{t}\dot{\mathbf{r}} + {}^{t+\Delta t}\dot{\mathbf{r}}\right), \tag{3.89}$$

$$ {}^{t+\Delta t}\dot{\mathbf{r}} = {}^{t}\dot{\mathbf{r}} + \frac{\Delta t}{2}\left({}^{t}\ddot{\mathbf{r}} + {}^{t+\Delta t}\ddot{\mathbf{r}}\right). \tag{3.90}$$

Using Eqs. (3.88) through (3.90), we obtain

$$ {}^{t+\Delta t}\ddot{\mathbf{r}}^{(k)} = \frac{4}{\Delta t^2}\left({}^{t+\Delta t}\mathbf{r}^{(k-1)} - {}^{t}\mathbf{r} + \delta\mathbf{r}^{(k)}\right) - \frac{4}{\Delta t}\,{}^{t}\dot{\mathbf{r}} - {}^{t}\ddot{\mathbf{r}}. \tag{3.91}$$

Substituting Eq. (3.91) into Eq. (3.87), we have

$$ {}^{t}\hat{\mathbf{K}}\,\delta\mathbf{r}^{(k)} = {}^{t+\Delta t}\mathbf{R}_0 - {}^{t+\Delta t}\mathbf{R}_I^{(k-1)} - \mathbf{M}\left[\frac{4}{\Delta t^2}\left({}^{t+\Delta t}\mathbf{r}^{(k-1)} - {}^{t}\mathbf{r}\right) - \frac{4}{\Delta t}\,{}^{t}\dot{\mathbf{r}} - {}^{t}\ddot{\mathbf{r}}\right], \tag{3.92}$$

where

$$ {}^{t}\hat{\mathbf{K}} = {}^{t}\mathbf{K} + \frac{4}{\Delta t^2}\mathbf{M}. \tag{3.93}$$

It is observed that the iterative equations in nonlinear dynamic analysis using implicit integration are of the same form as the equations in nonlinear static analysis, except that both the coefficient matrix and the force vector contain contributions from the inertia of the system. The dynamic response is usually smoother than the static response, due to the effect of inertia forces. Therefore, convergence of the iteration is expected to be more rapid than in static analysis. The convergence can be improved by decreasing Δt.

3.3.2 Mode Superposition

Formulation of the Incremental Equations

Consider the equations of motion [Eq. (3.1)] at time t for a system subjected to dynamic forces

$$ \mathbf{M}\,{}^{t}\ddot{\mathbf{r}} + \mathbf{C}\,{}^{t}\dot{\mathbf{r}} + {}^{t}\mathbf{F}_R = {}^{t}\mathbf{R}, \tag{3.94}$$

where superscript t denotes the time and ${}^t\mathbf{F}_R$ is the resisting (elastic or inelastic) force vector. Writing the equations at time $t + \Delta t$

$$\mathbf{M}\;{}^{t+\Delta t}\ddot{\mathbf{r}} + \mathbf{C}\;{}^{t+\Delta t}\dot{\mathbf{r}} + {}^{t+\Delta t}\mathbf{F}_R = {}^{t+\Delta t}\mathbf{R}, \tag{3.95}$$

and subtracting Eq. (3.94) from Eq. (3.95), we obtain the incremental equations

$$\mathbf{M}\;{}^t\Delta\ddot{\mathbf{r}} + \mathbf{C}\;{}^t\Delta\dot{\mathbf{r}} + {}^t\Delta\mathbf{F}_R = {}^t\Delta\mathbf{R}, \tag{3.96}$$

where

$${}^t\Delta\mathbf{r} = {}^{t+\Delta t}\mathbf{r} - {}^t\mathbf{r} \qquad {}^t\Delta\dot{\mathbf{r}} = {}^{t+\Delta t}\dot{\mathbf{r}} - {}^t\dot{\mathbf{r}} \qquad {}^t\Delta\ddot{\mathbf{r}} = {}^{t+\Delta t}\ddot{\mathbf{r}} - {}^t\ddot{\mathbf{r}}, \tag{3.97}$$

$${}^t\Delta\mathbf{F}_R = {}^{t+\Delta t}\mathbf{F}_R - {}^t\mathbf{F}_R \qquad {}^t\Delta\mathbf{R} = {}^{t+\Delta t}\mathbf{R} - {}^t\mathbf{R}.$$

Using the first-order Taylor series approximations of the resisting force we obtain

$${}^t\Delta\mathbf{F}_R = {}^{t+\Delta t}\mathbf{F}_R - {}^t\mathbf{F}_R \cong \frac{\partial\,{}^t\mathbf{F}_s}{\partial\mathbf{r}}\;{}^t\Delta\mathbf{r}. \tag{3.98}$$

The tangent stiffness matrix at time t, ${}^t\mathbf{K}$, is given by

$${}^t\mathbf{K} = \frac{\partial\,{}^t\mathbf{F}_R}{\partial\mathbf{r}}. \tag{3.99}$$

Thus, Eq. (3.98) can be expressed as

$${}^t\Delta\mathbf{F}_R = {}^t\mathbf{K}\;{}^t\Delta\mathbf{r}. \tag{3.100}$$

Substituting Eq. (3.100) into Eq. (3.96) yields

$$\mathbf{M}\;{}^t\Delta\ddot{\mathbf{r}} + \mathbf{C}\;{}^t\Delta\dot{\mathbf{r}} + {}^t\mathbf{K}\;{}^t\Delta\mathbf{r} = {}^t\Delta\mathbf{R}. \tag{3.101}$$

Assuming that the displacements ${}^t\mathbf{r}$ are known and solving Eq. (3.101) for ${}^t\Delta\mathbf{r}$, the displacements ${}^{t+\Delta t}\mathbf{r}$ are calculated by [see Eq. (3.97)]

$${}^{t+\Delta t}\mathbf{r} = {}^t\mathbf{r} + {}^t\Delta\mathbf{r}. \tag{3.102}$$

Simplified Solution Procedure

In linear dynamic analysis, the mode superposition approach can be more effective than direct integration in cases where only the lowest mode shapes may be considered and the integration must be carried out for many time steps. In nonlinear dynamic analysis, the structure properties change during the solution process and as a result the eigenpairs also change. A major difficulty in the solution process is the need to repeat the eigenproblem solution many times. The complete mode superposition analysis of nonlinear dynamic response is generally effective only when the solution can be obtained without updating the stiffness matrix too frequently, and only a few mode shapes are considered.

Assume the transformation from the nodal displacements to the generalized displacements

$$ {}^t\Delta\mathbf{r} = {}^t\boldsymbol{\Phi}\ {}^t\Delta\mathbf{Z} \qquad {}^t\Delta\dot{\mathbf{r}} = {}^t\boldsymbol{\Phi}\ {}^t\Delta\dot{\mathbf{Z}} \qquad {}^t\Delta\ddot{\mathbf{r}} = {}^t\boldsymbol{\Phi}\ {}^t\Delta\ddot{\mathbf{Z}}, \tag{3.103} $$

where ${}^t\Delta\mathbf{Z}$ is a vector of generalized displacements and ${}^t\boldsymbol{\Phi}$ is the matrix of eigenvectors at time t. The eigenpairs ${}^t\boldsymbol{\Phi}_i$, ${}^t\lambda_i$ are obtained by solving the eigenproblem

$$ {}^t\mathbf{K}\,{}^t\boldsymbol{\Phi}_i = {}^t\lambda_i\,\mathbf{M}\,{}^t\boldsymbol{\Phi}_i. \tag{3.104} $$

In the presentation that follows we assume proportional damping such that classical modal analysis can be used. Considering p mode shapes, where $p << n$ (n is the number of degrees of freedom), substituting Eqs. (3.103) into Eq. (3.101) and pre-multiplying the resulting equations by ${}^t\boldsymbol{\Phi}^T$, we obtain the decoupled equations of motion

$$ \mathbf{I}\ {}^t\Delta\ddot{\mathbf{Z}} + {}^t\mathbf{C}_d\ {}^t\Delta\dot{\mathbf{Z}} + {}^t\boldsymbol{\Omega}^2\ {}^t\Delta\mathbf{Z} = {}^t\boldsymbol{\Phi}^T\ {}^t\Delta\mathbf{R}. \tag{3.105} $$

In these equations the identity matrix $\mathbf{I} = {}^t\boldsymbol{\Phi}^T\mathbf{M}\,{}^t\boldsymbol{\Phi}$ is the mass matrix in normalized coordinates. The damping matrix $\mathbf{C}_d = {}^t\boldsymbol{\Phi}^T\mathbf{C}\,{}^t\boldsymbol{\Phi}$ and the stiffness matrix ${}^t\boldsymbol{\Omega}^2 = {}^t\boldsymbol{\Phi}^T\ {}^t\mathbf{K}\ {}^t\boldsymbol{\Phi}$ in these coordinates are diagonal low-order matrices, dependent on time and given by [see Eqs. (3.48), (3.49)]

$$ {}^t\mathbf{C}_d = {}^t\boldsymbol{\Phi}^T\,\mathbf{C}\,{}^t\boldsymbol{\Phi} = \begin{bmatrix} 2\ {}^t\omega_1\zeta_1 & & \\ & \ddots & \\ & & 2\ {}^t\omega_p\zeta_p \end{bmatrix}, \tag{3.106} $$

$$ {}^t\boldsymbol{\Omega}^2 = {}^t\boldsymbol{\Phi}^T\ {}^t\mathbf{K}\,{}^t\boldsymbol{\Phi} = \begin{bmatrix} {}^t\omega_1^2 & & \\ & \ddots & \\ & & {}^t\omega_p^2 \end{bmatrix}, \tag{3.107} $$

where ${}^t\omega_i$ are the free vibration frequencies (${}^t\omega_i^2 = {}^t\lambda_i$) and ζ_i are the damping ratios, usually estimated by experimental data.

To illustrate a simplified solution procedure, assume material nonlinearity and approximate bi-linear relations. Starting with initial analysis at time $t = 0$, the stiffness matrix $\mathbf{K}_0$ is first assembled and factorized. We find the p dominant eigenpairs $\mathbf{\Phi}_{0i}$, λ_{0i} from the initial eigenproblem

$$\mathbf{K}_0 \mathbf{\Phi}_{0i} = \lambda_{0i} \mathbf{M} \mathbf{\Phi}_{0i} . \tag{3.108}$$

The generalized displacements $\Delta\mathbf{Z}$, velocities $\Delta\dot{\mathbf{Z}}$ and accelerations $\Delta\ddot{\mathbf{Z}}$ are obtained by solving decoupled equations Eq. (3.105). The Newmark time integration method, for example, can be used for this purpose. The incremental displacements, velocities and accelerations $(\Delta\mathbf{r}, \Delta\dot{\mathbf{r}}, \Delta\ddot{\mathbf{r}})$ are then determined by Eq. (3.103), and the element forces are calculated and checked for material nonlinearity. If any element is in the nonlinear region, we reduce the time steps, to find the transition point of the material properties. For small steps the out-of-balance forces are also small. If the force in any element has reached the elastic limit point we use a smaller time step. When the force approaches the unloading region we use a similar procedure and find the point where the velocity (or change in forces) is zero.

The following steps are repeated for each time interval.

- Assemble the updated stiffness matrix ${}^t\mathbf{K}$ and solve the updated eigenproblem [Eq. (3.104)] to find the p eigenpairs ${}^t\lambda_i$, ${}^t\mathbf{\Phi}_i$ $(i = 1, ..., p)$. Find the diagonal low-order matrices ${}^t\mathbf{C}_d$, ${}^t\mathbf{\Omega}^2$ [Eqs. (3.106), (3.107)].
- Introduce and solve the p decoupled equations in the generalized coordinates [Eq. (3.105)], evaluate ${}^t\Delta\mathbf{Z}$, ${}^t\Delta\dot{\mathbf{Z}}$, ${}^t\Delta\ddot{\mathbf{Z}}$, and calculate ${}^t\Delta\mathbf{r}$, ${}^t\Delta\dot{\mathbf{r}}$, ${}^t\Delta\ddot{\mathbf{r}}$ by Eq. (3.103). Evaluate ${}^{t+\Delta t}\mathbf{r}$, ${}^{t+\Delta t}\dot{\mathbf{r}}$, ${}^{t+\Delta t}\ddot{\mathbf{r}}$ by [see Eq. (3.97)]

$${}^{t+\Delta t}\mathbf{r} = {}^t\mathbf{r} + {}^t\Delta\mathbf{r} \qquad {}^{t+\Delta t}\dot{\mathbf{r}} = {}^t\dot{\mathbf{r}} + {}^t\Delta\dot{\mathbf{r}} \qquad {}^{t+\Delta t}\ddot{\mathbf{r}} = {}^t\ddot{\mathbf{r}} + {}^t\Delta\ddot{\mathbf{r}} . \tag{3.109}$$

- Evaluate the member forces and check the properties of the members. If the stiffness coefficients of all members do not change, then start the calculations for the next time interval from the previous step with the new initial displacements, velocities and accelerations. If the stiffness coefficients of any member change (the force reached the elastic limit point or the unloading point), reduce the time step size Δt and repeat the calculations from the first step.

During the solution process it is necessary to update the generalized coordinates $\mathbf{Z}$. Using the transformation

$$ {}^{t}\mathbf{r} = {}^{t}\boldsymbol{\Phi}\, {}^{t}\mathbf{Z} \,, \tag{3.110} $$

pre-multiplying Eq. (3.110) by ${}^{t}\boldsymbol{\Phi}^{T}\mathbf{M}$ and noting that ${}^{t}\boldsymbol{\Phi}^{T}\mathbf{M}\, {}^{t}\boldsymbol{\Phi} = \mathbf{I}$, we obtain the required expression

$$ {}^{t}\mathbf{Z} = {}^{t}\boldsymbol{\Phi}^{T}\mathbf{M}\, {}^{t}\mathbf{r} \,. \tag{3.111} $$

References

1. Bathe KJ (1996) Finite element procedures. Prentice Hall, NJ
2. Chopra AK (2001) Dynamics of structures. Prentice Hall, NJ
3. Clough RW, Penzien JP (1993) Dynamics of structures. McGraw-Hill, NY
4. Belitchko T, Hughes TJR (eds 1983) Computational methods in transient analysis. North Holland, NY
5. Houbolt JC (1950) A recurrence matrix solution for the dynamic response of elastic aircraft. J of Aeronautical Sciences 17:540-550
6. Wilson EL, Farhoomand I, Bathe KJ (1973) Nonlinear analysis of complex structures. Int J Earthquake Engrg and Struct Dynamics 1:241-252
7. Newmark NM (1959) A method of computation for structural dynamics. ASCE J of Engrg Mech Div 85:67-94
8. Noor AK (1994) Recent advances and applications of reduction methods. Appl Mech Rev 47:125-146
9. Noor AK, Peters JM (1980) Reduced basis technique for nonlinear analysis of structures. AIAA J 18:455-462

4 Reanalysis of Structures

Repeated analysis, or reanalysis, is needed in various problems of structural analysis, design and optimization. In general, the structural response cannot be expressed explicitly in terms of the structure properties, and structural analysis involves solution of a set of simultaneous equations. Reanalysis methods are intended to analyze efficiently structures that are modified due to various changes in their properties. The object is to evaluate the structural response (e.g. displacements, forces and stresses) for such changes without solving the complete set of modified simultaneous equations. The solution procedures usually use the response of the original structure. Some common problems, where multiple repeated analyses are needed, are described in the following.

- In structural optimization the solution is iterative and consists of repeated analyses followed by redesign steps. The high computational cost involved in repeated analyses of large-scale problems is one of the main obstacles in the solution process. In many problems the analysis part will require most of the computational effort, therefore only methods that do not involve numerous time-consuming implicit analyses might prove useful. Reanalysis methods, intended to reduce the computational cost, have been motivated by some typical difficulties involved in the solution process. The number of design variables is usually large, and various failure modes under each of several load conditions are often considered. The constraints are implicit functions of the design variables, and evaluation of the constraint values for any assumed design requires the solution of a set of simultaneous analysis equations.
- In structural damage analysis it is necessary to analyze the structure for various changes due to deterioration, poor maintenance, damage, or accidents. In general many hypothetical damage scenarios, describing various types of damage, should be considered. These include partial or complete damage in various elements of the structure and changes in the support conditions. Numerous analyses are required to assess the adequacy of redundancy and to evaluate various hypothetical damage scenarios for different types of damage.

- In the design of the construction stages of complex structures, it might be necessary to analyze repeatedly structures that are modified during the construction. The modified structures are subjected to different loading conditions. The changes in the structure may include additional elements and different support conditions.
- Nonlinear analysis of structures is usually carried out in an iterative process. The solution can be performed by different methods but, in general, a set of updated linear equations must be solved repeatedly. Similarly, many of the vibration (or eigenproblem) solution techniques are based on matrix iteration methods. To calculate the mode shapes it is necessary to solve repeatedly a set of updated analysis equations.
- Reanalysis methods might prove useful in other applications such as probabilistic analysis, controlled structures, smart structures, adaptive structures, and for conceptual design problems.

One basic question that may arise is whether we really need efficient reanalysis methods, in view of the significant increase in computer processing power, memory and storage space. In this regard, it has been noted [1] that the rapid developments in computer technology have not eliminated computational cost and time constraints on the use of structural optimization for design. This is due to the constant increase in the required fidelity, and hence complexity, of analysis models. It seems that analysis models of acceptable accuracy have required an overnight run throughout the last decades.

The two main components of complexity of the problem are related to the complexity of the analysis model and the analysis procedure. The model complexity is a function of various parameters such as the number of degrees of freedom of the finite element model and the topology of the structure (which determines the bandwidth of the stiffness matrix). In terms of analysis complexity, linear elastic analysis is the simplest. More complex analysis such as non-linear elastic or dynamic response use linearization algorithms that require linear analysis as a repeated step. History-dependent nonlinear analysis and nonlinear dynamic analysis are currently the extremes of analysis complexity. These types of analysis may require numerous linear analysis equivalents. The analysis complexity can be considered as the number of matrix inversions performed. Reanalysis methods are intended to overcome the difficulties involved in the above complexities.

4.1 Design Variables

A structural system can be described by a set of quantities, some of which are viewed as variables during the design process. Those quantities defining a structural system that are fixed during the design are called pre-assigned parameters. Those quantities that are not pre-assigned are called design variables. The pre-assigned parameters together with the design variables completely describe a design. Quantities are designated as pre-assigned parameters for a variety of reasons. It may be that the designer is not free to choose certain parameters, or it may be known from experience that a particular value of the parameter produces good results. From a physical point of view, the design variables that are varied during the design process may represent:

- the mechanical or physical properties of the material;
- the topology of the structure, i.e., the pattern of connection of members (number and orientation of elements and joints);
- the geometry or the shape of the structure (e.g. coordinates of joints);
- the cross-sectional dimensions or the sizes of elements.

From a mathematical point of view, it is important to distinguish between continuous and discrete design variables. In cases of discrete variables with a large number of values uniformly distributed over a given interval, use of a continuous variable representation is often satisfactory, followed by selection of the nearest available discrete value. When a strictly discrete design variable is handled in this way, it is categorized as pseudo-discrete. However, it should be recognized that situations arise when it is essential to employ discrete or integer variables. An example for an integer variable is the number of elements in the structure.

Material selection presents a special problem with conventional materials, as they have discrete properties. That is, a choice is to be made from a discrete set of variables. Application of high-performance composite materials in structural components has encouraged further consideration of material properties as design variables. For example, in fiber composites the volume fraction of fibers or the modulus of elasticity in the longitudinal direction of carbon fibers could be considered as design variables.

Topological optimization is perhaps the most challenging class of problems in structural optimization because there are numerous possible topologies, which are difficult to classify and quantify. At the same time, topological optimization is of considerable importance because it leads to significant material savings. The topology of the structure can be optimized automatically in many cases when elements are allowed to reach zero size. This permits elimination of some uneconomical elements during the optimization process. In other cases, however, it may be necessary to represent some design variables as integer variables and to declare the existence or absence of a structural element. An example of an integer topological variable is a truss member joining two nodes, which is limited to the values 1 (the member exists), or 0 (the member is absent). Other examples of integer topological variables include the number of spans in a bridge, the number of columns supporting a roof system, or the number of elements in a grillage system.

Geometrical variables may represent, for example, the coordinates of joints in a truss or in a frame. Other examples for this class of variable include the location of supports in a bridge, the length of spans in a continuous beam, and the height of a shell structure. In general, the geometry of the structure is represented by continuous variables.

Cross-sectional dimensions are the simplest design variables. The cross-sectional areas of truss members, the moments of inertia of flexural members, or the thickness of a plate are some examples of this class of design variables. In certain cases a single design variable is adequate to describe the cross section, but a more detailed design with several design variables for each cross section may be necessary. For example, if the axial buckling of members is considered, the cross-sectional dimensions which define the area and the moment of inertia can be taken as design variables. In practical design, cross-sectional variables may be restricted to some discrete values, e.g. the areas of commercial steel section shapes.

It should be noted that a change in the cross sections or in the geometry might affect the topology. For example, the topology will be changed due to zero areas during sizing modifications or the coalescence of joints during geometrical modifications. In addition, the geometry might be affected by topological changes due to elimination of members and joints.

Changes in the design often affect only the numerical values of the coefficients of the analysis equations. However, in some cases of topological changes, members and joints are deleted or added and the structural model is allowed to vary during the design process. We may distinguish between the following cases of topological changes, considered later in this text:

- Deletion and addition of members, where the number of Degrees of Freedom (DOF) is unchanged (Fig. 4.1*a*). In this case the number of analysis equations is also unchanged and only the numerical values of the coefficients of the equations are modified.
- Deletion and addition of members, and deletion of some joints, where the number of DOF is decreased (Fig. 4.1*b*). In this case it is necessary

to change the analysis model such that the deleted DOF are not included in the modified analysis equations.
- Deletion and addition of members, and addition of some joints, where the number of DOF is increased (Fig. 4.1*c*). In this case it is necessary to augment the analysis model such that the new degrees of freedom are included in the modified analysis equations.

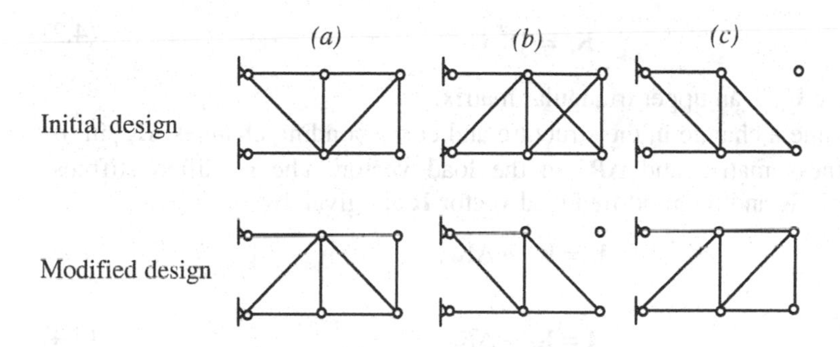

Fig. 4.1. Types of topological changes

The resulting modified structures may be classified as follows:

- *Stable structures*, where the modified response can be evaluated by solving the modified equilibrium equations.
- *Conditionally Unstable structures*, where the forces in the structure satisfy equilibrium conditions only for some specific loadings. That is, the structure can carry only these specific loading conditions, and it is unstable for other loading conditions.
- *Unstable structures*, where the structure or part of it is unstable for a general loading condition. In this case the modified equilibrium equations cannot be solved and a collapse of the structure will occur.

4.2 Formulation of Static Reanalysis

4.2.1 Linear Static Reanalysis

Linear static reanalysis is encountered in numerous analysis, design and optimization problems. The formulation presented in this section is general, and covers a wide range of problems. Assuming the displacement method of analysis, we can state a typical reanalysis problem as follows:

- Given an initial structure, the corresponding stiffness matrix $\mathbf{K}_0$ and the load vector $\mathbf{R}_0$, the displacements $\mathbf{r}_0$ are computed by solving the equilibrium equations [Eq. (1.2)]

$$\mathbf{K}_0 \, \mathbf{r}_0 = \mathbf{R}_0. \tag{4.1}$$

  The symmetric positive-definite stiffness matrix $\mathbf{K}_0$ is usually given from the initial analysis in the decomposed form [Eq. 1.23)]

$$\mathbf{K}_0 = \mathbf{U}_0^T \, \mathbf{U}_0, \tag{4.2}$$

  where $\mathbf{U}_0$ is an upper triangular matrix.
- Assume a change in the structure and corresponding changes $\Delta\mathbf{K}_0$ in the stiffness matrix and $\Delta\mathbf{R}_0$ in the load vector. The modified stiffness matrix $\mathbf{K}$ and the modified load vector $\mathbf{R}$ are given by

$$\mathbf{K} = \mathbf{K}_0 + \Delta\mathbf{K}_0, \tag{4.3}$$

$$\mathbf{R} = \mathbf{R}_0 + \Delta\mathbf{R}_0. \tag{4.4}$$

  In general, the elements of the stiffness matrix $\mathbf{K}$ are some explicit functions of the design variables. The changes $\Delta\mathbf{K}_0$ are functions of the members' cross-sections, the material properties, the geometry and the topology of the structure. In general, the changes $\Delta\mathbf{R}_0$ are also functions of the design variables. However, the elements of the load vector $\mathbf{R}$ are often assumed to be independent of these variables, that is, $\Delta\mathbf{R}_0 = \mathbf{0}$.
- The object is to estimate the modified displacements $\mathbf{r}$ due to the changes in the structure, without solving the complete set of modified analysis equations

$$(\mathbf{K}_0 + \Delta\mathbf{K}_0) \, \mathbf{r} = \mathbf{R}. \tag{4.5}$$

Once the displacements have been evaluated, the forces $\mathbf{N}$ and the stresses $\boldsymbol{\sigma}$ are readily calculated [Eqs. (1.3), (1.4)]. Thus, reanalysis methods essentially replace the formal solution of the implicit equations (4.5).

In this formulation the initial stiffness matrix $\mathbf{K}_0$, the load-vector $\mathbf{R}_0$ and the corresponding displacements $\mathbf{r}_0$ are given from initial analysis of the structure. But the present reanalysis formulation might prove useful also in cases where results of exact analysis are not available, that is, the initial displacements $\mathbf{r}_0$ are unknown. The object in this case is to evaluate $\mathbf{r}$ for the given stiffness matrix $\mathbf{K} = \mathbf{K}_0 + \Delta\mathbf{K}_0$, where matrices $\mathbf{K}_0$ and $\Delta\mathbf{K}_0$ can be chosen such that the solution is simple and effective. For example, it is possible to choose $\mathbf{K}_0$ as a diagonal matrix $\mathbf{K}_d$ consisting of the diagonal elements of $\mathbf{K}$. In this case, $\Delta\mathbf{K}_0$ and $\mathbf{R}$ are given by

$$\Delta\mathbf{K}_0 = \mathbf{K} - \mathbf{K}_d, \tag{4.6}$$

$$\mathbf{R} = \mathbf{R}_0. \tag{4.7}$$

The initial displacements are given explicitly by $\mathbf{r}_0 = (\mathbf{K}_d)^{-1}\mathbf{R}$, and the requested displacements $\mathbf{r}$ are calculated by Eq. (4.5). It will be shown later that despite this poor selection of $\mathbf{r}_0$, accurate results can be achieved by the reanalysis approach presented in this text.

4.2.2 Nonlinear Static Reanalysis

General Formulation

Consider the equations used in the Newton-Raphson method for the kth iteration cycle, obtained by linearizing the response about the conditions at time $t+\Delta t$, iteration k-1 [Eq. (1.83)]

$$^{t+\Delta t}\mathbf{K}^{(k\text{-}1)}\,\delta\mathbf{r}^{(k)} = \delta\mathbf{R}^{(k\text{-}1)}, \tag{4.8}$$

where $^{t+\Delta t}\mathbf{K}^{(k\text{-}1)}$ is the current tangent stiffness matrix. The out-of-balance load vector $\delta\mathbf{R}^{(k)}$ is given by

$$\delta\mathbf{R}^{(k\text{-}1)} = {}^{t+\Delta t}\mathbf{R}_0 - {}^{t+\Delta t}\mathbf{R}_I^{(k-1)}, \tag{4.9}$$

and $\delta\mathbf{r}^{(k)}$ is the vector of incremental displacements due to $\delta\mathbf{R}^{(k\text{-}1)}$.

For simplicity of presentation, consider the notation

$$\mathbf{K}_T = {}^{t+\Delta t}\mathbf{K}^{(k\text{-}1)} \qquad \delta\mathbf{R} = \delta\mathbf{R}^{(k\text{-}1)} \qquad \delta\mathbf{r} = \delta\mathbf{r}^{(k)}. \tag{4.10}$$

Substituting Eqs. (4.10) into Eq. (4.8), the vector $\delta\mathbf{r}$ is calculated at each iteration cycle by solving the set of equations

$$\mathbf{K}_T\,\delta\mathbf{r} = \delta\mathbf{R}. \tag{4.11}$$

The tangent stiffness matrix $\mathbf{K}_T$ can be expressed in terms of $\mathbf{K}_{ref}$ and the matrix of changes $\Delta\mathbf{K}$ as

$$\mathbf{K}_T = \mathbf{K}_{ref} + \Delta\mathbf{K}, \tag{4.12}$$

where $\mathbf{K}_{ref}$ is a reference stiffness matrix, which is the tangent stiffness matrix calculated at some previous step. Matrix $\mathbf{K}_{ref}$ might represent, for example, the tangent stiffness matrix at the end of the previous increment, the elastic stiffness matrix, or another choice, depending on the solution procedure discussed later in Sect. 5.4. Substituting Eq. (4.12) into Eq. (4.11), we obtain the set of equations to be solved at each iteration cycle

$$\mathbf{K}_T\,\delta\mathbf{r} = (\mathbf{K}_{ref} + \Delta\mathbf{K})\,\delta\mathbf{r} = \delta\mathbf{R}. \tag{4.13}$$

The definition of matrix $\Delta\mathbf{K}$ depends on the type of problem to be solved. We may distinguish between the following two problems of nonlinear analysis and reanalysis:

- The general case of *nonlinear reanalysis of a modified structure*, where matrix $\Delta\mathbf{K}$ is expressed in terms of the following two types of changes

$$\Delta\mathbf{K} = \Delta\mathbf{K}_{NL} + \Delta\mathbf{K}_0. \tag{4.14}$$

  Matrix $\Delta\mathbf{K}_{NL}$ represents the changes in the stiffness matrix due to the nonlinear behavior and it is usually calculated at each iteration cycle. Matrix $\Delta\mathbf{K}_0$ represents the changes in the stiffness matrix due to design considerations and it is constant for any given modified design. It should be noted that the present formulation is suitable also for situations where $\Delta\mathbf{K}$ is not calculated explicitly. Rather, we can calculate $\Delta\mathbf{K}$ at each iteration cycle from [Eq. (4.12)]

$$\Delta\mathbf{K} = \mathbf{K}_T - \mathbf{K}_{ref}. \tag{4.15}$$

- The case of *nonlinear analysis of the original structure*, where $\Delta\mathbf{K}_0 = 0$ and

$$\Delta\mathbf{K} = \Delta\mathbf{K}_{NL}. \tag{4.16}$$

  That is, matrix $\Delta\mathbf{K}$ represents only changes due to the nonlinear behavior of the structure. As noted earlier, $\mathbf{K}_{ref}$ is the tangent stiffness matrix calculated at some previous step.

Particular Formulations

Various formulations may be viewed as particular cases of the general formulation. Considering first the particular case of *geometric nonlinear analysis* of the original structure, we start with the initial displacements $\mathbf{r}_0$, computed by [Eq. (1.56)]

$$\mathbf{K}_0\,\mathbf{r}_0 = \mathbf{R}_0. \tag{4.17}$$

The modified equations, which must be solved at each iteration cycle, are expressed as

$$(\mathbf{K}_{ref} + \Delta\mathbf{K}^{(k-1)})\,\delta\mathbf{r}^{(k)} = \delta\mathbf{R}^{(k-1)}. \tag{4.18}$$

In this case, $\mathbf{K}_{ref}$ can be chosen as the elastic stiffness matrix

$$\mathbf{K}_{ref} = \mathbf{K}_0 \tag{4.19}$$

and the resulting matrix of stiffness changes $\Delta\mathbf{K}^{(k\text{-}1)}$ is the geometric stiffness matrix $\mathbf{K}_G^{(k-1)}$, that is

$$\Delta\mathbf{K}^{(k\text{-}1)} = \mathbf{K}_T^{(k-1)} - \mathbf{K}_0 = \mathbf{K}_G^{(k-1)}. \tag{4.20}$$

Considering the problem of *geometric nonlinear reanalysis* of a modified structure, it is still possible to choose $\mathbf{K}_{ref}$ as the elastic stiffness matrix [Eq. (4.19)]. In this case, the resulting matrix of stiffness changes $\Delta\mathbf{K}^{(k)}$ is

$$\Delta\mathbf{K}^{(k\text{-}1)} = \mathbf{K}_T^{(k-1)} - \mathbf{K}_0 = \Delta\mathbf{K}_0 + \mathbf{K}_G^{(k-1)}, \tag{4.21}$$

where $\Delta\mathbf{K}_0$ represents changes in the elastic stiffness matrix due to changes in the design. The modified equations, solved at each iteration cycle, are

$$(\mathbf{K}_0 + \Delta\mathbf{K}^{(k\text{-}1)})\, \delta\mathbf{r}^{(k)} = \delta\mathbf{R}^{(k\text{-}1)}. \tag{4.22}$$

Problems of material nonlinearity can be formulated in a similar way. Consider for illustrative purposes the particular case of *plastic analysis* presented in Sect. 1.4.2. The structure is first analyzed for the initial stiffness matrix $\mathbf{K}_0$ and load vector $\mathbf{R}_0$. The load vector is then increased to obtain $\mathbf{R}_1 = \lambda_1\, \mathbf{R}_0$, which is the load that causes the yield stress to be reached at the first section. The modified stiffness matrix,

$$\mathbf{K}_2 = \mathbf{K}_0 + \Delta\mathbf{K}_1, \tag{4.23}$$

is then determined, accounting for the reduction $\Delta\mathbf{K}_1$ in the stiffness due to yield of the first section. The additional load vector $\mathbf{R}_2$ that causes the yield stress to be reached at the second section $\mathbf{R}_2 = \lambda_2\, \mathbf{R}_0$ is determined from the modified equations $\mathbf{K}_2\, \mathbf{r}_2 = \mathbf{R}_2$. We proceed with the modified matrix

$$\mathbf{K}_3 = \mathbf{K}_0 + \Delta\mathbf{K}_1 + \Delta\mathbf{K}_2, \tag{4.24}$$

considering the reduced stiffness due to yield at the first and the second sections. These steps are repeated until collapse. Given the initial $\mathbf{K}_0$, $\mathbf{R}_0$, $\mathbf{r}_0$ and defining,

$$\Delta\mathbf{K} = \sum_{i=1}^{m-1} \Delta\mathbf{K}_i\,, \tag{4.25}$$

$$\mathbf{R}_m = \mathbf{R}_0 \lambda_m\,, \tag{4.26}$$

plastic analysis at the *m*th load stage involves solution of the modified equations

$$(\mathbf{K}_0 + \Delta\mathbf{K})\mathbf{r}_m = \mathbf{R}_m. \tag{4.27)}$$

In summary, various nonlinear analysis and reanalysis problems can be formulated in the form of linear reanalysis. The mathematical expressions of the formulations presented in Sects. 4.2.1, 4.2.2 are shown in Table 4.1.

Table 4.1. Formulation of static reanalysis problems

Problem	Reanalysis Equations	Matrix of Changes	Right-hand side
Linear Analysis	(4.5) $(\mathbf{K}_0+\Delta\mathbf{K}_0)\ \mathbf{r} = \mathbf{R}$	(4.6) $\Delta\mathbf{K}_0=\mathbf{K}-\mathbf{K}_d$	(4.7) $\mathbf{R}=\mathbf{R}_0$
Linear Reanalysis	(4.5) $(\mathbf{K}_0+\Delta\mathbf{K}_0)\mathbf{r}=\mathbf{R}$	(4.3) $\Delta\mathbf{K}_0=\mathbf{K}-\mathbf{K}_0$	(4.4) $\mathbf{R}=\mathbf{R}_0+\Delta\mathbf{R}_0$
General Nonlinear Analysis	(4.10), (4.13) $(\mathbf{K}_{ref}+\Delta\mathbf{K})\delta\mathbf{r}=\delta\mathbf{R}$	(4.16) $\Delta\mathbf{K} = \Delta\mathbf{K}_{NL}$	(4.9), (4.10) $\delta\mathbf{R}={}^{t+\Delta t}\mathbf{R}_0-{}^{t+\Delta t}\mathbf{R}_I^{(k\text{-}1)}$
General Nonlinear Reanalysis	(4.10), (4.13) $(\mathbf{K}_{ref}+\Delta\mathbf{K})\delta\mathbf{r}=\delta\mathbf{R}$	(4.14) $\Delta\mathbf{K} = \Delta\mathbf{K}_{NL} + \Delta\mathbf{K}_0$	$\delta\mathbf{R}={}^{t+\Delta t}\mathbf{R}_0-{}^{t+\Delta t}\mathbf{R}_I^{(k\text{-}1)}$
Geometric Nonlinear Analysis	(4.18) $(\mathbf{K}_{ref}+\Delta\mathbf{K}^{(k\text{-}1)})\delta\mathbf{r}^{(k)}=\delta\mathbf{R}^{(k\text{-}1)}$	(4.20) $\Delta\mathbf{K}^{(k\text{-}1)}=\mathbf{K}_G^{(k\text{-}1)}$	(4.9) $\delta\mathbf{R}^{(k\text{-}1)}={}^{t+\Delta t}\mathbf{R}_0-{}^{t+\Delta t}\mathbf{R}_I^{(k\text{-}1)}$
Geometric Nonlinear Reanalysis	(4.22) $(\mathbf{K}_0+\Delta\mathbf{K}^{(k\text{-}1)})\delta\mathbf{r}^{(k)}=\delta\mathbf{R}^{(k\text{-}1)}$	(4.21) $\Delta\mathbf{K}^{(k\text{-}1)}=\Delta\mathbf{K}_0+\mathbf{K}_G^{(k\text{-}1)}$	(4.9) $\delta\mathbf{R}^{(k\text{-}1)}={}^{t+\Delta t}\mathbf{R}_0-{}^{t+\Delta t}\mathbf{R}_I^{(k\text{-}1)}$
Plastic Analysis	(4.27) $(\mathbf{K}_0+\Delta\mathbf{K})\mathbf{r}_m=\mathbf{R}_m$	(4.25) $\Delta\mathbf{K} = \sum_{i=1}^{m-1}\Delta\mathbf{K}_i$	(4.26) $\mathbf{R}_m=\mathbf{R}_0\lambda_m$

4.3 Formulation of Vibration Reanalysis

4.3.1 Eigenproblem Reanalysis

In a typical vibration analysis, the following eigenproblem is solved for an initial structure [see Eq. (2.8)]

$$\mathbf{K}_0\,\boldsymbol{\Phi}_0 = \lambda_0\,\mathbf{M}_0\,\boldsymbol{\Phi}_0\,, \tag{4.28}$$

where $\mathbf{K}_0$, $\mathbf{M}_0$ are the stiffness and mass matrices, and $\lambda_0, \boldsymbol{\Phi}_0$ represent the ith eigenpair (for simplicity, the subscript i is omitted). Assume a change in the structure and corresponding changes $\Delta\mathbf{K}_0$, $\Delta\mathbf{M}_0$ in the stiffness and mass matrices such that the modified matrices are expressed as

$$\mathbf{K} = \mathbf{K}_0 + \Delta\mathbf{K}_0, \tag{4.29}$$

$$\mathbf{M} = \mathbf{M}_0 + \Delta\mathbf{M}_0. \tag{4.30}$$

The modified analysis equations are given by

$$\mathbf{K}\,\boldsymbol{\Phi} = \lambda\,\mathbf{M}\,\boldsymbol{\Phi}\,. \tag{4.31}$$

Denoting the right hand side vector of Eq. (4.31) as

$$\mathbf{R} = \lambda\,\mathbf{M}\,\boldsymbol{\Phi}\,, \tag{4.32}$$

and substituting Eqs. (4.29), (4.32) into Eq. (4.31) we find

$$(\mathbf{K}_0 + \Delta\mathbf{K}_0)\,\boldsymbol{\Phi} = \mathbf{R}. \tag{4.33}$$

It can be observed that the modified equations (4.5) and (4.33) are of similar form. However, the difference is that the terms of $\mathbf{R}$ in the latter case [Eq. (4.32)] are functions of the unknown eigenpair $\lambda, \boldsymbol{\Phi}$.

4.3.2 Iterative Procedures

It has been noted in Chap. 2 that in various procedures of *eigenproblem analysis*, a linear set of equations is solved iteratively. In such cases it is possible to state the eigenproblem reanalysis as a linear reanalysis problem. In this section only the inverse vector iteration (Sect. 2.3.1), the vector iteration with shifts (Sect. 2.3.2) and the subspace iteration (Sect. 2.8) procedures are considered. Similar formulations can be introduced for other iterative procedures.

The basic step in the inverse iteration is the solution of Eq. (2.46) – a set of algebraic equations. The initial trial vector $\mathbf{r}^{(0)} = \boldsymbol{\Phi}_0$ is known from solution of the initial eigenproblem [Eq. (4.28)]. For the changes $\Delta\mathbf{K}_0$, $\Delta\mathbf{M}_0$ [Eqs. (4.29), (4.30)] the reanalysis problem, to be solved at the kth iteration cycle, is [Eq. (2.46)]

$$(\mathbf{K}_0 + \Delta\mathbf{K}_0)\,\bar{\mathbf{r}}^{(k)} = \mathbf{R}^{(k\text{-}1)}, \tag{4.34}$$

where the vector $\mathbf{R}^{(k\text{-}1)}$ is defined as

$$\mathbf{R}^{(k\text{-}1)} = \mathbf{M}\,\mathbf{r}^{(k\text{-}1)}. \tag{4.35}$$

The basic step in inverse iteration with shifts is the solution of Eq. (2.58). Considering again the known initial eigenvector $\mathbf{r}^{(0)} = \boldsymbol{\Phi}_0$, the reanalysis problem solved at the kth iteration cycle is [see Eq. (2.58)]

$$(\mathbf{K}_0 + \Delta\mathbf{K}^{(k\text{-}1)})\,\bar{\mathbf{r}}^{(k)} = \mathbf{R}^{(k\text{-}1)}, \tag{4.36}$$

where $\mathbf{R}^{(k\text{-}1)}$ is given by Eq. (4.35) and $\Delta\mathbf{K}^{(k\text{-}1)}$ is defined as

$$\Delta\mathbf{K}^{(k\text{-}1)} = \Delta\mathbf{K}_0 - \lambda^{(k\text{-}1)}\,\mathbf{M}. \tag{4.37}$$

It is observed that both $\Delta\mathbf{K}^{(k\text{-}1)}$ and $\mathbf{R}^{(k\text{-}1)}$ are changed at each iteration cycle. Again, Eqs. (4.5) and (4.36) are of similar form, therefore eigenproblem reanalysis can be stated in the form of linear reanalysis. For any change in the design, $\Delta\mathbf{K}^{(k\text{-}1)}$ and $\mathbf{R}^{(k\text{-}1)}$ are revised accordingly.

The basic step in the subspace iteration procedure is the solution of Eq. (2.102). The reanalysis problem, to solved at the kth iteration cycle for the changes $\Delta\mathbf{K}_0$, $\Delta\mathbf{M}_0$ [Eqs. (4.29), (4.30)], can be stated as

$$(\mathbf{K}_0 + \Delta\mathbf{K}_0)\bar{\mathbf{r}}^{(k+1)} = \mathbf{R}^{(k)}, \tag{4.38}$$

where the vector $\mathbf{R}^{(k)}$ is defined as

$$\mathbf{R}^{(k)} = \mathbf{M}\,\mathbf{r}^{(k)}. \tag{4.39}$$

It is observed that all the modified equations presented in this section [Eqs. (4.33), (4.34), (4.36), (4.38)] and the linear reanalysis problem [Eq. (4.5)] are of similar form. The various vibration reanalysis formulations are summarized in Table 4.2.

Table 4.2. Formulation of vibration reanalysis problems

Problem	Equations	Matrix of Changes	Right-hand side
Eigenproblem Reanalysis	(4.33) $(\mathbf{K}_0 + \Delta\mathbf{K}_0)\,\mathbf{\Phi} = \mathbf{R}$	(4.29) $\Delta\mathbf{K}_0 = \mathbf{K} - \mathbf{K}_0$	(4.32) $\mathbf{R} = \lambda\,\mathbf{M}\,\mathbf{\Phi}$
Inverse Iteration	(4.34) $(\mathbf{K}_0 + \Delta\mathbf{K}_0)\,\bar{\mathbf{r}}^{(k)} = \mathbf{R}^{(k\text{-}1)}$	(4.29) $\Delta\mathbf{K}_0 = \mathbf{K} - \mathbf{K}_0$	(4.35) $\mathbf{R}^{(k\text{-}1)} = \mathbf{M}\,\mathbf{r}^{(k\text{-}1)}$
Inverse Iteration with Shifts	(4.36) $(\mathbf{K}_0+\Delta\mathbf{K}^{(k\text{-}1)})\,\bar{\mathbf{r}}^{(k)} = \mathbf{R}^{(k\text{-}1)}$	(4.37) $\Delta\mathbf{K}^{(k\text{-}1)} = \Delta\mathbf{K}_0 - \lambda^{(k\text{-}1)}\mathbf{M}$	(4.35) $\mathbf{R}^{(k\text{-}1)} = \mathbf{M}\,\mathbf{r}^{(k\text{-}1)}$
Subspace Iteration	(4.38) $(\mathbf{K}_0 + \Delta\mathbf{K}_0)\bar{\mathbf{r}}^{(k+1)} = \mathbf{R}^{(k)}$	(4.29) $\Delta\mathbf{K}_0 = \mathbf{K} - \mathbf{K}_0$	(4.39) $\mathbf{R}^{(k)} = \mathbf{M}\,\mathbf{r}^{(k)}$

4.4 Formulation of Dynamic Reanalysis

4.4.1 Linear Dynamic Reanalysis

The equations of motion for multiple degrees of freedom system subjected to external dynamic forces are [Eq. (3.1)]

$$\mathbf{M}\ddot{\mathbf{r}}(t) + \mathbf{C}\dot{\mathbf{r}}(t) + \mathbf{K}\mathbf{r}(t) = \mathbf{R}(t)\,, \tag{4.40}$$

where $\mathbf{M}$ is the mass matrix, $\mathbf{C}$ is the damping matrix and $\mathbf{K}$ is the stiffness matrix. The displacement vector $\mathbf{r}(t)$, the velocity vector $\dot{\mathbf{r}}(t)$, the acceleration vector $\ddot{\mathbf{r}}(t)$, and the load vector $\mathbf{R}(t)$ are functions of the time t.

Given is an initial structure represented by the initial values $\mathbf{K}_0$, $\mathbf{M}_0$. Assuming a change in the design, and corresponding changes $\Delta\mathbf{K}_0$, $\Delta\mathbf{M}_0$, the modified matrices are given by

$$\mathbf{K} = \mathbf{K}_0 + \Delta\mathbf{K}_0 \qquad \mathbf{M} = \mathbf{M}_0 + \Delta\mathbf{M}_0. \tag{4.41}$$

The problem of dynamic reanalysis is to estimate efficiently and accurately $\mathbf{r}(t)$, $\dot{\mathbf{r}}(t)$, $\ddot{\mathbf{r}}(t)$ of the modified structure, without solving the complete set of Eq. (4.40). It has been noted that the common procedures for solving these equations can be divided into two methods of solution:

- *Direct integration*, where Eqs. (4.40) are integrated using a numerical step-by-step procedure. It is assumed that the equations are satisfied only at discrete time intervals, and the variation of displacements, velocities and accelerations within each time interval has a certain form.
- *Mode superposition*, where the equilibrium equations are transformed into a form in which the step-by-step solution is less costly. This method may be more effective if the integration must be carried out for many time steps. In addition, the effectiveness of the method depends on the number of modes that must be considered. If only a few modes may need to be considered, the mode superposition procedure can be much more effective than direct integration.

In this section some common formulations of linear dynamic reanalysis are presented. Other formulations can be introduced in a similar way.

Solution by Direct Integration

Using the Newmark implicit integration method [Eqs. (3.12) – (3.21)], we solve at each time step the equilibrium equations [Eqs. (3.17), (3.19)]

$$\hat{\mathbf{K}}\,{}^{t+\Delta t}\mathbf{r} = {}^{t+\Delta t}\hat{\mathbf{R}}\,, \tag{4.42}$$

where the effective stiffness matrix $\hat{\mathbf{K}}$ and the effective load vector ${}^{t+\Delta t}\hat{\mathbf{R}}$ are given by [Eqs. (3.16), (3.18)]

$$\hat{\mathbf{K}} = \mathbf{K} + a_0\mathbf{M} + a_1\mathbf{C}\,, \tag{4.43}$$

$${}^{t+\Delta t}\hat{\mathbf{R}} = {}^{t+\Delta t}\mathbf{R} + \mathbf{M}(a_0{}^t\mathbf{r} + a_2{}^t\dot{\mathbf{r}} + a_3{}^t\ddot{\mathbf{r}}) + \mathbf{C}(a_1{}^t\mathbf{r} + a_4{}^t\dot{\mathbf{r}} + a_5{}^t\ddot{\mathbf{r}})\,. \tag{4.44}$$

It is observed that any change in the structure results in corresponding changes in $\hat{\mathbf{K}}$ and ${}^{t+\Delta t}\hat{\mathbf{R}}$. Thus, Eq. (4.42) can be expressed in terms of the initial matrix $\hat{\mathbf{K}}_0(\mathbf{K}_0,\mathbf{M}_0,\mathbf{C})$ and the matrix of changes $\Delta\hat{\mathbf{K}}$ as

$$(\hat{\mathbf{K}}_0 + \Delta\hat{\mathbf{K}})\,{}^{t+\Delta t}\mathbf{r} = {}^{t+\Delta t}\hat{\mathbf{R}}\,, \tag{4.45}$$

where $\Delta\hat{\mathbf{K}}$ is given by

$$\Delta\hat{\mathbf{K}} = \hat{\mathbf{K}}(\mathbf{K},\mathbf{M},\mathbf{C}) - \hat{\mathbf{K}}_0(\mathbf{K}_0,\mathbf{M}_0,\mathbf{C})\,. \tag{4.46}$$

The effective load vector ${}^{t+\Delta t}\hat{\mathbf{R}}$ is changed for each change in **M**, **C** and at each time step. It is observed that the modified analysis equations (4.5) and Eq. (4.45) are of similar form.

Solution by Mode Superposition

Computation of the dynamic response by modal analysis requires first to determine the eigenpairs λ_i, $\mathbf{\Phi}_i$ by solving the eigenproblem

$$\mathbf{K}\mathbf{\Phi}_i = \lambda_i\,\mathbf{M}\mathbf{\Phi}_i \qquad i = 1, \ldots, n. \tag{4.47}$$

For the case of proportional damping, we then solve the decoupled equations [Eq. (3.46)]

$$\ddot{Z}_i(t) + 2\omega_i\zeta_i\dot{Z}_i(t) + \omega_i^2 Z_i(t) = T_i(t)\,, \tag{4.48}$$

where Z_i are the components of the modal coordinates, ζ_i are the damping ratios and $T_i(t)$ are given by

$$T_i(t) = \mathbf{\Phi}_i^T\mathbf{R}(t)\,. \tag{4.49}$$

Finally, we compute the nodal displacements by [Eq. (3.38)]

$$\mathbf{r} = \sum_{i=1}^{n}\mathbf{\Phi}_i Z_i = \mathbf{\Phi}\mathbf{Z}\,, \tag{4.50}$$

where $\mathbf{\Phi}$ is the modal matrix, made up of the n mode shapes, and $\mathbf{Z}$ is the vector of modal coordinates.

As noted earlier, a significant part of the computational effort is involved in repeated solutions of the eigenproblem. The reanalysis formulations presented in Sect. 4.3 are suitable for this type of problem.

4.4.2 Nonlinear Dynamic Reanalysis

Solution by Direct Integration

Considering the common trapezoidal rule, which is Newmark's method with $\delta = 0.5$ and $\alpha = 0.25$, neglecting the effect of damping, and using the modified Newton-Raphson iteration, the modified equilibrium equations to be solved at each iteration cycle are [Eq. (3.87)]

$$\mathbf{M}\,{}^{t+\Delta t}\ddot{\mathbf{r}}^{(k)} + {}^{t}\mathbf{K}\,\delta\mathbf{r}^{(k)} = {}^{t+\Delta t}\mathbf{R}_0 - {}^{t+\Delta t}\mathbf{R}_I^{(k-1)}, \tag{4.51}$$

where [Eq. (3.88)]

$$\delta\mathbf{r}^{(k)} = {}^{t+\Delta t}\mathbf{r}^{(k)} - {}^{t+\Delta t}\mathbf{r}^{(k-1)}, \tag{4.52}$$

and ${}^{t}\mathbf{K}$ is the stiffness matrix considered for the solution of Eq. (4.51). Using the trapezoidal rule of time integration, we obtain [Eqs. (3.92), (3.93)]

$${}^{t}\hat{\mathbf{K}}\,\delta\mathbf{r}^{(k)} = {}^{t+\Delta t}\hat{\mathbf{R}}^{(k)}, \tag{4.53}$$

where

$${}^{t}\hat{\mathbf{K}} = {}^{t}\mathbf{K} + \frac{4}{\Delta t^2}\mathbf{M}, \tag{4.54}$$

$${}^{t+\Delta t}\hat{\mathbf{R}}^{(k)} = {}^{t+\Delta t}\mathbf{R}_0 - {}^{t+\Delta t}\mathbf{R}_I^{(k-1)} - \mathbf{M}\left[\frac{4}{\Delta t^2}\left({}^{t+\Delta t}\mathbf{r}^{(k-1)} - {}^{t}\mathbf{r}\right) - \frac{4}{\Delta t}\,{}^{t}\dot{\mathbf{r}} - {}^{t}\ddot{\mathbf{r}}\right]. \tag{4.55}$$

It is observed that the effective stiffness matrix ${}^{t}\hat{\mathbf{K}}$ might change at each time step whereas the effective load vector ${}^{t+\Delta t}\hat{\mathbf{R}}^{(k)}$ – at each iteration cycle. The iterative equations are similar to the nonlinear static equations, except that ${}^{t}\hat{\mathbf{K}}$ and ${}^{t+\Delta t}\hat{\mathbf{R}}^{(k)}$ contain contributions from the inertia of the system. Thus, Eq. (4.53) can be expressed in the form

$$(\hat{\mathbf{K}}_0 + {}^{t}\Delta\hat{\mathbf{K}})\,\delta\mathbf{r}^{(k)} = {}^{t+\Delta t}\hat{\mathbf{R}}^{(k)}, \tag{4.56}$$

where subscript 0 denotes the initial matrix and ${}^{t}\Delta\hat{\mathbf{K}}$ is given by

$$ {}^{t}\Delta\hat{\mathbf{K}} = {}^{t}\hat{\mathbf{K}}({}^{t}\mathbf{K},\mathbf{M}) - \hat{\mathbf{K}}_0(\mathbf{K}_0,\mathbf{M}_0) . \tag{4.57} $$

Again, the modified analysis equations (4.5) and (4.56) are of similar form.

Solution by Mode Superposition

In nonlinear dynamic analysis, the eigenpairs ${}^{t}\mathbf{\Phi}_i$, ${}^{t}\lambda_i$ at time t are first obtained by solving the eigenproblem [Eq. (3.104)]

$$ {}^{t}\mathbf{K}\,{}^{t}\mathbf{\Phi}_i = {}^{t}\lambda_i\,\mathbf{M}\,{}^{t}\mathbf{\Phi}_i , \tag{4.58} $$

where ${}^{t}\mathbf{K}$ is the tangent stiffness matrix at time t, given by [Eq. (3.99)]

$$ {}^{t}\mathbf{K} = \frac{\partial\, {}^{t}\mathbf{F}_R}{\partial \mathbf{r}} . \tag{4.59} $$

For proportional damping, we solve the uncoupled equations (3.105)

$$ \mathbf{I}\ {}^{t}\Delta\ddot{\mathbf{Z}} + {}^{t}\mathbf{C}_d\ {}^{t}\Delta\dot{\mathbf{Z}} + {}^{t}\mathbf{\Omega}^2\ {}^{t}\Delta\mathbf{Z} = {}^{t}\mathbf{\Phi}^T\ {}^{t}\Delta\mathbf{R} , \tag{4.60} $$

where ${}^{t}\Delta\mathbf{Z}$ are the generalized displacements and ${}^{t}\mathbf{\Phi}$ is the matrix of eigenvectors. The identity matrix $\mathbf{I} = {}^{t}\mathbf{\Phi}^T\mathbf{M}\,{}^{t}\mathbf{\Phi}$ is the mass matrix in normalized coordinates. The damping matrix $\mathbf{C}_d = {}^{t}\mathbf{\Phi}^T\mathbf{C}\,{}^{t}\mathbf{\Phi}$ and the stiffness matrix ${}^{t}\mathbf{\Omega}^2 = {}^{t}\mathbf{\Phi}^T\ {}^{t}\mathbf{K}\ {}^{t}\mathbf{\Phi}$ in these coordinates are diagonal matrices. The nodal displacements are computed by [Eqs. (3.102), (3.103)]

$$ {}^{t}\Delta\mathbf{r} = {}^{t}\mathbf{\Phi}\ {}^{t}\Delta\mathbf{Z} , \tag{4.61} $$

$$ {}^{t+\Delta t}\mathbf{r} = {}^{t}\mathbf{r} + {}^{t}\Delta\mathbf{r}. \tag{4.62} $$

As noted earlier, a significant part of the computational effort involves repeated eigenproblem solutions in the nonlinear region. The nonlinear eigenproblem [Eq. (4.58)] can be expressed as

$$ (\mathbf{K}_0 + {}^{t}\Delta\mathbf{K})\,{}^{t}\mathbf{\Phi}_i = {}^{t}\mathbf{R}_i \qquad i = 1, \ldots, n, \tag{4.63} $$

where

$$ {}^{t}\mathbf{R}_i = {}^{t}\lambda_i\ \mathbf{M}\,{}^{t}\mathbf{\Phi}_i . \tag{4.64} $$

The matrix of changes ${}^{t}\Delta\mathbf{K}$ consists of the following two parts

$$ {}^{t}\Delta\mathbf{K} = \Delta\mathbf{K}_0 + {}^{t}\Delta\mathbf{K}_{NL}, \tag{4.65}$$

where matrix $\Delta\mathbf{K}_0$, representing the changes in the elastic stiffness matrix due to changes in the design, is constant for any given design, whereas matrix ${}^{t}\Delta\mathbf{K}_{NL} = {}^{t}\mathbf{K} - (\mathbf{K}_0+\Delta\mathbf{K}_0)$, representing the changes in stiffness due to the nonlinear behavior, might change at each time step.

The mathematical expressions of the problems presented in this section are summarized in Table 4.3.

Table 4.3. Formulation of dynamic reanalysis problems

Problem	Equations	Matrix of Changes	Right-hand side
Linear Direct Integration	(4.45) $(\hat{\mathbf{K}}_0 + \Delta\hat{\mathbf{K}})\,{}^{t+\Delta t}\mathbf{r} = {}^{t+\Delta t}\hat{\mathbf{R}}$	(4.46) $\Delta\hat{\mathbf{K}} = \hat{\mathbf{K}}(\mathbf{K},\mathbf{M},\mathbf{C}) - \hat{\mathbf{K}}_0(\mathbf{K}_0,\mathbf{M}_0,\mathbf{C})$	(4.44) ${}^{t+\Delta t}\hat{\mathbf{R}}({}^{t+\Delta t}\mathbf{R}, \mathbf{M}, {}^{t}\mathbf{r}, {}^{t}\dot{\mathbf{r}}, {}^{t}\ddot{\mathbf{r}})$
Linear Mode Superposition	(4.33) $(\mathbf{K}_0 + \Delta\mathbf{K}_0)\,\boldsymbol{\Phi} = \mathbf{R}$	(4.29) $\Delta\mathbf{K}_0 = \mathbf{K} - \mathbf{K}_0$	(4.32) $\mathbf{R} = \lambda\,\mathbf{M}\,\boldsymbol{\Phi}$
Nonlinear Direct Integration	(4.56) $(\hat{\mathbf{K}}_0 + {}^{t}\Delta\hat{\mathbf{K}})\,\delta\mathbf{r}^{(k)} = {}^{t+\Delta t}\hat{\mathbf{R}}^{(k)}$	(4.57) ${}^{t}\Delta\hat{\mathbf{K}} = {}^{t}\hat{\mathbf{K}}({}^{t}\mathbf{K},\mathbf{M},\mathbf{C}) - \hat{\mathbf{K}}_0(\mathbf{K}_0,\mathbf{M}_0,\mathbf{C})$	(4.55) ${}^{t+\Delta t}\hat{\mathbf{R}}^{(k)}({}^{t+\Delta t}\mathbf{R}_0, {}^{t+\Delta t}\mathbf{R}_I^{(k-1)}, \mathbf{M}, {}^{t+\Delta t}\mathbf{r}^{(k-1)}, {}^{t}\mathbf{r}, {}^{t}\dot{\mathbf{r}}, {}^{t}\ddot{\mathbf{r}})$
Nonlinear Mode Superposition	(4.63) $(\mathbf{K}_0 + {}^{t}\Delta\mathbf{K})\,{}^{t}\boldsymbol{\Phi}_i = {}^{t}\mathbf{R}_i$	(4.65) ${}^{t}\Delta\mathbf{K} = \Delta\mathbf{K}_0 + {}^{t}\Delta\mathbf{K}_{NL}$	(4.64) ${}^{t}\mathbf{R}_i = {}^{t}\lambda_i\,\mathbf{M}\,{}^{t}\boldsymbol{\Phi}_i$

4.5 Reanalysis Methods

Several comprehensive reviews on reanalysis methods have been published in the past [e.g. 2–4]. The various methods may be divided into the following two general categories, described in this section:

- Direct methods, giving exact closed-form solutions and might be efficient only when a relatively small part of the structure is changed.
- Approximate methods, giving approximate solutions, with the accuracy being dependent on the type of changes and the specifiec method used. These methods are usually suitable for changes in large parts of the structure.

4.5.1 Direct Methods

Direct reanalysis methods are efficient for low-rank changes in the stiffness matrix. In particular, these methods are applicable to situations where a relatively small proportion of the structure is changed and the changes in the stiffness matrix can be represented by a small sub-matrix (e.g. when the cross sections of only a few members are changed). Direct methods are inefficient when the sub-matrix of changes in the stiffness matrix is large.

Direct methods are usually based on the Sherman-Morrison [5] and Woodbury [6] formulae for the update of the inverse of a matrix. Surveys on these methods are given elsewhere [7–9]. A comprehensive historical survey of the origin of these formulae is presented in [8]. It has been shown [9] that various reanalysis methods may be viewed as variants of these formulae. Several methods for calculating the modified response due to changes in the structure were proposed in the late 1960s and the early 1970s. Most of these improved methods are based on the Sherman-Morrison identity [e.g. 10–12]. Direct methods are described later in Chap. 8. The *Combined Approximations* (CA) approach, introduced in this text, provides exact solutions under certain conditions. It is shown in Sects. 8.2.2 and 8.2.3 that in such cases exact solutions achieved by the CA approach and the Sherman-Morrison-Woodbury formulae are equivalent.

Other direct methods are the *Virtual Distortion Method* [VDM, 13–14] and the *Theorems of Structural Variation* [TSV, 15–20]. The two methods, called load-based methods and may be viewed as variants of the Woodbury formula [9], require collinear loads to be applied to the modified members, to compute an influence matrix. In the VDM, a reduced set of equations is then solved for a set of scalar multipliers of the influence vectors. In the TSV approach the modified displacements and forces are expressed in terms of the original values and the values due to unit loadings.

4.5.2 Approximate Methods

Approximate reanalysis methods are suitable for cases of changes in large parts or all of the structure. These methods have been used extensively in structural optimization to reduce the number of exact analyses and the overall computational cost during the solution process. Reduction of the computational cost allows the solution of large-scale problems.

In general, the following factors are considered in choosing an approximate reanalysis method for a specific application:

- The accuracy of the calculations (the quality of the approximations).
- The computational effort involved (the efficiency of the method).

- The ease-of-implementation.

The implementation effort is weighted against the performance of the algorithms as reflected in their computational efficiency and accuracy. The quality of the results and the efficiency of the calculations are usually conflicting factors. That is, better approximations are often achieved at the expense of more computational effort. The different levels of analysis range from inexpensive and inaccurate to costly and accurate. The common approximations can be divided into the following classes [4, 21]:

- *Local approximations* (called also *single-point approximations*), such as the first-order Taylor series expansion or the binomial series expansion about a given design point. Local approximations are based on information calculated at a single point. These methods are very efficient but they are effective only for small changes in the design variables. For large changes in the design the accuracy of the approximations often deteriorates and the results may become meaningless. That is, the approximations are valid only in the vicinity of a design point. To improve the quality of the results, reciprocal cross-sectional areas have been assumed as design variables [22, 23]. A hybrid form of the direct and reciprocal approximations, which is more conservative than either, can also be introduced [24]. This approximation has the advantage of being convex [25], but it has been found that the hybrid approximation tends to be less accurate than either the direct or the reciprocal approximation. More accurate convex approximations can be introduced by the method of moving asymptotes [26], but the quality of the results might be dependent on the selection of these asymptotes. Another possibility to improve the quality of the results is to consider second-order approximations [27, 28] but this considerably increases the computational effort.
- *Global approximations* (called also *multipoint approximations*), such as *polynomial fitting*, *response surface* or *reduced basis* methods [29–33]. These approximations are obtained by analyzing the structure at a number of design points, and they are valid for the whole design space (or, at least, large regions of it). In response surface methods [e.g. 30, 31], the response functions are replaced by simple functions (polynomials), which are fitted to data computed at a set of selected design points. So far, the use of response surface methods has been limited to problems with a relatively small number of design variables. In reduced basis methods [32, 33] the response of a large system, which was originally described by a large number of degrees of freedom, is approximated by a linear combination of a few pre-selected basis

vectors. The problem is then stated in terms of a small number of unknown coefficients of the basis vectors. The approach is most effective when highly accurate approximations can be introduced by the reduced and much smaller system of equations. A basic question in using reduced basis methods is the choice of an appropriate set of the basis vectors. Response vectors of previously analyzed designs could be used, but an ad hoc or intuitive choice may not lead to satisfactory approximations. In addition, calculation of the basis vectors requires several exact analyses of the structure for the basis designs, which involves extensive computational effort. In summary, global approximations may require much computational effort, particularly in problems with large numbers of design variables.

- *Combined approximations*. In the next chapters of this text we develop a third class of approximations, called Combined Approximations (CA). In this approach we attempt to give global qualities to local approximations. This can be achieved by considering the terms of local approximations as basis vectors in a global expression. Specifically, the binomial series terms are used as basis vectors in reduced basis approximations. The advantage is that the efficiency of local approximations and the improved quality of global approximations are combined to obtain an effective solution procedure. That is, the above choice of basis vectors provides accurate results efficiently. Various means can be used to improve both the accuracy of the results and the efficiency of the calculations. The approach is general, providing various options and possibilities in applications. The main developments of the approach are reviewed in the next subsection.

4.5.3 The Combined Approximations Approach

Initially, the main objective in developing the CA approach was to simplify design optimization procedures for practical structures. Later, it was found that the approach might prove useful not only in structural optimization but also in various analysis and design tasks. In particular, solutions for the following classes of problems have been developed:

- *Linear and nonlinear static analysis*. Solutions of various analysis and reanalysis problems have been demonstrated, including topological and geometrical changes, geometric and material nonlinearity, accurate and exact solutions. Linear and nonlinear static reanalysis are presented in Chap. 5.

- *Vibration analysis*. Effective solution procedures for linear and nonlinear eigenproblems have been developed. Various procedures for vibration reanalysis are introduced in Chap. 6.
- *Linear and nonlinear dynamic analysis*. Solutions of various analysis and reanalysis problems of structures subjected to dynamic loadings are discussed in Chap. 7.
- *Sensitivity analysis*. Repeated sensitivity analysis for various problems is demonstrated in Chap. 9. Solutions of linear, nonlinear, static and dynamic problems are presented.

Efficiency and accuracy considerations for the above classes of problems are discussed in Chap. 10. The main developments of the CA approach are described in the next subsections.

Early Developments

Early developments related to linear static reanalysis problems are presented in [34–39]. Several studies in the early 1980s [34, 35] showed that improved local approximations can be achieved by scaling of the initial design such that the changes in the design variables are reduced. The advantage is that the solution is still based on results of a single exact analysis. It was found that scaling procedures may significantly improve the accuracy of the results with little computational effort. Moreover, scaling might prove useful for various types of design variables and response functions. Several criteria for selecting the scaling multiplier have been proposed, based on geometrical [34] and mathematical [35] considerations. In the early 1990s it was shown [36–38] that scaling of both the initial design and the modified approximate displacements can be expressed in a reduced basis form, using transformations of variables. Extending the concept of scaling to include also the approximate displacements, in addition to the initial design, significantly improved the results. It was found [39] that accurate approximations of displacements, stresses and forces can be achieved by the CA approach, for very large changes in the design variables, by considering only first-order approximations. Reanalysis of static problems by the CA approach is developed in Chap. 5.

Geometrical and Topological Changes

Following the early stages of developments, it was found that the CA approach is very effective for geometrical and topological changes [40–52]. In the early 1990s, it has been shown that the approach provides accurate solutions for topological changes [40]. It was found [41] that exact solutions can be achieved for cross-sectional variables if for a changed member only one basis vector is considered. For simultaneous changes in several members, exact solutions are achieved if for each changed member a corresponding basis vector is considered. This result is valid also for all types of topological changes in the structure, including elimination and addition of members and joints [42, 43]. Exact and accurate solutions for all types of topological and geometrical changes have been demonstrated [44–47]. In the early 2000s the solution approach was improved to obtain more accurate results [48–52]. It was found [48] that a preconditioned conjugate gradient method and the CA procedure provide theoretically identical results. Improved solution procedures for the challenging problem of increasing the number of degrees of freedom were developed [51, 52]. Topological changes are discussed in Sect. 5.3. Direct solutions for topological and geometrical changes are demonstrated in Sect. 8.3.

Direct Solutions

It was found in the early 2000s [9] that exact solutions achieved by the CA approach and the Sherman-Morrison-Woodbury formulae are equivalent. The two solution procedures show that the change in nodal displacements due to a change in the cross section area of a truss member is a multiple of the response to a pair of collinear forces acting at the ends of the member. It has been shown also that this result can be generalized to any structural member such as a frame element or a plate element. Direct solutions by the CA approach were developed in the early 1900s [40, 41]. Such solutions have been demonstrated also for various cases of topological and geometrical changes [42, 46, 47]. Direct reanalysis is discussed in Chap. 8.

Vibration, Dynamic and Nonlinear Reanalysis

In the late 1990s, eigenvalue reanalysis of damaged structures [53] and nonlinear reanalysis [54, 55] by the CA approach were first presented. It was found that by using a Gram-Schmidt orthonormalization procedure a new set of basis vectors can be generated such that the reduced set of analysis equations becomes uncoupled. For any assumed number of basis vectors, the results obtained by considering either the original set of basis vectors or the new set of uncoupled basis vectors, are identical. The advantage in using the latter vectors is that all expressions for evaluating the displacements become explicit functions of the parameters of the structure. As a result, additional vectors can be considered without modifying the calculations that have been already carried out. In addition, the uncoupled system is more well-conditioned. Several procedures, based on the CA approach, have been developed recently for solving various nonlinear analysis and reanalysis problems [56]. These procedures are used to evaluate the modified displacements at each iteration cycle.

Effective procedures for eigenproblem reanalysis have been developed in the early 2000s [57–61]. Significant improvements in the accuracy of the results and the efficiency of the calculations for vibration reanalysis were reported. These studies formed the basis for developing improved solution procedures for linear-dynamic [62] and nonlinear-dynamic [63, 64] reanalysis. Vibration reanalysis is discussed in Chap. 6 and dynamic reanalysis is developed in Chap. 7.

Repeated Sensitivity Analysis

Various approximations that might be adequate for structural reanalysis are not sufficiently accurate for repeated sensitivity analysis. It has been shown [65–72] that the CA approach can be used also for effective approximations of the response derivatives for designs where results of exact analysis are not available. Accurate results have been achieved by either the direct method or the adjoint-variable method for calculations of analytical derivatives [65, 66] and finite-difference derivatives [67] of static and vibration response. Methods for calculations of repeated sensitivity analysis of structures subjected to linear [68] and nonlinear [69] dynamic response have been developed. Improved efficiency has been demonstrated in the solution of various problems [70, 71]. It was found [72] that improved accuracy of shape sensitivity calculations can be achieved using refined basis vectors. Repeated sensitivity analysis is presented in Chap. 9.

Computational Considerations

Various considerations related to the convergence properties of the solution process and to the accuracy of the results have been studied [48, 73, 74]. It has been noted [48] that a preconditioned conjugate gradient method and the CA approach provide theoretically identical results. As a result, some convergence criteria and error expressions developed for conjugate gradient methods can be used for the CA approach. The approach has been successfully applied to both low-rank and moderately high-rank modifications to structures [75]. Accurate results were reported for large scale systems [76] and for probabilistic analysis [77]. Efficient procedures have been developed for numerous repeated calculations of the structural response in large scale structures [78] and for nonlinear reliability analysis of structural systems [79]. In a recent study [80] a methodology was developed to integrate the CA method with the commercial program MSC-

NASTRAN for finite element analysis. A modified CA method has been developed and integrated with other methods to achieve efficient probabilistic vibration analysis of complex structures [81] and efficient reanalysis of large finite element models [82]. Various computational advantages of the CA approach have been discussed in several studies [83–86].

References

1. Venkataraman S, Haftka RT (2004) Structural optimization complexity: what has Moore's law done for us. Struct and Multidis Opt 28:375-387
2. Arora JS (1976) Survey of structural reanalysis techniques. J Struct Div ASCE 102:783-802
3. Abu Kasim AM, Topping BHV (1987) Static reanalysis: a review. J Struct Div ASCE 113:1029-1045
4. Barthelemy JFM, Haftka RT (1993) Approximation concepts for optimum structural design - a review. Struct and Multidis Opt 5:129-144
5. Sherman J, Morrison WJ (1949) Adjustment of an inverse matrix corresponding to changes in the elements of a given column or a given row of the original matrix. Ann. Math. Statist. 20:621
6. Woodbury M (1950) Inverting modified matrices. Memorandum Report 42 Statistical Research Group, Princeton University, Princeton, NJ
7. Householder AS (1957) A survey of some closed form methods for inverting matrices. SIAM J 3:155 -169
8. Hager WW (1989) Updating the inverse of a matrix. SIAM Rev 31:221-239
9. Akgun MA, Garcelon JH, Haftka RT (2001) Fast exact linear and nonlinear structural reanalysis and the Sherman-Morrison-Woodbury formulas. Int J Num Meth Engrg 50:1587-1606
10. Sack RL, Carpenter WC, Hatch GL (1967) Modification of elements in displacement method. AIAA J 5:1708-1710
11. Argyris JH, Roy JR (1972) General treatment of structural modifications. J Struct Div ASCE 98:462-492
12. Kirsch U (1981) Optimum structural design. McGraw Hill, New York
13. Holnicki-Szulc J (1995) Structural analysis design and control by the virtual distortion method. Wiley, England
14. Makode PV, Ramirez MR, Corotis RB (1996) Reanalysis of rigid frame structures by the virtual distortion method. Struct and Multidis Opt 11:71-79
15. Majid KI (1974) Optimum design of structures. Newnes-Butterworths, London
16. Majid KI, Saka MP, Celik T (1978) The theorem of structural variation generlized for rigidly jointed frames. Proc Inst Civ Engineers 65:839-856
17. Topping BHV (1983) The application of the theorems of structural variation to finite element problems. Int J Num Meth Engrg 19:141-151
18. Atrek E (1985) Theorems of structural variation: a simplification. Int J Num Meth Engrg 21:481-485

19. Topping BHV, Kassim AMA (1987) The use and the efficiency of the theorems of structural variation to finite element analysis. Int J Num Meth Engrg 24:1901-1920
20. Saka MP (1991) Finite element applications of the theorems of structural variation. Computers and Structures 41:519-530
21. Kirsch U (1993) Structural Optimization, fundamentals and applications. Springer-Verlag, Berlin
22. Fuchs MB (1980) Linearized homogeneous constraints in structural design. Int J Mech Sci 22:333-400
23. Schmit LA, Farshi B (1974) Some approximation concepts for structural synthesis. AIAA J 11:489-494
24. Starnes JHJr, Haftka RT (1979) Preliminary design of composite wings for buckling stress and displacement constraints. J Aircraft 16:564-570
25. Fleury C, Braibant V (1986) Structural optimization: a new dual method using mixed variables. Int J Num Meth Engrg 23:409-428
26. Svanberg K (1987) The method of moving asymptotes - a new method for structural optimization Int J Num Meth Engrg 24:359-373
27. Fleury C (1989) Efficient approximation concepts using second-order information. Int J Num Meth Engrg 28:2041-2058
28. Fleury C (1989) First- and second-order convex approximation strategies in structural optimization. Struct Opt 1:3-10
29. Haftka RT, Nachlas JA, Watson LT, Rizzo T, Desai R (1989) Two-point constraint approximation in structural optimization. Comp Meth Appl Mech Engrg 60:289-301
30. Unal R, Lepsch R, Engelund W, Stanley D (1996) Approximation model building and multidisciplinary design optimization using response surface methods. Proc 6th AIAA/NASA/ISSMO Symp on multidisciplinary analysis and optimization. Bellevue, WA
31. Sobieszczanski-Sobieski J, Haftka RT (1997) Multidisciplinary aerospace design optimization: survey and recent developments. Struct and Multidis Opt 14:1-23
32. Fox RL, Miura H (1971) An approximate analysis technique for design calculations. AIAA J 9:177-179.
33. Noor AK (1994) Recent advances and applications of reduction methods. Appl Mech Rev 47:125-146
34. Kirsch U (1984) Approximate behavior models for optimum structural design. In: Atrek E et al (eds) New directions in optimum structural design. John Wiley &Sons, NY
35. Kirsch U, Toledano G (1983) Approximate reanalysis for modifications of structural geometry. Computers & Structures 16:269-279
36. Kirsch U (1991) Reduced basis approximations of structural displacements for optimal design. AIAA J 29:1751-1758
37. Kirsch U Eisenberger M (1991) Approximate interactive design of large Structures, Comp Sys in Engrg 2:67-74
38. Kirsch, U (1993) Approximate reanalysis methods. In: Kamat PM (ed) Structural optimization: status and promise. AIAA

39. Kirsch U (1995) Improved stiffness-based first-order approximations for structural optimization. AIAA J 33:143-150.
40. Kirsch U (1993) Efficient reanalysis for topological optimization. Struct and Multidis Opt 6:143-150
41. Kirsch U, Liu S (1995) Exact structural reanalysis by a first-order reduced basis approach. Struct and Multidis Opt 10:153-158
42. Kirsch U, Liu S (1997) Structural reanalysis for general layout modifications. AIAA J 35:382-388
43. Chen S, Huang C, Liu Z (1998) Structural approximate reanalysis for topological modifications by finite element systems. AIAA J 36:1760-1762
44. Kirsch U, Moses F (1998) An improved reanalysis method for grillage-type structures. Computers & Structures 68:79-88
45. Kirsch U, Moses F (1999) Effective reanalysis for damaged structures. In: Frangopol DM (ed) Case studies in optimal design. ASCE
46. Kirsch U, Papalambros PY (2001) Structural reanalysis for topological modifications. Struct and Multidis Opt 21:333-344
47. Kirsch U, Papalambros PY (2001) Exact and accurate reanalysis of structures for geometrical changes. Engrg with Comp 17:363-372
48. Kirsch U, Kocvara M, Zowe J (2002) Accurate reanalysis of structures by a preconditioned conjugate gradient method. Int J Num Meth Engrg 55:233-251
49. Wu BS, Li ZG (2001) Approximate reanalysis for modifications of structural layout. Engineering Structures 23:1590-1596
50. Chen SH, Yang ZJ (2004) A universal method for structural static reanalysis of topological modifications. Int J Num Meth Engrg 61:673–686
51. Wu BS, Lim CW, Li ZG (2004) A finite element algorithm for reanalysis of structures with added degrees of freedom. Finite Elements in Analysis and Design 40:1791-1801
52. Wu BS, Li ZG (2006) Static reanalysis of structures with added degrees of freedom. Comm Num Meth Engrg 22:269-281
53. Aktas A, Moses F (1998) Reduced basis eigenvalue solutions for damaged structures. Mech of Struct and Mach 26:63-79
54. Leu LJ, Huang CW (1998) A reduced basis method for geometric nonlinear analysis of structures. IASS J 39:71-75
55. Kirsch U (1999) Efficient-accurate reanalysis for structural optimization. AIAA J 37:1663-1669
56 Amir O, Kirsch U, Sheinman I (in press) Efficient nonlinear reanalysis of skeletal structures using combined approximations. Int J Num Meth Engrg
57. Chen SH, Yang, XW (2000) Extended Kirsch combined method for eigenvalue reanalysis. AIAA J 38:927-930
58. Chen SH, Yang XW, Lian HD (2000) Comparison of several eigenvalue reanalysis methods for modified structures. Struct and Multidis Opt 20:253-259
59. Kirsch U (2003) Approximate vibration reanalysis of structures. AIAA J 41:504-511
60. Rong F et al (2003) Structural modal reanalysis for topological modifications with extended Kirsch method. Comp Meth Appl Mech Engrg 192:697-707

61. Kirsch U, Bogomolni M (2004) Procedures for approximate eigenproblem reanalysis of structures. Int J Num Meth Engrg 60:1969-1986
62. Kirsch U, Bogomolni M, Sheinman I (2007) Efficient dynamic reanalysis of structures. ASCE J Struct Engrg 133:440-448
63. Kirsch U, Bogomolni M (2007) Nonlinear and dynamic structural analysis using linear reanalysis. Computers & Structures 85:566-578
64. Kirsch U, Bogomolni M, Sheinman I (2006) Nonlinear dynamic reanalysis for structural optimization. Comp Meth Appl Mech Engrg 195:4420-4432
65. Kirsch U (1994) Effective sensitivity analysis for structural optimization. Comp Meth Appl Mechanics Engrg 117:143-156
66. Kirsch U, Papalambros PY (2001) Accurate displacement derivatives for structural optimization using approximate reanalysis. Comp Meth Appl Mech Engrg 190:3945-3956
67. Kirsch U, Bogomolni M, van Keulen F (2005) Efficient finite-difference design-sensitivities. AIAA J 43:399-405
68. Bogomolni M, Kirsch U, Sheinman I (2006) Efficient design-sensitivities of structures subjected to dynamic loading. Int J Sol & Str 43:5485-5500
69. Bogomolni M, Kirsch U, Sheinman I (2006) Nonlinear-dynamic sensitivities of structures using combined approximations. AIAA J 44: 2675-2772
70. Kirsch U, Bogomolni M, Sheinman I (2007) Efficient procedures for repeated calculations of structural sensitivities. Engrg Opt 39:307-325
71. Kirsch U, Bogomolni M, Sheinman I (2007) Efficient structural optimization using reanalysis and sensitivity reanalysis. Engrg with Computers 23:229-239
72. van Keulen F, Vervenne K, Cerulli C (2004) Improved combined approach for shape modifications. In proc 10th AIAA/ISSMO Multidisciplinary Analysis and Optimization Conference, Albany, New York, AIAA 2004-4378
73. Kirsch U, Papalambros PY (2001) Exact and accurate solutions in the approximate reanalysis of structures. AIAA J 39:2198-2205
74. Kirsch U, Bogomolni M (2004) Error evaluation in approximate reanalysis of structures. Struct and Multidis Opt 28:77-86
75. Garcelon JH, Haftka RT, Scotti SJ (2000) Approximations in optimization of damage tolerant structures. AIAA J 38:517-524
76. Heiserer D, Baier H (2001) Applied reanalysis techniques for large scale structural mechanics multidisciplinary optimization in automotive industry. Proc 4th World Congress of Struct and Multidis Opt, Dalian, China
77. Keertti A, Nikolaidis E, Ghiocel D, Kirsch U (2004) Combined approximations for efficient probabilistic analysis of structures. AIAA J 42:1321-1330
78. Kirsch U, Bogomolni M, Sheinman I (2006) Efficient procedures for repeated calculations of the structural response. Struct and Multidis Opt 32:435-446
79. Khurana P (2005) Efficient deterministic and probabilistic analysis of nonlinear plate structures using combined approximation. MS thesis, Mechanical Industrial and Manufacturing Engineering dept, the Univ of Toledo, USA
80. Kadupukotla CR (2006) An efficient implementation of the combined approximations method and integration of this method into a commercial FEA software. MS thesis, Mechanical Industrial and Manufacturing Engineering dept, the Univ of Toledo, USA

81. Zhang G, Mourelatos ZP, Nikolaidis E (2007) An efficient reanalysis methodology for probabilistic vibration of complex structures. In proc 48th AIAA/ASME/ASCE/AHS/ASC Structures, Structural Dynamics, and Materials Conference, Honolulu, Hawaii, AIAA 2007-1963
82. Zhang G, Mourelatos ZP, Nikolaidis E (2007) An efficient reanalysis methodology for vibration of large-scale structures. In proc SAE Noise and Vibration Conference and Exhibition, St Charles, IL, USA
83. Kirsch U (2000) Combined approximations – a general reanalysis approach for structural optimization. Struct and Multidis Opt 20:97-106
84. Kirsch U (2002) Design-oriented analysis of structures. Kluwer Academic Publishers, Dordrecht
85. Kirsch U (2003) Design-oriented analysis of structures – a unified approach. ASCE J Engrg Mech 129:264-272
86. Kirsch U (2003) A unified reanalysis approach for structural analysis, design and optimization. Struct and Multidis Opt 25:67-85

5. Static Reanalysis

It has been noted that the overall effectiveness of reanalysis depends to a large extent on the numerical procedures used for the solution of the equilibrium equations. The accuracy of the analysis can, in general, be improved if a more refined model is used. Because of the tendency to employ more and more refined models to approximate the actual structure, the cost of an analysis and its practical feasibility depend to a considerable degree on the algorithms available for the solution of the resulting equations. The time required for solving the equilibrium equations of large-scale systems can be a high percentage of the total solution time, particularly in nonlinear analysis, when the solution must be repeated many times. An analysis may not be possible if the solution procedures are too costly or unstable.

In this section approximate solution procedures, for static analysis and reanalysis by the CA approach, are developed. The solution is based on results of a single exact analysis and the integration of several concepts and methods, most of them presented in previous chapters. These include matrix factorization, series expansion, reduced basis, and Gram-Schmidt orthogonalization. In the approach presented, the terms of the binomial series are used as high quality basis vectors in a reduced basis expression. The advantage is that efficient local approximations (series expansion) and accurate global approximations (the reduced basis method) are combined to achieve an effective solution procedure. Due to the nature of the selected basis vectors, high accuracy can be achieved by considering only a few vectors. Yet, the accuracy of the results can be improved at the expense of more computational effort, by considering additional vectors.

Determination of the basis vectors is introduced in Sect. 5.1. Linear reanalysis using coupled and uncoupled basis vectors is developed in Sect. 5.2, and topological changes, where members and joints are deleted or added, are presented in Sect. 5.3. Nonlinear analysis and reanalysis are demonstrated in Sect. 5.4.

The main advantages of the CA approach, which have been studied in terms of several criteria, are as follows [1–4]:

- *Generality*. Various analysis models (linear, nonlinear, elastic, plastic, static, dynamic), different types of structures (e.g. trusses, frames, grillages, continuum structures) and all types of changes in the structure (changes in cross-sections, geometry, topology and material properties) may be considered. The changes can be of different extent, varying from changes in only a few elements to all elements of the structure. As well, changes in the structural model itself may be considered.

- *Accuracy*. Accurate approximations can be achieved for significant changes in large-scale structures. Various means can be used to improve the accuracy of the results, including consideration of high-order terms and Gram-schmidt orthogonalization procedures. Moreover, exact solutions can be achieved in certain cases.
- *Efficiency*. Similar to local approximations, the calculations are based on results of a single exact analysis. The number of algebraic operations and the total CPU effort are usually much smaller than those needed to carry out complete analysis of modified structures.
- *Flexibility*. The efficiency of the calculations and the accuracy of the results can be controlled by the level of simplification or sophistication considered and the amount of information used. Depending on the problem to be solved, various simplified and efficient versions of the approach may be considered. On the other hand, various sophisticated means may be used to improve the accuracy of the approximations at the expense of more computational effort.
- *Ease of implementation*. The solution steps are straightforward, the approach can be readily used with general finite element systems, and the calculations are based on simple analytical expressions.

5.1 Determination of the Basis Vectors

5.1.1 The Binomial Series

In the CA approach presented in this chapter the first few terms of the binomial series expansion are used as basis vectors. Consider the modified analysis equations $(\mathbf{K}_0 + \Delta\mathbf{K})\,\mathbf{r} = \mathbf{R}$ [Eq. (4.5)], rearranged to read

$$\mathbf{K}_0\,\mathbf{r} = \mathbf{R} - \Delta\mathbf{K}\,\mathbf{r}. \qquad (5.1)$$

Writing these equations as the recurrence relation

$$\mathbf{K}_0\,\mathbf{r}^{(k+1)} = \mathbf{R} - \Delta\mathbf{K}\,\mathbf{r}^{(k)}, \qquad (5.2)$$

where $\mathbf{r}^{(k+1)}$ is the value of $\mathbf{r}$ after the kth iteration cycle, and assuming the initial value $\mathbf{r}^{(0)} = \mathbf{0}$, we obtain the binomial series

$$\mathbf{r} = (\mathbf{I} - \mathbf{B} + \mathbf{B}^2 - \dots)\,\mathbf{r}_1, \qquad (5.3)$$

where $\mathbf{I}$ is an identity matrix. The matrix $\mathbf{B}$ and the vector $\mathbf{r}_1$, which is the first term of the series, are defined as

$$\mathbf{B} = \mathbf{K}_0^{-1} \Delta\mathbf{K}, \tag{5.4}$$

$$\mathbf{r}_1 = \mathbf{K}_0^{-1} \mathbf{R}, \tag{5.5}$$

The additional terms of the series are given by the recurrence relation

$$\mathbf{r}_i = -\mathbf{B}\, \mathbf{r}_{i\text{-}1} \qquad (i = 2, ..., s). \tag{5.6}$$

The series of Eq. (5.3) can be obtained in an alternative way as follows. Pre-multiplying the modified equations $(\mathbf{K}_0+\Delta\mathbf{K})\, \mathbf{r} = \mathbf{R}$ by $\mathbf{K}_0^{-1}$ and substituting Eqs. (5.4), (5.5) yields

$$(\mathbf{I} + \mathbf{B})\, \mathbf{r} = \mathbf{r}_1. \tag{5.7}$$

Pre-multiplying Eq. (5.7) by $(\mathbf{I} + \mathbf{B})^{-1}$ and expanding

$$(\mathbf{I} + \mathbf{B})^{-1} \cong \mathbf{I} - \mathbf{B} + \mathbf{B}^2 - ..., \tag{5.8}$$

gives the series (5.3). Convergence of the series is discussed in Sect. 5.1.3.

5.1.2 Calculation of the Basis Vectors

The positive-definite stiffness matrix $\mathbf{K}_0$ is usually given from initial analysis in the decomposed form $\mathbf{K}_0 = \mathbf{U}_0^T\, \mathbf{U}_0$ [Eq. (4.2)]. If the load vector is unchanged ($\Delta\mathbf{R} = \mathbf{0}$, $\mathbf{R} = \mathbf{R}_0$) then the first term $\mathbf{r}_1 = \mathbf{r}_0$ is already given from initial analysis. It is shown in the following that calculation of the series terms by Eq. (5.6) involves only forward and backward substitutions.

The vector $\mathbf{r}_2$ is calculated by [Eqs. (4.2), (5.4), (5.6)]

$$\mathbf{K}_0\, \mathbf{r}_2 = \mathbf{U}_0^T\, \mathbf{U}_0\, \mathbf{r}_2 = -\Delta\mathbf{K}\, \mathbf{r}_1. \tag{5.9}$$

We first solve for the vector of unknowns $\mathbf{t}$ by the forward substitution

$$\mathbf{U}_0^T\, \mathbf{t} = -\Delta\mathbf{K}\, \mathbf{r}_1. \tag{5.10}$$

The vector $\mathbf{r}_2$ is then calculated by the backward substitution

$$\mathbf{U}_0\, \mathbf{r}_2 = \mathbf{t}. \tag{5.11}$$

Similarly, the vector $\mathbf{r}_i$ is calculated by

$$\mathbf{K}_0\, \mathbf{r}_i = -\Delta\mathbf{K}\, \mathbf{r}_{i\text{-}1}. \tag{5.12}$$

In some particular cases, repeated calculation of the series terms involves almost no computational effort. Consider for example the common case of approximations along the line defined by

$$\mathbf{K} = \mathbf{K}_0 + \Delta\mathbf{K} = \mathbf{K}_0 + \alpha\, \Delta\mathbf{K}_0, \tag{5.13}$$

where the elements of matrix $\Delta\mathbf{K}_0$ are constant, representing a given direction of movement, the scalar α is a step size variable and

$$\Delta\mathbf{K} = \alpha\, \Delta\mathbf{K}_0. \tag{5.14}$$

The expression of Eq. (5.14) is obtained also if the matrix of changes $\Delta\mathbf{K}$ represents a change of rank-one in the stiffness matrix. For $\mathbf{B}_0 = \mathbf{K}_0^{-1}\Delta\mathbf{K}_0$ and $\mathbf{R} = \mathbf{R}_0$, the terms of the binomial series [Eqs. (5.5), (5.6)] become

$$\mathbf{r}_1 = \mathbf{r}_0, \tag{5.15}$$

$$\mathbf{r}_i = \alpha^{i-1}\, \mathbf{r}_{0i} \quad (i = 2, \ldots, s), \tag{5.16}$$

where the elements of vectors $\mathbf{r}_{0i}$ are constant, defined by

$$\mathbf{r}_{0i} = -\,\mathbf{B}_0\, \mathbf{r}_{0i\text{-}1} \quad (i = 2, \ldots, s). \tag{5.17}$$

That is, once the vectors $\mathbf{r}_{0i}$ are calculated, then for any assumed α it is necessary to calculate only the products $\alpha^{i-1}\, \mathbf{r}_{0i}$ $(i = 2, \ldots, s)$. Moreover, it will be shown later in Sect. 5.2.1 that, in the CA approach, multiplication of any basis vector by a scalar does not change the results. That is, the constant vectors $\mathbf{r}_{0i}$ can be used as basis vectors for any value of the scalar α.

The terms of the binomial series are equivalent to those of the Taylor series for homogeneous displacement functions. Both the Taylor series and the binomial series are based on information of a single design. As a result, the accuracy of the results might be insufficient for large changes in the design, where problems of slow convergence or divergence of the series may be encountered. Several methods have been proposed to improve the series convergence. These include the Jacobi iteration, block Gauss-Seidel iteration, dynamic acceleration and scaling of the initial design [5, 6, 7]. One advantage of using the binomial series is that, unlike the Taylor series, calculation of derivatives is not required and high-order terms can readily be calculated. This makes the method more attractive in various applications where derivatives are not available or difficult to calculate.

The accuracy of the binomial series approximations can be significantly improved by considering its terms as basis vectors in a reduced basis expression, as will be shown later in this chapter.

5.1.3 Convergence of the Series

The series of basis vectors [Eq. (5.3)] converges if and only if [8]

$$\lim_{k\to\infty} \mathbf{B}^k = \mathbf{0}, \tag{5.18}$$

which in turn holds if and only if

$$\rho(\mathbf{B}) < 1, \tag{5.19}$$

where $\rho(\mathbf{B})$ is the spectral radius (the largest eigenvalue) of matrix $\mathbf{B}$. A sufficient criterion for the convergence of the series is

$$\|\mathbf{B}\| \leq 1, \tag{5.20}$$

where $\|\mathbf{B}\|$ is the norm of $\mathbf{B}$.

To evaluate the errors involved in the binomial series approximations, we see that the sum of the additional terms in the series of Eq. (5.3), beyond the first s terms,

$$\Delta\mathbf{r} = \mathbf{B}^s\,\mathbf{r}_0 - \mathbf{B}^{s+1}\,\mathbf{r}_0 + \mathbf{B}^{s+2}\,\mathbf{r}_0 - ..., \tag{5.21}$$

can be expressed as

$$\Delta\mathbf{r} = \mathbf{B}^s(\mathbf{I} - \mathbf{B} + \mathbf{B}^2 - ...)\,\mathbf{r}_0. \tag{5.22}$$

This expression is bounded from above by [9]

$$\Delta\mathbf{r} \leq \|\,\mathbf{B}\,\|^s \frac{1}{1-\|\,\mathbf{B}\,\|} \|\,\mathbf{r}_0\,\|. \tag{5.23}$$

It is observed that for large changes $\Delta\mathbf{K}$ (and large elements of $\mathbf{B}$) this bound becomes very large and the series diverges. It will be shown that even in such cases the CA approach provides accurate results.

5.2 Linear Reanalysis

5.2.1 Coupled Basis Vectors

Linear elastic reanalysis by the CA approach, which has been discussed in various studies [e.g. 1–4], is described in this section. We assume that the binomial series terms developed in Sect. 5.1.1 are used as basis vectors in the reduced basis expression presented in Sect. 3.2.1 [Eq. (3.72)]. The displacements $\mathbf{r}$ of a new design are approximated by a linear combination of pre-selected s linearly independent basis vectors, $\mathbf{r}_1$, $\mathbf{r}_2$, ..., $\mathbf{r}_s$

$$\mathbf{r} = y_1\mathbf{r}_1 + y_2\mathbf{r}_2 + \ldots + y_s\mathbf{r}_s = \mathbf{r}_B\,\mathbf{y}, \tag{5.24}$$

where s is much smaller than the number of degrees of freedom n, $\mathbf{r}_B$ is the $n{\times}s$ matrix of the basis vectors and $\mathbf{y}$ is a vector of unknown coefficients

$$\mathbf{r}_B = [\mathbf{r}_1, \mathbf{r}_2, \ldots, \mathbf{r}_s], \tag{5.25}$$

$$\mathbf{y}^T = \{y_1, y_2, \ldots, y_s\}. \tag{5.26}$$

The modified analysis equations are approximated by a smaller system of equations in the new unknowns $\mathbf{y}$. Substituting Eqs. (5.24) into the modified equations $\mathbf{K}\,\mathbf{r}$=$\mathbf{R}$ and pre-multiplying by $\mathbf{r}_B^T$ gives the $s{\times}s$ system

$$\mathbf{r}_B^T\mathbf{K}\mathbf{r}_B\mathbf{y} = \mathbf{r}_B^T\mathbf{R}\,. \tag{5.27}$$

Introducing the notation

$$\mathbf{K}_R = \mathbf{r}_B^T\mathbf{K}\mathbf{r}_B \qquad \mathbf{R}_R = \mathbf{r}_B^T\mathbf{R}\,, \tag{5.28}$$

and substituting Eqs. (5.28) into Eq. (5.27), we obtain

$$\mathbf{K}_R\,\mathbf{y} = \mathbf{R}_R. \tag{5.29}$$

The term *coupled basis vectors* indicates that the basis vectors are coupled by a set of simultaneous equations [Eqs. (5.28), (5.29)].

Given the initial stiffness matrix $\mathbf{K}_0$ in the decomposed form $\mathbf{K}_0 = \mathbf{U}_0^T\,\mathbf{U}_0$ [Eq. (4.2)], the initial load vector $\mathbf{R}_0$, and the initial displacement vector $\mathbf{r}_0$, calculation of the modified displacements $\mathbf{r}$ by the CA approach for any assumed changes $\Delta\mathbf{K}_0$ and $\Delta\mathbf{R}_0$ involves the following steps:

- Calculate the modified $\mathbf{K}$ and $\mathbf{R}$ by Eqs. (4.3) and (4.4)

$$\mathbf{K} = \mathbf{K}_0 + \Delta\mathbf{K}_0, \tag{5.30}$$

$$\mathbf{R} = \mathbf{R}_0 + \Delta\mathbf{R}_0.$$

  Since the initial values $\mathbf{K}_0$ and $\mathbf{R}_0$ are already given, this step involves only calculation of $\Delta\mathbf{K}_0$ and $\Delta\mathbf{R}_0$.

- Calculate the basis vectors by Eqs. (5.5), (5.6)

$$\mathbf{r}_1 = \mathbf{K}_0^{-1}\,\mathbf{R}, \tag{5.31}$$

$$\mathbf{r}_i = -\mathbf{B}\,\mathbf{r}_{i\text{-}1} \qquad (i = 2, \ldots, s), \tag{5.32}$$

  where $\mathbf{B}$ is defined by Eq. (5.4). It has been shown in Sect. 5.1.2 that calculation of the basis vectors involves only forward and backward substitutions when $\mathbf{K}_0$ is given from the initial analysis in the decomposed form of Eq. (4.2).

For large changes in the design, the elements of the basis vectors may become very large, due to large $\Delta\mathbf{K}$ values. To overcome numerical round off errors, it is possible to normalize any basis vector $\mathbf{r}_i$ by dividing it by an arbitrary reference element of the vector (say, the first element r_{1i}) to obtain the normalized vector $\mathbf{r}_{Ni}$

$$\mathbf{r}_{Ni} = \mathbf{r}_i / r_{1i}. \tag{5.33}$$

This operation scales the first element of the vector to unity. It does not change the final solution, as shown below [Eq. (5.36)].

- Calculate the reduced stiffness matrix and load vector by Eqs. (5.28)

$$\mathbf{K}_R = \mathbf{r}_B^T \mathbf{K} \mathbf{r}_B \qquad \mathbf{R}_R = \mathbf{r}_B^T \mathbf{R}\,. \tag{5.34}$$

This calculation is straightforward.

- Calculate the vector of coefficients $\mathbf{y}$ by solving Eq. (5.29)

$$\mathbf{K}_R\,\mathbf{y} = \mathbf{R}_R. \tag{5.35}$$

Since the number of basis vectors s is much smaller than the number of degrees of freedom n, it is necessary to solve only the smaller $s{\times}s$ system of Eq. (5.35) for $\mathbf{y}$ instead of computing the exact solution by solving the large $n{\times}n$ system $\mathbf{K}\,\mathbf{r} = \mathbf{R}$.

- Evaluate the final displacements by Eq. (5.24)

$$\mathbf{r} = \mathbf{r}_B\,\mathbf{y}. \tag{5.36}$$

Equation (5.36) shows that the transformation of Eq. (5.33) does not change the final solution (but only the corresponding scalars y_i).

The above solution procedure is most effective in many cases where high accuracy is achieved with only a small number of basis vectors.

Example 5.1

The object of this example is to demonstrate the accuracy of the results achieved by the CA approach for very large changes in the design. Consider the ten-bar truss shown in Fig. 5.1. Assuming arbitrary units, the design variables $\mathbf{X}$ are the member cross-sectional areas, the initial cross section areas $\mathbf{X}_0$ are all unity, the modulus of elasticity is 30000 and the eight analysis unknowns are the horizontal and vertical displacements at joints 1, 2, 3 and 4, respectively. The stress constraints are $-\mathbf{25.0} \le \boldsymbol{\sigma} \le \mathbf{25.0}$ and the minimum size constraints are $\mathbf{0.001} \le \mathbf{X}$. Assuming the weight as an objective function, the optimal design is

$$\mathbf{X}_{opt}^T = \{8.0, 0.001, 8.0, 4.0, 0.001, 0.001, 5.667, 5.667, 5.667, 0.001\}. \qquad (a)$$

The line from the initial design to the optimal design is given by

$$\mathbf{X} = \mathbf{X}_0 + \alpha\, \Delta\mathbf{X}_0, \qquad (b)$$

where $\Delta\mathbf{X}_0 = \mathbf{X}_{opt} - \mathbf{X}_0$ is given by

$$\Delta\mathbf{X}_0^T = \{7.0, -0.999, 7.0, 3.0, -0.999, -0.999, 4.667, 4.667, 4.667, -0.999\} \qquad (c)$$

and α is a step-size variable. For $\alpha = 1.0$ (the optimum) the changes in the design are most significant: members 1 and 3 are increased by 700%, member 4 is increased by 300%, members 7, 8, 9 are increased by 467%, and the topology is changed by effectively eliminating members 2, 5, 6 and 10, and hence joint 2 (displacements 3 and 4).

To illustrate results for various magnitudes of change in the design variables, the following three cases of change are considered:

- Large change in the design (up to –10% and +70%), $\alpha = 0.1$.
- Very large change in the design (up to –50% and +350%), $\alpha = 0.5$.
- Most significant change in the design (up to –99% and +700%), $\alpha = 1.0$.

The Results obtained by the CA approach, summarized in Table 5.1, show that larger step sizes require more basis vectors to achieve certain accuracy. For two-digit accuracy, only 3 vectors are needed for $\alpha = 0.1$, whereas 4 vectors are needed for $\alpha = 0.5$ and 5 vectors are needed for $\alpha = 1.0$.

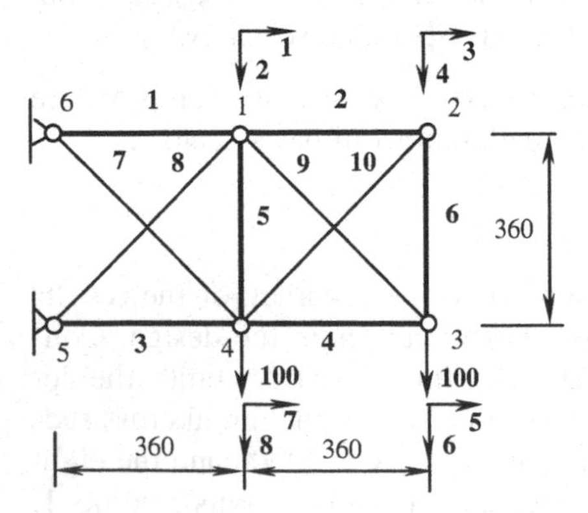

Fig. 5.1. Ten-bar truss

Table 5.1. Approximations of displacements by the CA method

Number of basis vectors	2	3	4	5	Exact Solution
$\alpha = 0.1$ (large)	1.36	1.37			1.37
	3.59	3.56			3.56
	1.76	1.77			1.77
	8.23	8.25			8.25
	-2.06	-2.10			-2.10
	8.62	8.65			8.65
	-1.44	-1.45			-1.45
	3.92	3.89			3.89
$\alpha = 0.5$ (very large)	0.50	0.52	0.52		0.52
	1.53	1.46	1.49		1.49
	0.71	0.76	0.77		0.77
	3.56	3.63	3.64		3.64
	-0.89	-0.98	-0.98		-0.98
	3.77	3.87	3.89		3.89
	-0.54	-0.55	-0.55		-0.55
	1.70	1.64	1.62		1.62
$\alpha = 1.0$ (most significant)	0.28	0.29	0.29	0.30	0.30
	0.90	0.84	0.88	0.90	0.90
	0.41	0.45	0.47	0.49	0.49
	2.10	2.17	2.19	2.21	2.21
	-0.53	-0.61	-0.62	-0.60	-0.60
	2.24	2.34	2.37	2.40	2.40
	-0.30	-0.31	-0.31	-0.30	-0.30
	1.01	0.95	0.93	0.90	0.90

Example 5.2

Originally, the CA approach was developed for reanalysis of structures where results from initial analysis are known. It is shown in this example how the approach can also be used in cases where the initial displacements are not available. The object is to find the unknown displacements $\mathbf{r}$, where the known $\mathbf{K}$ is expressed in the form $\mathbf{K} = \mathbf{K}_0 + \Delta\mathbf{K}_0$, matrices $\mathbf{K}_0$, $\Delta\mathbf{K}_0$ are to be defined, and $\mathbf{r}_0$ is unknown. Since results of previous analysis are not available, one simple approach is to choose $\mathbf{K}_0$ as a diagonal matrix $\mathbf{K}_d$ consisting of the diagonal elements of $\mathbf{K}$ [see Eq. (4.6)]. As a result the displacements corresponding to the initial stiffness matrix $\mathbf{K}_0$ are uncoupled, given directly by $\mathbf{r}_0 = (\mathbf{K}_d)^{-1}\mathbf{R}_0$. In this case $\mathbf{K}_d$ might represent several substructures, which are completely different from the actual structure represented by $\mathbf{K}$, as is shown in this example. Since the initial and the modified displacements are of different nature, the CA procedure may require a large number of basis vectors to achieve accurate results.

Consider again the ten-bar truss shown in Fig. 5.1, subjected to a single loading condition of two loads. The modulus of elasticity is 30 000 and the 8 analysis unknowns are the horizontal and the vertical displacements at joints 1, 2, 3 and 4, respectively. The object is to analyze the structure for cross-sectional areas equal to unity. With the initial structure as represented by $\mathbf{K}_d$, we have four different substructures with cross sections corresponding to the stiffness coefficients, as shown in Fig. 5.2. The initial displacements calculated by $\mathbf{r}_0 = (\mathbf{K}_d)^{-1}\mathbf{R}_0$ are given by

$$\mathbf{r}_0^T = \{0, 0, 0, 0, 0, 0.89, 0, 0.70\}. \qquad (a)$$

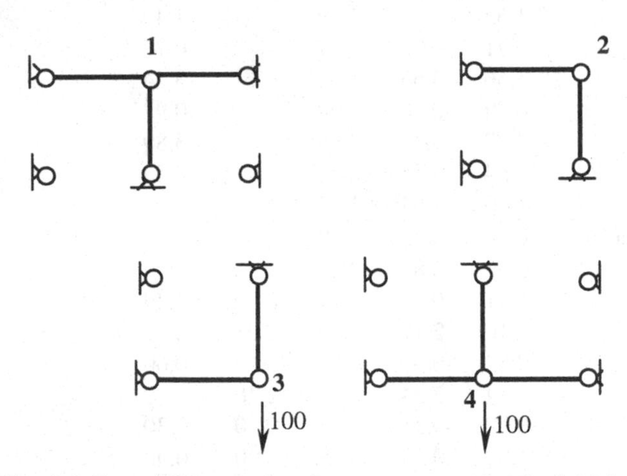

Fig. 5.2. Four different substructures, representing the initial solution

Table 5.2. Complete analysis by the CA approach

Number of basis vectors	2	3	4	5	6	7	Exact solution
r	0.45	0.45	2.25	2.32	2.30	2.34	2.34
	2.30	4.84	5.50	5.64	5.56	5.58	5.58
	-0.71	0.11	2.45	2.75	2.77	2.83	2.83
	3.24	6.92	11.72	12.50	12.65	12.65	12.65
	-0.89	-0.96	-3.06	-3.31	-3.24	-3.17	-3.17
	3.96	7.41	11.88	13.10	13.13	13.13	13.13
	0.00	-1.21	-2.58	-2.54	-2.50	-2.46	-2.46
	3.14	5.17	6.33	5.99	6.00	6.01	6.01

The displacements obtained by considering various numbers of basis vectors are shown in Table 5.2. The results show that despite the very poor initial displacements, the CA procedure converges to the exact solution.

5.2.2 Uncoupled Basis Vectors

The Gram-Schmidt orthogonalization procedure, presented in Sect. 2.3.3, can be used to generate a new set of basis vectors such that the reduced set of analysis equations [Eq. (5.35)] becomes uncoupled with respect to **K**. This procedure has been used for nonlinear analysis [10] and linear reanalysis of structures [11]. For any assumed number of basis vectors, the results obtained by considering the reduced set of equations, for either the original set of basis vectors or the new set of uncoupled vectors, are identical. The advantage in using the latter vectors is that all expressions for evaluating the displacements become explicit functions of the design variables. As a result, additional vectors can be considered without modifying the calculations that already were carried out for previous terms. In addition, the results obtained by the uncoupled basis vectors are more well-conditioned, particularly in problems of nonlinear analysis.

It can be observed from Eq. (5.34) that the elements K_{Rij} of the reduced stiffness matrix $\mathbf{K}_R$ are given by

$$K_{Rij} = \mathbf{r}_i^T \mathbf{K}\, \mathbf{r}_j\,. \tag{5.37}$$

The object is to transform the reduced system of Eq. (5.35) into an uncoupled set of equations. This can be done by generating a set of new vectors $\mathbf{V}_i$ ($i = 1, 2, \ldots, s$), from the original vectors $\mathbf{r}_i$, such that for any two vectors $\mathbf{V}_i$ and $\mathbf{V}_j$

$$\mathbf{V}_i^T \mathbf{K} \mathbf{V}_j = \delta_{ij}\,, \tag{5.38}$$

where δ_{ij} is the kronecker delta, for which

$$\delta_{ij} = 0\ (i \neq j) \qquad \delta_{ii} = 1\,. \tag{5.39}$$

The vectors $\mathbf{V}_i$ and $\mathbf{V}_j$ are orthogonal with respect to **K** if the condition of Eq. (5.38) is satisfied. The new basis vectors, which are linear combinations of the original vectors, are generated as follows.

The first normalized vector $\mathbf{V}_1$ is chosen as

$$\mathbf{V}_1 = \frac{\mathbf{r}_1}{|\,\mathbf{r}_1^T \mathbf{K} \mathbf{r}_1\,|^{1/2}}. \tag{5.40}$$

To generate the second normalized vector $\mathbf{V}_2$ we first define the non-normalized vector $\overline{\mathbf{V}}_2$, which is a linear combination of $\mathbf{V}_1$ and $\mathbf{r}_2$, by

$$\overline{\mathbf{V}}_2 = \mathbf{r}_2 - \alpha\,\mathbf{V}_1, \tag{5.41}$$

where α is chosen such that the orthogonality condition of Eq.(5.38),

$$\overline{\mathbf{V}}_2^T \mathbf{K} \mathbf{V}_1 = 0, \tag{5.42}$$

is satisfied. Substituting Eq. (5.41) into the condition of Eq. (5.42) yields

$$\overline{\mathbf{V}}_2^T \mathbf{K} \mathbf{V}_1 = \mathbf{r}_2^T \mathbf{K} \mathbf{V}_1 - \alpha \mathbf{V}_1^T \mathbf{K} \mathbf{V}_1 = 0. \tag{5.43}$$

Since $\mathbf{V}_1^T \mathbf{K} \mathbf{V}_1 = 1$ [Eqs. (5.38), (5.40)], then Eq. (5.43) becomes

$$\alpha = \mathbf{r}_2^T \mathbf{K} \mathbf{V}_1. \tag{5.44}$$

Substituting Eq. (5.44) into Eq. (5.41) gives

$$\overline{\mathbf{V}}_2 = \mathbf{r}_2 - \left(\mathbf{r}_2^T \mathbf{K} \mathbf{V}_1\right)\mathbf{V}_1. \tag{5.45}$$

Finally, normalizing $\overline{\mathbf{V}}_2$ we obtain the second normalized vector

$$\mathbf{V}_2 = \frac{\overline{\mathbf{V}}_2}{|\,\overline{\mathbf{V}}_2^T \mathbf{K} \overline{\mathbf{V}}_2\,|^{1/2}}. \tag{5.46}$$

Additional basis vectors are generated in a similar way. The resulting general expressions for all $i = 2, \ldots, s$ vectors are

$$\overline{\mathbf{V}}_i = \mathbf{r}_i - \sum_{j=1}^{i-1}\left(\mathbf{r}_i^T \mathbf{K} \mathbf{V}_j\right)\mathbf{V}_j, \tag{5.47}$$

$$\mathbf{V}_i = \frac{\overline{\mathbf{V}}_i}{|\,\overline{\mathbf{V}}_i^T \mathbf{K} \overline{\mathbf{V}}_i\,|^{1/2}},$$

where $\overline{\mathbf{V}}_i$ and $\mathbf{V}_i$ are the ith non-normalized and normalized vectors.

Now, we will show how the new basis vectors $\mathbf{V}_i$ are used to evaluate the displacements. It is observed [Eqs. (5.34), (5.38)] that, for the basis vectors $\mathbf{V}_i$ ($i = 1, 2, \ldots, s$), the diagonal elements of the new reduced stiffness matrix equal unity and all other elements equal zero. That is, the new reduced stiffness matrix is the identity matrix **I**. Define the matrix of new basis vectors $\mathbf{V}_B$ and the vector of new coefficients **z** by

$$\mathbf{V}_B = [\mathbf{V}_1 , \mathbf{V}_2 , \ldots , \mathbf{V}_s], \tag{5.48}$$

$$\mathbf{z}^T = \{z_1, z_2, \ldots , z_s\}.$$

Rather than the reduced system of Eq. (5.35), we have [see Eq. (5.34)]

$$\mathbf{I}\,\mathbf{z} = \mathbf{z} = \mathbf{V}_B^T\mathbf{R}\,, \tag{5.49}$$

where **z** is the vector of new coefficients, which can be determined directly by Eq. (5.49). Since this system is uncoupled, the final displacements are given by the explicit expression [see Eq. (5.36)]

$$\mathbf{r} = \mathbf{V}_B\,\mathbf{z} = \mathbf{V}_B\left(\mathbf{V}_B^T\,\mathbf{R}\right). \tag{5.50}$$

The displacements calculated by Eq. (5.50) can be expressed as an additively separable quadratic function of the basis vectors $\mathbf{V}_i$ by

$$\mathbf{r} = \sum_{i=1}^{s}\mathbf{V}_i\left(\mathbf{V}_i^T\mathbf{R}\right). \tag{5.51}$$

One advantage in using the new vectors is that all expressions for evaluating the displacements are explicit functions of the original basis vectors and, therefore, might be explicit functions of the design variables. This can be seen from the following expressions:

- The stiffness matrices $\Delta\mathbf{K}$, **K** and load vectors $\Delta\mathbf{R}$, **R** are often explicit functions of the design variables.
- The vectors $\mathbf{r}_i$ are explicit functions of $\Delta\mathbf{K}$ and **R** [Eqs. (5.31), (5.32)].
- The vectors $\mathbf{V}_i$ are explicit functions of $\mathbf{r}_i$ and **K** [Eqs. (5.40), (5.47)].
- The final displacements **r** are explicit functions of $\mathbf{V}_i$ and **R** [Eq. (5.51)].

Another advantage is that calculation of any new basis vector $\mathbf{V}_i$ leads to an additional term in the displacements expression [Eq. (5.51)]. As a result, additional vectors can be considered without modifying the calculations that already were carried out.

It was found that the normalized vectors $\mathbf{V}_i$ are of similar magnitude, whereas the values of the z_i coefficients of the vector **z**, and therefore the corresponding terms of the series of Eq. (5.51), are gradually decreased. That is, transformation of the binomial series terms [Eqs. (5.31), (5.32)] into the terms of the CA series [Eq. (5.51)] provides accurate solutions even in cases where the binomial series diverges.

In summary, calculation of the displacements for any assumed changes $\Delta\mathbf{K}$, $\Delta\mathbf{R}$, using the presented procedure, involves the following steps:

- Generate the original basis vectors $\mathbf{r}_i$ by the procedure described in Sect. 5.2.1.
- Generate the normalized basis vectors $\mathbf{V}_i$. The first vector is [Eq. (5.40)]

$$\mathbf{V}_1 = \frac{\mathbf{r}_1}{|\mathbf{r}_1^T \mathbf{K}\, \mathbf{r}_1|^{1/2}}. \tag{5.52}$$

Additional vectors $\mathbf{V}_i$ $(i = 2, ..., s)$ are generated by [Eqs. (5.47)]

$$\overline{\mathbf{V}}_i = \mathbf{r}_i - \sum_{j=1}^{i-1} \left(\mathbf{r}_i^T \mathbf{K} \mathbf{V}_j\right) \mathbf{V}_j, \tag{5.53}$$

$$\mathbf{V}_i = \frac{\overline{\mathbf{V}}_i}{|\overline{\mathbf{V}}_i^T \mathbf{K} \overline{\mathbf{V}}_i|^{1/2}},$$

where $\overline{\mathbf{V}}_i$, $\mathbf{V}_i$ are the ith non-normalized and normalized vectors.

- Evaluate the displacements $\mathbf{r}$ by [Eq. (5.51)]

$$\mathbf{r} = \sum_{i=1}^{s} \mathbf{V}_i \left(\mathbf{V}_i^T \mathbf{R}\right). \tag{5.54}$$

The accuracy of the results for any specific number of basis vectors, s, can be evaluated by several methods described later in Sect. 10.2. If the accuracy is insufficient, additional basis vectors are introduced and the updated displacements are evaluated. Equations (5.53) and (5.54) show that additional vectors can be considered without modifying the calculations that were carried out already.

Example 5.3

Consider again the ten-bar truss shown in Fig. 5.1. The displacements obtained by the CA approach, with the original basis vectors $\mathbf{r}_i$ for various numbers of basis vectors, are summarized in Table 5.1. Considering the case of most significant changes in the design ($\alpha = 1.0$, up to –99% and +700%), Fig. 5.3*a* shows how the norm of the terms $\mathbf{r}_i$ of the series of basis vectors (the binomial series) increases and the series diverges. Figure 5.3*b* shows that the norms of the uncoupled basis vectors $\mathbf{V}_i$ are of similar magnitude and Figs. 5.3*c*, 5.3*d* show that the z_i coefficients of the vector $\mathbf{z}$ and the norm of the CA terms, $\mathbf{V}_i z_i$, gradually decrease and the series converges as the number of basis vectors is increased.

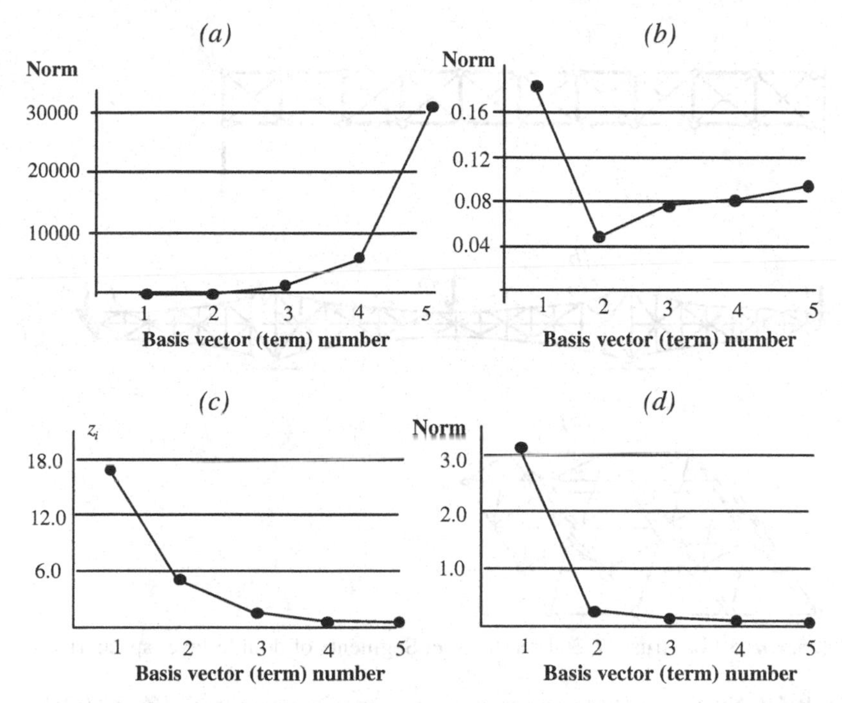

Fig. 5.3. Results for $\alpha = 1$ ***a***. Norm of the original basis vectors $\mathbf{r}_i$ ***b***. Norm of the uncoupled basis vectors $\mathbf{V}_i$ ***c***. Values of coefficients z_i ***d***. Norm of the terms $\mathbf{V}_i\, z_i$

Example 5.4

The accuracy of low-order approximations (small number of basis vectors) might be insufficient for large changes in large-scale problems. In such cases a larger number of basis vectors might be needed to improve the accuracy of the approximations. Results obtained by high-order approximations are presented in this example. The object is to illustrate the accuracy achieved by the CA approach for structures having different numbers of DOF and solved with various numbers of basis vectors. In all structures the design variables are the member cross-sectional areas and the initial areas equal unity. Many cases of random changes in the cross-sections were assumed for each structure. The structures considered are as follows:

- The 50-bar cantilever truss shown in Fig. 5.4*a*.
- The 204-bar bridge truss shown in Fig. 5.4*b*.
- 356-bar and 968-bar rectangular space trusses made up of the double-layer segments shown in Fig. 5.4*c*, subjected to uniformly distributed loads and supported along the four edges.

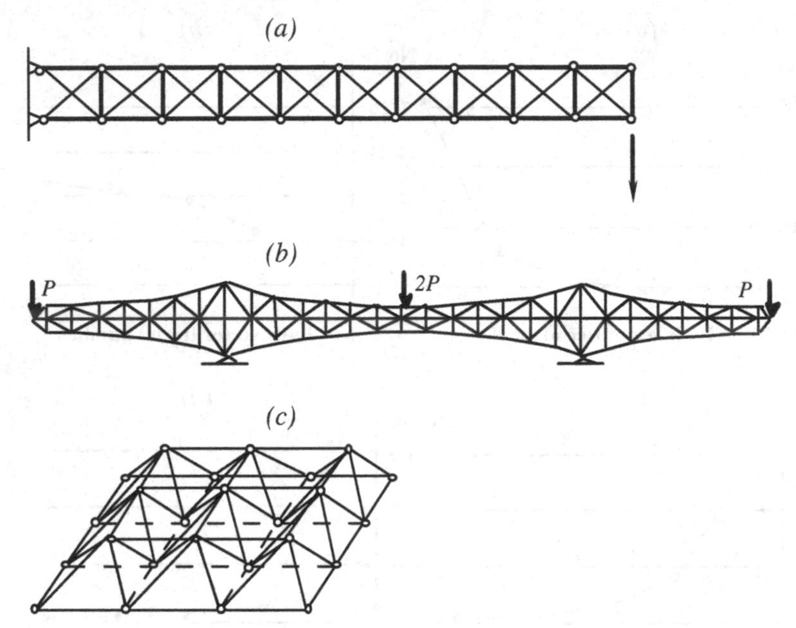

Fig. 5.4. ***a.*** 50-bar truss ***b.*** 204-bar truss ***c.*** Segments of double-layer space truss

Table 5.3. Numbers of basis vectors needed for maximum errors of 1% and 0.1%

Structure	Maximum error 1%	Maximum error 0.1%
50-bar truss	5-6	6-7
204-bar truss	6-7	8-9
356-bar truss	8-9	9-10
968-bar truss	10-11	11-12

The numbers of basis vectors needed for the various structures to limit maximum errors to 1% and 0.1% are shown in Table 5.3. It is observed that the difference in the numbers of basis vectors required for the two cases of errors is small. In addition, the number of vectors is not significantly increased with the size of the structure.

Example 5.5

To illustrate changes in the geometry of the structure, consider the 130-bar truss shown in Fig. 5.5, having 10-stories and 3-bays, subjected to 10 horizontal loads of 10.0. Two types of changes are considered:

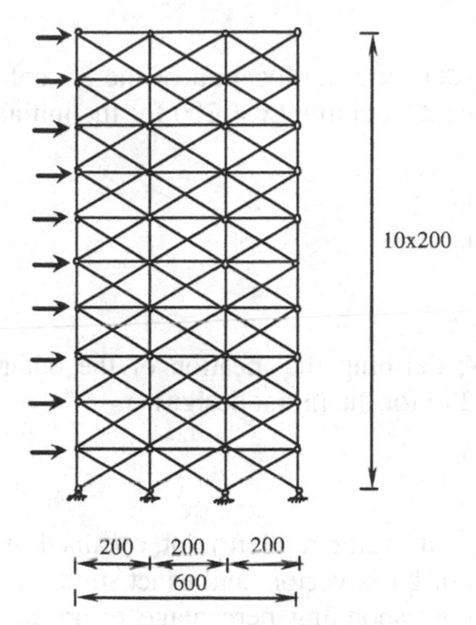

Fig. 5.5. Initial geometry, 130-bar truss

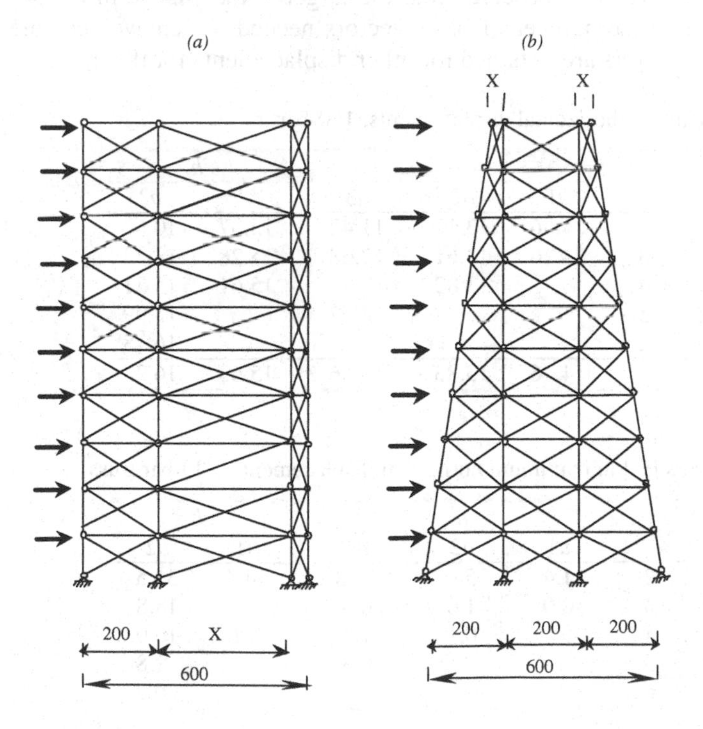

Fig. 5.6. Modified geometries, 130-bar truss

Type a (Fig. 5.6*a*)
The following modified values of X, defining the location of the 3rd column from the left relative to that of the 2nd column ($X = 200$ for the initial design):

Case $a1$: $X = 250$ (increase of 25%).
Case $a2$: $X = 300$ (increase of 50%).
Case $a3$: $X = 350$ (increase of 75%).

Type b (Fig. 5.6*b*)
The following modified values of X, defining the location of the outer joints relative to the inner joints ($X = 200$ for the initial design):

Case $b1$: $X = 100$ (decrease of 50%).
Case $b2$: $X = 50$ (decrease of 75%).

The maximum horizontal displacements at the top left joint, obtained by the CA approach for various numbers of basis vectors and exact solutions, are summarized in Table 5.4. The corresponding percentage errors are shown in Table 5.5. It is observed that the larger is the change in the geometry the more the number of basis vectors needed to achieve accurate results. Similar errors are obtained for other displacements for the truss.

Table 5.4. Maximum horizontal displacements, 130-bar truss

Change type			*a*		*b*	
Case		*a*1	*a*2	*a*3	*b*1	*b*2
Basis vectors	2	14.06	13.12	11.42	13.37	10.67
	3	14.26	13.61	12.68	15.28	13.22
	4		13.82	13.33	15.61	15.31
	5			13.52		15.99
	6					16.18
Exact solution		14.26	13.83	13.56	15.63	16.28

Table 5.5. Errors [%] in maximum horizontal displacements, 130-bar truss

Change type			*a*		*b*	
Case		*a*1	*a*2	*a*3	*b*1	*b*2
Basis vectors	2	1.4	5.1	15.8	14.4	34.5
	3	0.0	1.6	6.5	2.2	18.8
	4		0.1	1.7	0.1	16.0
	5			0.3		1.8
	6					0.7

5.3 Topological Changes

Developing a reanalysis approach for topological changes is most challenging, because the structural model is changed and the resulting response might be significantly different from the original response. Various approximate reanalysis methods are not suitable for such changes and provide inadequate or meaningless results. It is shown in this section that accurate approximations can be achieved efficiently by the CA approach for significant changes in the topology.

In cases where the number of degrees of freedom (DOF) is unchanged, the general solution procedure presented in Sect. 5.2 can be used for various topological changes. This procedure can be used also in cases where members are deleted or added [12–14]. In this section the two basic cases of decreasing and increasing the number of DOF are considered. It is shown later in Chap. 8 that exact solutions can be obtained efficiently by the CA approach for low-rank topological changes in the stiffness matrix.

5.3.1 Number of DOF is Decreased

This case is encountered in many topological optimization problems where various members and joints are deleted from a ground structure consisting of numerous members and joints. As a result, the number of DOF and the corresponding number of analysis equations are decreased.

The sizes of the stiffness matrix and the load vector are decreased according to the number of joints deleted from the structure. The modified stiffness matrix and the modified load vector can be expressed as

$$\mathbf{K} = \mathbf{K}_0 + \Delta\mathbf{K}_0 = \begin{bmatrix} \mathbf{K}_r & \mathbf{0} \\ \mathbf{0} & \mathbf{0} \end{bmatrix}, \tag{5.55}$$

$$\mathbf{R} = \mathbf{R}_0 + \Delta\mathbf{R}_0 = \begin{Bmatrix} \mathbf{R}_r \\ \mathbf{0} \end{Bmatrix}, \tag{5.56}$$

where $\mathbf{K}$ and $\mathbf{R}$ are the modified stiffness matrix and the modified load vector, respectively, of the complete set of equations, including the original degrees of freedom; and $\mathbf{K}_r$ and $\mathbf{R}_r$ are the stiffness sub-matrix and the load sub-vector, respectively, of the modified structure with the reduced number of DOF. Since some analysis equations become zero identities, stiffness analysis of the complete set of modified equations cannot be carried out. The reduced set of equations to be solved is

$$\mathbf{K}_r \, \mathbf{r}_r = \mathbf{R}_r, \tag{5.57}$$

where $\mathbf{r}_r$ is now a reduced vector of modified displacements. Despite the reduction in the size of the stiffness matrix, the number of modified analysis equations is often large and efficient reanalysis might prove useful.

It should be noted that the resulting structure represented by Eq. (5.57) might be conditionally unstable, in which case the modified stiffness matrix is singular and exact analysis cannot be carried out. The reduced stiffness matrix $\mathbf{K}_R$ used by the CA approach is often not singular even in cases where the modified stiffness matrix $\mathbf{K}_r$ is singular. Therefore, approximate reanalysis by the CA approach with a reduced number of unknowns may provide accurate results even when the modified structure is conditionally unstable. If the number of basis vectors considered is smaller than the number of DOF of the modified structure, the usual CA procedure might provide accurate results efficiently.

Example 5.6

Consider the nineteen-bar tower truss shown in Fig. 5.7 subjected to a single loading condition of two concentrated loads. The modulus of elasticity is 10000 and the twelve analysis unknowns are the horizontal and the vertical displacements at joints 2, 3, 4, 6, 7, and 8, respectively. With initial cross-sectional areas of unity and an initial nineteen-bar topology, the following three cases are solved by considering 0.001 (practically zero) cross-section areas for the eliminated members (Fig. 5.8):

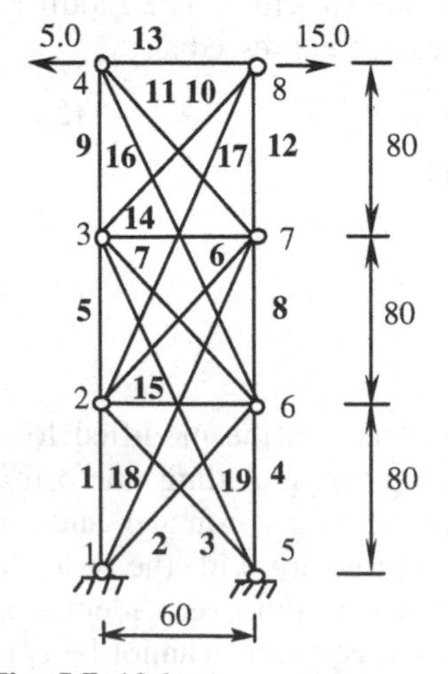

Fig. 5.7. 19-bar tower truss

a. Elimination of six members (Fig. 5.8*a*).
b. Elimination of seven members (Fig. 5.8*b*).
c. Elimination of nine members (Fig. 5.8*c*).

The results shown in Table 5.6 indicate that high accuracy is achieved by the CA procedure with only 3 basis vectors (CA3) for these large topological changes.

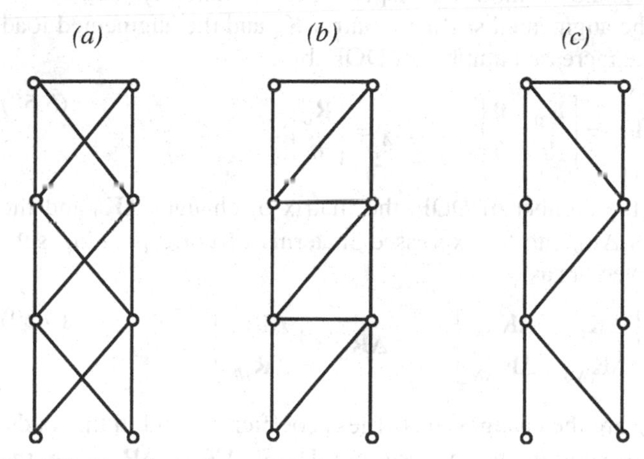

Fig. 5.8. Modified truss topologies

Table 5.6. CA3 and Exact modified displacements, 19-bar truss

	Case *a*		Case *b*		Case *c*	
Number	CA3	Exact	CA3	Exact	CA3	Exact
1	0.49	0.48	0.78	0.76	0.71	0.70
2	0.27	0.26	0.21	0.21	0.32	0.32
3	1.57	1.58	2.07	2.09	*	*
4	0.43	0.43	0.33	0.32	*	*
5	2.88	2.88	3.61	3.62	3.53	3.53
6	0.48	0.48	0.35	0.32	0.54	0.53
7	0.50	0.50	0.72	0.70	*	*
8	-0.27	-0.27	-0.32	-0.32	*	*
9	1.54	1.54	2.00	2.03	1.97	1.98
10	-0.42	-0.43	-0.51	-0.53	-0.42	-0.43
11	2.93	2.93	3.64	3.65	3.62	3.62
12	-0.48	-0.48	-0.63	-0.64	-0.42	-0.43

* These displacements correspond to joints that are practically eliminated

5.3.2 Number of DOF is Increased

If some members and joints are added to the structure, the number of DOF is increased, the number of analysis equations is changed, and the sizes of the stiffness matrix and the load vector are increased accordingly. Several procedures have been proposed to solve this problem [e.g. 15–18]. The procedure presented in this section, which is based on the CA approach, can be used to calculate the modified displacements efficiently [16].

Let us define the augmented stiffness matrix $\mathbf{K}_A$ and the augmented load vector $\mathbf{R}_A$, with the increased number of DOF, by

$$\mathbf{K}_A = \begin{bmatrix} \mathbf{K}_0 & \mathbf{0} \\ \mathbf{0} & \mathbf{0} \end{bmatrix} \qquad \mathbf{R}_A = \begin{Bmatrix} \mathbf{R}_0 \\ \mathbf{0} \end{Bmatrix}. \tag{5.58}$$

Upon increasing the number of DOF, the matrix of changes $\Delta\mathbf{K}_0$ and the vector of changes $\Delta\mathbf{R}_0$ can be expressed in terms of corresponding sub-matrices and sub-vectors as

$$\Delta\mathbf{K}_0 = \begin{bmatrix} \Delta\mathbf{K}_{00} & \Delta\mathbf{K}_{0N} \\ \Delta\mathbf{K}_{N0} & \Delta\mathbf{K}_{NN} \end{bmatrix} \qquad \Delta\mathbf{R}_0 = \begin{Bmatrix} \Delta\mathbf{R}_{00} \\ \Delta\mathbf{R}_{NN} \end{Bmatrix}, \tag{5.59}$$

where $\Delta\mathbf{K}_{00}$, $\Delta\mathbf{R}_{00}$ are the changes in stiffness coefficients and in the loads, respectively, corresponding to the original DOF; $\Delta\mathbf{K}_{NN}$, $\Delta\mathbf{R}_{NN}$ are the changes corresponding to the new DOF; and $\Delta\mathbf{K}_{0N}$, $\Delta\mathbf{K}_{N0}$ are the changes in the coefficients corresponding to both the original and the new DOF.

The modified stiffness matrix and the modified load vector are given by

$$\mathbf{K} = \mathbf{K}_A + \Delta\mathbf{K}_0 \qquad \mathbf{R} = \mathbf{R}_A + \Delta\mathbf{R}_0, \tag{5.60}$$

where the new degrees of freedom are included in the set of modified analysis equations. The number of added DOF is usually small, compared with the original number of DOF.

Upon increasing the number of DOF, it is necessary first to establish a Modified Initial Analysis (MIA), such that the new degrees of freedom are included in the analysis model. For the augmented stiffness matrix and the augmented load vector [Eqs. (5.58)], the MIA model can be selected such that reanalysis will be convenient. Once the MIA is established, it is then possible to analyze structures modified due to addition or deletion of members, keeping the number of degrees of freedom unchanged.

The MIA is established as follows. The matrix of changes in the stiffness $\Delta\mathbf{K}_0$ is expressed first as a sum of the matrices $\Delta\mathbf{K}_A$ and $\Delta\mathbf{K}_N$ by

$$\Delta\mathbf{K}_0 = \Delta\mathbf{K}_A + \Delta\mathbf{K}_N. \tag{5.61}$$

These two matrices are defined in such a way that the modified initial analysis is easy to carry out. The *modified initial stiffness matrix* $\mathbf{K}_M$ is expressed as

$$\mathbf{K}_M = \mathbf{K}_A + \Delta\mathbf{K}_A. \tag{5.62}$$

Matrices $\Delta\mathbf{K}_A$ and $\Delta\mathbf{K}_N$ are defined as

$$\Delta\mathbf{K}_A = \alpha\begin{bmatrix} 0 & \Delta\mathbf{K}_{0N} \\ \Delta\mathbf{K}_{N0} & \Delta\mathbf{K}_{NN} \end{bmatrix}, \tag{5.63}$$

$$\Delta\mathbf{K}_N = \begin{bmatrix} \Delta\mathbf{K}_{00} & (1-\alpha)\Delta\mathbf{K}_{0N} \\ (1-\alpha)\Delta\mathbf{K}_{N0} & (1-\alpha)\Delta\mathbf{K}_{NN} \end{bmatrix},$$

where α is a scalar multiplier to be selected $(0 < \alpha \leq 1)$. Substituting the expressions of $\mathbf{K}_A$ and $\Delta\mathbf{K}_A$ [Eqs. (5.58), (5.63)] into Eq. (5.62) yields

$$\mathbf{K}_M = \begin{bmatrix} \mathbf{K}_0 & \alpha\Delta\mathbf{K}_{0N} \\ \alpha\Delta\mathbf{K}_{N0} & \alpha\Delta\mathbf{K}_{NN} \end{bmatrix}. \tag{5.64}$$

The rationale of this selection is that, once the decomposed form of Eq. (4.2) is available, factorization of the modified initial stiffness matrix $\mathbf{K}_M$,

$$\mathbf{K}_M = \mathbf{U}_M^T\,\mathbf{U}_M, \tag{5.65}$$

is straightforward. Specifically, matrix $\mathbf{U}_M$ can be expressed as

$$\mathbf{U}_M = \begin{bmatrix} \mathbf{U}_0 & \mathbf{U}_{0N} \\ \mathbf{0} & \mathbf{U}_{NN} \end{bmatrix}, \tag{5.66}$$

where the elements of matrix $\mathbf{U}_0$ are already given. That is, the rows and columns corresponding to the original degrees of freedom are unchanged and only rows and columns corresponding to the new degrees of freedom are calculated. In general the number of added DOF is small, and the factorization of Eq. (5.65) involves a small computational effort.

Concerning the selected value of α, it is observed that $\alpha = 1$ yields

$$\mathbf{K}_M = \begin{bmatrix} \mathbf{K}_0 & \Delta\mathbf{K}_{0N} \\ \Delta\mathbf{K}_{N0} & \Delta\mathbf{K}_{NN} \end{bmatrix}. \tag{5.67}$$

One drawback of this selection is that matrix $\mathbf{K}_M$ is not necessarily positive definite and the factorization of Eq. (5.65) might not be possible. In such cases we can use the symmetric factorization

$$\mathbf{K}_M = \mathbf{L}_M \mathbf{D}_M \mathbf{L}_M^T , \tag{5.68}$$

where $\mathbf{L}_M$ is a lower triangular matrix and $\mathbf{D}_M$ is a diagonal matrix. However, matrix $\mathbf{K}_M$ might not represent a real structure and the accuracy of the approximations might deteriorate.

In the presented procedure this difficulty is overcome by selecting a small α value such that matrix $\mathbf{K}_M$ [Eq. (5.64)] is a good approximation of the matrix $\mathbf{K}_A + \alpha\Delta\mathbf{K}_0$ [since $\alpha\, \Delta\mathbf{K}_{00} << \mathbf{K}_0$, see Eqs. (5.58), (5.59)].

In summary, the solution procedure involves the following two stages.

- The modified initial analysis (MIA) is established. Assuming a small α value, we calculate and factorize the matrix $\mathbf{K}_M$ [Eqs. (5.64), (5.65)]. Since the decomposed form of Eq. (4.2) is available, this operation involves a small computational effort. The modified initial displacements $\mathbf{r}_M$ are then calculated by solving

$$\mathbf{K}_M\, \mathbf{r}_M = \mathbf{R}. \tag{5.69}$$

  This calculation involves only forward and backward substitutions.
- Once the vector $\mathbf{r}_M$ has been determined, the displacements due to the remaining changes in the stiffness matrix $\Delta\mathbf{K}_N$ [Eq. (5.63)] are calculated by solving the modified equations

$$\mathbf{K}\, \mathbf{r} = (\mathbf{K}_M + \Delta\mathbf{K}_N)\, \mathbf{r} = \mathbf{R}. \tag{5.70}$$

  The CA approach described in Sect. 5.2 can be used for this purpose, with $\mathbf{r}_M$, $\mathbf{K}_M$, $\mathbf{U}_M$, $\mathbf{R}$ replacing $\mathbf{r}_0$, $\mathbf{K}_0$, $\mathbf{U}_0$, $\mathbf{R}_0$ as initial values.

Example 5.7

To illustrate reanalysis for the case of addition of members and joints, consider the initial six-bar truss shown in Fig. 5.9*a*. The six analysis unknowns are the horizontal and the vertical displacements at joints 1, 2 and 3, respectively. The initial values $\mathbf{r}_0$, $\mathbf{U}_0$ are given by

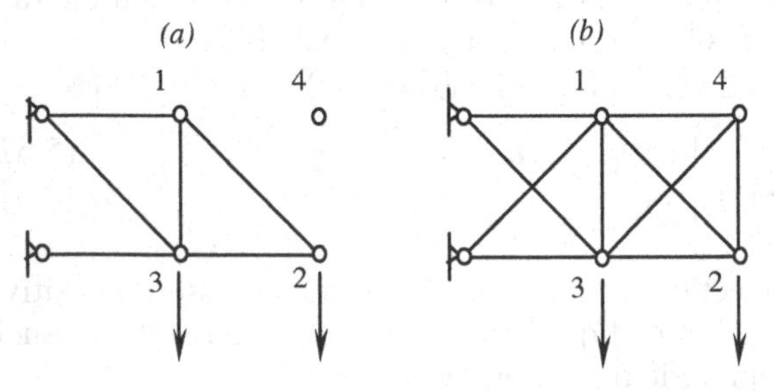

Fig. 5.9. ***a***. Six-bar truss, initial topology ***b***. Modified topology

$$\mathbf{r}_0^T = \{1.20, 11.59, -4.80, 20.98, -3.60, 10.39\}, \tag{a}$$

$$\mathbf{U}_0 = \begin{bmatrix} 10.62 & 2.77 & -2.77 & -2.77 & 0 & 0 \\ & 10.25 & -2.12 & -2.12 & 0 & -8.13 \\ & & 10.03 & 1.72 & -8.31 & -1.72 \\ & & & 3.78 & 3.78 & -3.78 \\ & & & & 10.62 & 2.77 \\ & & & & & 4.67 \end{bmatrix}. \tag{b}$$

Assume addition of one joint and four members to obtain the ten-bar truss shown in Fig. 5.9*b*. Reanalysis is carried out in the following two stages:

- The Modified Initial Analysis (MIA) is carried out. Selecting $\alpha = 0.001$, the initial decomposed stiffness matrix is given by $\mathbf{U}_M$ [Eq. (5.66)], where $\mathbf{U}_0$ is already given as above. Thus, we calculate only the submatrices $\mathbf{U}_{0N}$, $\mathbf{U}_{NN}$ to obtain

$$\mathbf{U}_{0N} = \begin{bmatrix} -0.0078 & 0 \\ 0.0021 & 0 \\ -0.0017 & 0 \\ -0.0038 & -0.0220 \\ -0.0028 & 0.0106 \\ 0.0080 & -0.0305 \end{bmatrix} \qquad \mathbf{U}_{NN} = \begin{bmatrix} 0.3356 & -0.0872 \\ 0 & 0.3220 \end{bmatrix}. \tag{c}$$

  For the $\mathbf{U}_M$ found from Eq. (5.66), calculation of the modified initial displacement vector $\mathbf{r}_M$ by Eq. (5.69) involves only forward and backward substitutions. The result is

$$\mathbf{r}_M^T = \{1.22, 11.74, -4.87, 21.28, -3.65, 10.52, 2.44, 20.06\}. \tag{d}$$

  It is observed that, due to the small change in stiffness the displacements of the original degrees of freedom have changed only slightly.

- The displacements due to the remaining changes in the stiffness matrix $\Delta\mathbf{K}_N$ are calculated. Employing the CA procedure described in Sect. 5.2, with $\mathbf{r}_M$, $\mathbf{K}_M$, $\mathbf{U}_M$, $\mathbf{R}$ replacing $\mathbf{r}_0$, $\mathbf{K}_0$, $\mathbf{U}_0$, $\mathbf{R}_0$, respectively, as initial values, we achieve the exact solution with only three basis vectors

$$\mathbf{r}^T = \{2.34, 5.58, -3.17, 13.13, -2.46, 6.01, 2.82, 12.65\}. \tag{e}$$

5.4 Nonlinear Analysis and Reanalysis

5.4.1 Problem Formulation

In the *full Newton-Raphson* procedure, the main computational effort is involved in calculations and factorizations of the tangent stiffness matrix. These calculations must be repeated at each iteration cycle. In the Newton-Raphson incremental-iterative procedure we assume that the solution at the time t is known, where time represents the load-level step. The incremental displacements $\delta\mathbf{r}^{(k)}$ at the current iteration cycle and time $t+\Delta t$ are calculated by solving the set of equations [Eq. (1.83)]

$$ {}^{t+\Delta t}\mathbf{K}^{(k-1)}\,\delta\mathbf{r}^{(k)} = {}^{t+\Delta t}\mathbf{R}_0 - {}^{t+\Delta t}\mathbf{R}_I^{(k-1)} = \delta\mathbf{R}^{(k-1)}. \tag{5.71} $$

The superscript $(k\text{-}1)$ denotes values calculated according to the displacements at the end of the previous iteration cycle, the superscript $(t+\Delta t)$ represents the current load level, ${}^{t+\Delta t}\mathbf{K}^{(k-1)}$ is the current tangent stiffness matrix, ${}^{t+\Delta t}\mathbf{R}_0$ is the vector of external forces, ${}^{t+\Delta t}\mathbf{R}_I^{(k-1)}$ is the vector of internal forces and $\delta\mathbf{R}^{(k-1)}$ is the out-of-balance (residual) force vector.

In the *modified Newton-Raphson* procedure, the tangent stiffness matrix is calculated and factorized only at the beginning of each load increment, in order to save calculation and factorization of a new matrix at each iteration cycle. Using this procedure, we obtain the following set of equations [Eq. (1.93)]

$$ {}^{\tau}\mathbf{K}\,\delta\mathbf{r}^{(k)} = {}^{t+\Delta t}\mathbf{R}_0 - {}^{t+\Delta t}\mathbf{R}_I^{(k-1)}, \tag{5.72} $$

where matrix ${}^{\tau}\mathbf{K}$ corresponds to equilibrium at the end of the previous increment. A requirement for the solution method is its ability to overcome various numerical problems associated with different types of behavior.

Consider the full Newton-Raphson procedure and assume, for simplicity of presentation, the notation,

$$ \mathbf{K}_T = {}^{t+\Delta t}\mathbf{K}^{(k-1)} \qquad \delta\mathbf{R} = \delta\mathbf{R}^{(k-1)} \qquad \delta\mathbf{r} = \delta\mathbf{r}^{(k)}, \tag{5.73} $$

where $\mathbf{K}_T$ is the tangent stiffness matrix, $\delta\mathbf{R}$ is the vector of unbalanced forces and $\delta\mathbf{r}$ is the vector of incremental displacements. Substituting Eqs. (5.73) into Eq. (5.71), the vector $\delta\mathbf{r}$ is calculated at each iteration cycle by solving the set of equations

$$ \mathbf{K}_T\,\delta\mathbf{r} = \delta\mathbf{R}. \tag{5.74} $$

Consider a positive-definite reference stiffness matrix $\mathbf{K}_{ref}$, which is the tangent stiffness matrix calculated at some previous step. Matrix $\mathbf{K}_{ref}$ might represent, for example, the tangent stiffness matrix at the end of the previous increment $\mathbf{K}_{ref} = {}^{\tau}\mathbf{K}$, the elastic stiffness matrix $\mathbf{K}_{ref} = \mathbf{K}_0$, or another choice. We assume that matrix $\mathbf{K}_{ref}$ is given in the decomposed form

$$\mathbf{K}_{ref} = \mathbf{U}_{ref}^T\,\mathbf{U}_{ref}, \tag{5.75}$$

where $\mathbf{U}_{ref}$ is an upper triangular matrix. Passing a limit point, the stiffness matrix may become non-positive-definite and Eq. (5.75) cannot be used. In such cases the $\mathbf{LDL}^T$ factorization can be considered, where $\mathbf{L}$ is a lower-triangular matrix and $\mathbf{D}$ is a diagonal matrix.

Expressing the current tangent stiffness matrix $\mathbf{K}_T$ in terms of $\mathbf{K}_{ref}$ and the matrix of changes $\Delta\mathbf{K}$

$$\mathbf{K}_T = \mathbf{K}_{ref} + \Delta\mathbf{K}, \tag{5.76}$$

and substituting Eq. (5.76) into Eq. (5.74), we obtain the set of equations to be solved at each iteration cycle

$$\mathbf{K}_T\,\delta\mathbf{r} = (\mathbf{K}_{ref} + \Delta\mathbf{K})\,\delta\mathbf{r} = \delta\mathbf{R}. \tag{5.77}$$

It has been noted in Sect. 4.2.2 that the definition of matrix $\Delta\mathbf{K}$ depends on the type of problem to be solved. Specifically, we distinguish between the following two problems of nonlinear analysis and reanalysis:

- The general case of *nonlinear reanalysis of a modified structure*, where matrix $\Delta\mathbf{K}$ is expressed in terms of the following two types of changes

$$\Delta\mathbf{K} = \Delta\mathbf{K}_{NL} + \Delta\mathbf{K}_0. \tag{5.78}$$

  Matrix $\Delta\mathbf{K}_{NL}$ represents the changes in the stiffness matrix due to the nonlinear behavior and it is usually calculated at each iteration cycle. Matrix $\Delta\mathbf{K}_0$ represents the changes in the stiffness matrix due to design considerations and it is constant for any given modified design. In nonlinear reanalysis of a modified structure we assume that $\mathbf{K}_{ref}$ is known from nonlinear analysis of the original structure. It should be noted that the present formulation is suitable also for situations where $\Delta\mathbf{K}$ is not calculated explicitly. Rather, we can calculate $\Delta\mathbf{K}$ at each iteration cycle from [Eq. (5.76)] $\Delta\mathbf{K} = \mathbf{K}_T - \mathbf{K}_{ref}$.
- The particular case of *nonlinear analysis of the original structure*, where $\Delta\mathbf{K}_0 = 0$ and $\Delta\mathbf{K} = \Delta\mathbf{K}_{NL}$. That is, matrix $\Delta\mathbf{K}$ represents only changes due to the nonlinear behavior of the structure. Choosing, for example, $\mathbf{K}_{ref} = \mathbf{K}_0$ then the resulting matrix of changes $\Delta\mathbf{K} = \mathbf{K}_T - \mathbf{K}_0$ is the geometric stiffness matrix $\mathbf{K}_G$.

5.4.2 Solution by Combined Approximations

Consider the general case of nonlinear reanalysis of a modified structure (a similar procedure is used for nonlinear analysis of the original structure). Given matrix $\mathbf{K}_{ref}$ in the decomposed form of Eq. (5.75), calculation of the incremental displacements $\delta\mathbf{r}$ for any $\mathbf{K}_T$ and $\delta\mathbf{R}$ at each iteration cycle by the CA approach involves the following steps:

- Calculate the matrix of changes $\Delta\mathbf{K} = \Delta\mathbf{K}_{NL} + \Delta\mathbf{K}_0 = \mathbf{K}_T - \mathbf{K}_{ref}$. Again, the elements of $\Delta\mathbf{K}_0$ are constant for any assumed change in the design whereas the elements of $\Delta\mathbf{K}_{NL}$ are changed at each iteration cycle.
- Calculate the the $n{\times}s$ matrix of basis vectors basis vectors $\mathbf{r}_B$

$$\mathbf{r}_B = [\mathbf{r}_1, \mathbf{r}_2, ..., \mathbf{r}_s], \tag{5.79}$$

where the basis vectors $\mathbf{r}_1, \mathbf{r}_2, ..., \mathbf{r}_s$ are calculated by

$$\mathbf{r}_1 = \mathbf{K}_{ref}^{-1}\, \delta\mathbf{R}, \tag{5.80}$$

$$\mathbf{r}_i = -\mathbf{B}\, \mathbf{r}_{i-1} \qquad (i = 2, ..., s), \tag{5.81}$$

and matrix $\mathbf{B}$ is defined as

$$\mathbf{B} = \mathbf{K}_{ref}^{-1}\, \Delta\mathbf{K}. \tag{5.82}$$

Again, since $\mathbf{K}_{ref}$ is given in a decomposed form [Eq. (5.75)], calculation of the basis vectors involves only forward and backward substitutions.

- Calculate the reduced stiffness matrix $\mathbf{K}_R$ and load vector $\mathbf{R}_R$ by

$$\mathbf{K}_R = \mathbf{r}_B^T \mathbf{K}_T \mathbf{r}_B \qquad \mathbf{R}_R = \mathbf{r}_B^T \delta\mathbf{R}\,. \tag{5.83}$$

- Calculate the unknown coefficient $\mathbf{y}$ by solving the reduced set

$$\mathbf{K}_R\, \mathbf{y} = \mathbf{R}_R. \tag{5.84}$$

Since the number of basis vectors s is much smaller than the number of degrees of freedom n, it is necessary to solve only the smaller $s{\times}s$ system in Eq. (5.84) for $\mathbf{y}$ instead of solving the large $n{\times}n$ system in Eq. (5.77). To improve the accuracy of the results, uncoupled basis vectors can be used (see Sect. 5.2.2), instead of solving Eq. (5.84)

- Evaluate the displacements $\delta\mathbf{r}$ in Eq. (5.77) by the linear combination

$$\delta\mathbf{r} = y_1\mathbf{r}_1 + y_2\mathbf{r}_2 + ... + y_s\mathbf{r}_s = \mathbf{r}_B\, \mathbf{y}. \tag{5.85}$$

The solution process can be repeated for various changes in the design as necessary, with the initial decomposed $\mathbf{K}_{ref}$ being unchanged.

Example 5.8

Consider again the two-bar truss presented in example 1.3 and shown in Fig. 5.10. The object is to illustrate the solution steps by the CA approach for the two cases discussed earlier in Sect. 5.4.1, namely:

- the particular case of *nonlinear analysis of the original structure*; and
- the general case of *nonlinear reanalysis of a modified structure.*

Since there are only two DOF, consideration of two basis vectors by the CA approach will provide the exact solution of the equations.

Nonlinear Analysis of the Original Structure. Choosing $\mathbf{K}_{ref}$ as the elastic stiffness matrix we first calculate the initial displacements $\mathbf{r}_0$ [Eq. (1.56)],

$$\mathbf{K}_{ref} = \mathbf{K}_0 = \begin{bmatrix} 5.0 & \\ & 10.0 \end{bmatrix} \qquad \mathbf{R}_0 = \begin{Bmatrix} 10.0 \\ 5.0 \end{Bmatrix} \qquad \mathbf{r}_0 = \begin{Bmatrix} 2.0 \\ 0.5 \end{Bmatrix}. \tag{a}$$

For the first iteration cycle we calculate $\delta\mathbf{R}$, $\mathbf{K}_T$ [Eqs. (1.62), (1.65)]

$$\mathbf{R}_I = \begin{Bmatrix} 11.4312 \\ 7.2508 \end{Bmatrix} \qquad \delta\mathbf{R} = \mathbf{R}_0 - \mathbf{R}_I = -\begin{Bmatrix} 1.4312 \\ 2.2508 \end{Bmatrix}, \tag{b}$$

$$\Delta\mathbf{K} = \Delta\mathbf{K}_{NL} = \mathbf{K}_G = \begin{bmatrix} 1.1 & 2.125 \\ 2.125 & 0.5047 \end{bmatrix}, \tag{c}$$

$$\mathbf{K}_T = \mathbf{K}_0 + \Delta\mathbf{K} = \begin{bmatrix} 6.1 & 2.125 \\ 2.125 & 10.5047 \end{bmatrix}.$$

The object now is to demonstrate solution of the modified equations [Eq. (5.77)] at the first iteration cycle by the CA approach. The matrix of basis vectors $\mathbf{r}_B$ is calculated by Eqs. (5.79) – (5.82)

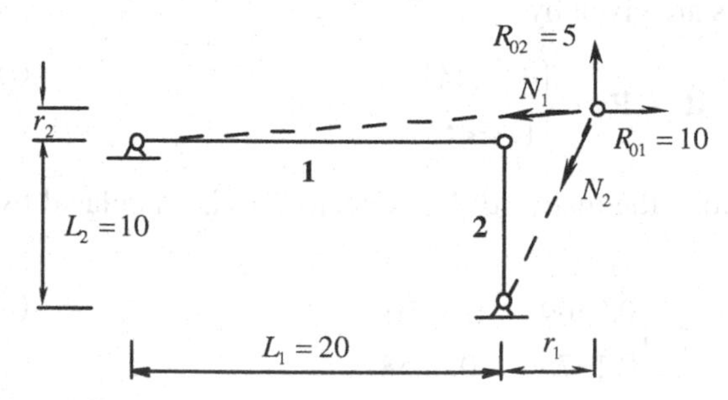

Fig. 5.10. Two-bar truss

$$\mathbf{r}_1 = \mathbf{K}_0^{-1}\,\delta\mathbf{R} = \begin{Bmatrix} -0.2862 \\ -0.2251 \end{Bmatrix} \qquad \mathbf{r}_2 = -\mathbf{B}\ \mathbf{r}_1 = \begin{Bmatrix} 0.1586 \\ 0.0722 \end{Bmatrix}, \tag{d}$$

$$\mathbf{r}_B = [\mathbf{r}_1, \mathbf{r}_2] = \begin{bmatrix} -0.2862 & 0.1586 \\ -0.2251 & 0.0722 \end{bmatrix}. \tag{e}$$

The reduced stiffness matrix $\mathbf{K}_R$ and load vector $\mathbf{R}_R$ are [Eqs. (5.83)]

$$\mathbf{K}_R = \mathbf{r}_B^T \mathbf{K}_T \mathbf{r}_B = \begin{bmatrix} 1.3058 & -0.5674 \\ -0.5674 & 0.2569 \end{bmatrix} \qquad \mathbf{R}_R = \mathbf{r}_B^T\,\delta\mathbf{R} = \begin{Bmatrix} 0.9163 \\ -0.3895 \end{Bmatrix}. \tag{f}$$

Finally, the multipliers $\mathbf{y}$ and the resulting incremental displacements $\delta\mathbf{r}$ for the first iteration cycle are [Eqs. (5.84), (5.85)]

$$\mathbf{y} = \mathbf{K}_R^{-1}\mathbf{R}_R = \begin{Bmatrix} 1.0665 \\ 0.8394 \end{Bmatrix} \qquad \delta\mathbf{r} = \mathbf{r}_B\mathbf{y} = \begin{Bmatrix} -0.1721 \\ -0.1794 \end{Bmatrix}. \tag{g}$$

As noted earlier, this is the exact solution of the modified equations for the first iteration cycle (see results of the first iteration in Table 1.1).

Nonlinear reanalysis of a modified structure. Assuming $\mathbf{K}_{ref} = \mathbf{K}_0$ and a reduction of 50% in the cross section areas, the tangent stiffness matrix for the first iteration cycle is calculated by [Eqs. (5.76), (5.78)]

$$\Delta\mathbf{K} = \Delta\mathbf{K}_{NL} + \Delta\mathbf{K}_0 = \begin{bmatrix} -1.95 & 1.0625 \\ 1.0625 & -4.7477 \end{bmatrix}, \tag{h}$$

$$\mathbf{K}_T = \mathbf{K}_{ref} + \Delta\mathbf{K} = \begin{bmatrix} 3.05 & 1.0625 \\ 1.0625 & 5.2523 \end{bmatrix}.$$

The out-of-balance forces are given by

$$\delta\mathbf{R} = \mathbf{R}_0 - \mathbf{R}_I = \begin{Bmatrix} 4.2844 \\ 1.3746 \end{Bmatrix}. \tag{i}$$

Assuming two basis vectors, the matrix of basis vectors $\mathbf{r}_B$ is calculated by Eqs. (5.79) – (5.82)

$$\mathbf{r}_B = [\mathbf{r}_1, \mathbf{r}_2] = \begin{bmatrix} 0.8569 & 0.3050 \\ 0.1375 & -0.0258 \end{bmatrix}. \tag{j}$$

Finally, the multipliers $\mathbf{y}$ and the resulting incremental displacements $\delta\mathbf{r}$ for the first iteration cycle are [Eqs. (5.84), (5.85)]

$$\mathbf{y} = \mathbf{K}_R^{-1}\mathbf{R}_R = \begin{Bmatrix} 0.4541 \\ 3.3578 \end{Bmatrix} \qquad \delta\mathbf{r} = \mathbf{r}_B\,\mathbf{y} = \begin{Bmatrix} 1.4131 \\ -0.0241 \end{Bmatrix}. \tag{k}$$

Again, since the number of basis vectors is equal to the number of DOF, this is the exact solution of the modified equations for the first iteration cycle. Using $\mathbf{r}^{(k)} = \mathbf{r}^{(k-1)} + \delta\mathbf{r}^{(k)}$, the solution converges very fast as follows

$$\text{Iteration 1} \qquad \delta\mathbf{r} = \begin{Bmatrix} 1.4131 \\ -0.0241 \end{Bmatrix} \qquad \mathbf{r} = \begin{Bmatrix} 3.4131 \\ 0.4759 \end{Bmatrix}.$$

$$\text{Iteration 2} \qquad \delta\mathbf{r} = \begin{Bmatrix} -0.0632 \\ -0.0707 \end{Bmatrix} \qquad \mathbf{r} = \begin{Bmatrix} 3.3499 \\ 0.4052 \end{Bmatrix}.$$

$$\text{Iteration 3} \qquad \delta\mathbf{r} = \begin{Bmatrix} -0.0010 \\ 0.0001 \end{Bmatrix} \qquad \mathbf{r} = \begin{Bmatrix} 3.3489 \\ 0.4053 \end{Bmatrix}.$$

5.4.3 Procedures for Analysis and Reanalysis

Nonlinear Analysis of the Original Structure

During nonlinear analysis of the original structure, the frequency of assembling and factorizing the tangent stiffness matrix depends on the solution scheme. In the full Newton-Raphson procedure we factorize the matrix at each iteration cycle, whereas in the modified Newton-Raphson procedure we factorize it at the beginning of each load increment. In this section, several solution procedures that are based on the CA approach are presented. The procedures differ in the strategies for choosing matrices $\mathbf{K}_{ref}$, $\Delta\mathbf{K}$ and $\mathbf{r}_B$. As a result, the expected efficiency and accuracy of the approximations is also different.

The matrix $\Delta\mathbf{K}$ is defined as [Eq. (5.76)]

$$\Delta\mathbf{K} = \mathbf{K}_T - \mathbf{K}_{ref}, \tag{5.86}$$

where $\mathbf{K}_T$ is the current tangent stiffness matrix and $\mathbf{K}_{ref}$ is a stiffness matrix from a certain solution step preceding the current step, given in the decomposed form of Eq. (5.75). We may start with the initial (elastic) stiffness matrix $\mathbf{K}_{ref} = \mathbf{K}_0$, and then consider various possibilities for updating $\mathbf{K}_{ref}$. Evidently, any strategy for this update directly influences the accuracy of the results and the efficiency of the solution process. Once matrix $\mathbf{K}_{ref}$ is chosen and factorized, the matrix of changes $\Delta\mathbf{K}$ and the matrix of basis vectors $\mathbf{r}_B$ are determined accordingly for subsequent iteration cycles. It should be noted that frequent updates of $\mathbf{K}_{ref}$ improves the accuracy of the results at the expense of additional computational effort, because each update involves factorization of the corresponding tangent stiffness matrix.

As to the nature of matrix $\Delta\mathbf{K}$, it has been noted that in nonlinear analysis of the original structure $\Delta\mathbf{K} = \Delta\mathbf{K}_{NL}$, that is, $\Delta\mathbf{K}$ represents only changes due to the nonlinear behavior of the structure. The accuracy of $\Delta\mathbf{K}$ depends on the accuracy of matrices $\mathbf{K}_T$ and $\mathbf{K}_{ref}$ [Eq. (5.86)]. Thus, improving the accuracy of both matrices might lead to more effective approximations in terms of accuracy and efficiency. In material nonlinearity the changes in the stiffness matrix are often of local nature. When yielding occurs in a structural element, only the stiffness coefficients related to that element are changed. In that sense, material nonlinearity is similar to linear reanalysis for local changes in the design. It has been shown [19] that in such cases exact or near exact solutions can be achieved with a small number of basis vectors. On the other hand, in geometric nonlinearity, the changes in the geometry are often of global nature.

Another factor that may significantly contribute to the computational effort is the update of matrix $\mathbf{r}_B$. It is observed [Eqs. (5.80), (5.81)] that for any given factorization of $\mathbf{K}_{ref}$ [Eq. (5.75)] the basis vectors depend on matrix $\Delta\mathbf{K}$ and on the residual force vector $\delta\mathbf{R}$. Since both $\Delta\mathbf{K}$ and $\delta\mathbf{R}$ are modified at each iteration cycle, it might be more efficient to update the basis vectors only after several iteration cycles.

Various possibilities for updating matrices $\mathbf{K}_{ref}$, $\mathbf{K}_T$, $\Delta\mathbf{K}$, $\mathbf{r}_B$ have been proposed and examined in terms of accuracy and efficiency [19], including the following strategies:

- *Strategy A*. Matrix $\mathbf{K}_{ref}$ is the tangent stiffness matrix at the beginning of an increment, matrices $\mathbf{K}_T$, $\Delta\mathbf{K}$, $\mathbf{r}_B$ are updated at each iteration cycle.
- *Strategy B*. Matrix $\mathbf{K}_{ref}$ is updated every few increments, matrices $\mathbf{K}_T$, $\Delta\mathbf{K}$, $\mathbf{r}_B$ are updated at each iteration cycle.
- *Strategy C*. Matrix $\mathbf{K}_{ref}$ is the initial stiffness matrix $\mathbf{K}_0$, matrices $\mathbf{K}_T$, $\Delta\mathbf{K}$ are updated at each increment and matrix $\mathbf{r}_B$ is updated at each iteration cycle.
- *Strategy D*. Matrix $\mathbf{K}_{ref}$ is the initial stiffness matrix $\mathbf{K}_0$, matrices $\mathbf{K}_T$, $\Delta\mathbf{K}$ are updated at each increment and matrix $\mathbf{r}_B$ is updated every few iteration cycles.

Solution by some of these strategies is demonstrated later in this section by numerical examples.

Nonlinear Reanalysis of Modified Structures

The procedures presented in this section are intended to reduce the number of matrix factorizations during nonlinear reanalysis. The object is to evaluate the response under incrementally applied loads after changes in the structure properties due to various design modifications. The presented procedures are suitable for multiple successive design modifications, which might be necessary in various applications (e.g. design optimization, structural damage analysis, probabilistic analysis). It is recognized that various approximations considered in nonlinear analysis may result in inaccurate or inefficient solutions, and even divergence or incorrect results. The object in this section is to demonstrate the performance of some reanalysis procedures in terms of both accuracy and efficiency. In the procedures presented we assume that the factorized matrix $\mathbf{K}_{ref}$ is calculated (once, or more) only for the original structure.

Using the CA approach, we start with nonlinear analysis of the original structure and calculate M factorized tangent stiffness matrices (M is for example the number of load increments) chosen as reference matrices ${}^{m}\mathbf{K}_{ref}$ ($m=1,2,...,M$). Having calculated ${}^{m}\mathbf{K}_{ref}$, the following two options for choosing the appropriate reference increment are considered [19]:

- *Option 1 – load resemblance.* The reference increment is related to the current load intensity, that is ${}^{t}\mathbf{R}_0 \approx {}^{m}\mathbf{R}_{ref}$. It is assumed in this case that the load distribution does not change significantly, thus choosing the appropriate increment is rather simple.
- *Option 2 – displacement resemblance.* The reference increment is related to the current displacement field, that is ${}^{t}\mathbf{r} \approx {}^{m}\mathbf{r}_{ref}$. The most similar displacement vector can be found according to a particular significant displacement or as the reference displacement vector having the smallest angle with the current displacement vector.

Noting that $\Delta\mathbf{K} = \Delta\mathbf{K}_{NL} + \Delta\mathbf{K}_0$ [Eq. (5.78)], the entries in $\Delta\mathbf{K}_{NL}$ for option 1 might be mainly functions of the displacements, because $\mathbf{K}_{ref}$ represents load resemblance. On the other end, the entries for option 2 might be mainly functions of the internal forces because $\mathbf{K}_{ref}$ represents displacement resemblance. As a result, option 1 is more suitable for cases where the nonlinearity is mostly dominated by forces whereas option 2 is suitable for cases where the nonlinearity is dominated by displacements.

Different numerical examples have been solved considering the various options and strategies presented. Some small scale examples are demonstrated in this section. The accuracy of the results is discussed in the next sub-section and in Sect. 10.2. The efficiency of the calculations for large-scale structures is examined in Sect. 10.1.

Accuracy Considerations

Various examples of nonlinear analysis have shown that high accuracy can be achieved by the procedures presented in this section for both material nonlinearity (MNL) and geometric nonlinearity (GNL). Convergence of the procedures is assumed when consideration of additional basis vectors does not improve the quality of the results.

The high accuracy achieved for MNL is due to the local nature of the changes in the structure. Very efficient results have been achieved in MNL examples, where the stiffness matrix is factorized only once during the solution process. In some examples the average size of the errors is reduced consistently when considering additional basis vectors. The high accuracy achieved for GNL is due to the smooth nature of the problem formulation.

In some examples, increase in the number of basis vectors considered does not improve the accuracy of the results. It was found that in such cases the basis vectors become close to linearly dependent as their number is increased. One possible solution is to calculate the angles between the basis vectors and avoiding those vectors for which their angle relative to a previous vector is close to zero. Another possible criterion is the condition number of the reduced stiffness matrix. When this number is high the matrix becomes ill-conditioned. Various topics related to accuracy considerations are discussed later in Sect. 10.2.

Example 5.9

Consider the simple frame shown in Fig. 5.11, having 30 co-rotational beam elements. Using [*kN*, *m*] units, the reference load is P=1000, the cross-section area is 0.007548, the moment of inertia is 1.29 10^{-4} and the modulus of elasticity is 206 850 000. The object is to demonstrate results for both geometric nonlinear analysis and reanalysis

Analysis of the original structure. Assuming automatic load incrementation up to a maximum load factor of λ = 20, we evaluate the nonlinear response using *strategies A and B* with 2–6 basis vectors (CA2–CA6). The results are compared with the Full Newton-Raphson (FNR) and the Modified Newton-Raphson (MNR) procedures. For strategy B the stiffness matrix is factorized at the beginning of every second increment.

All the above procedures provide identical results for λ = 20 (Fig. 5.12). The horizontal displacement at the top left corner is 2.0921 and the moment at the left base is 34855. The resulting numbers of factorizations, increments and iteration cycles are shown in Table 5.7. The following observations are made:

- For strategy A, the numbers of matrix factorizations required by the CA approach are 21%–34% of the numbers required by the FNR procedure. For the more efficient strategy B, these numbers are only 15%–21%.
- Using automatic incrementation, the numbers of increments required by the CA approach are significantly smaller than those required by the MNR procedure and similar to those required by the FNR procedure.
- Increasing the number of basis vectors, the numbers of CA iteration cycles are reduced, due to the more accurate problem formulation.

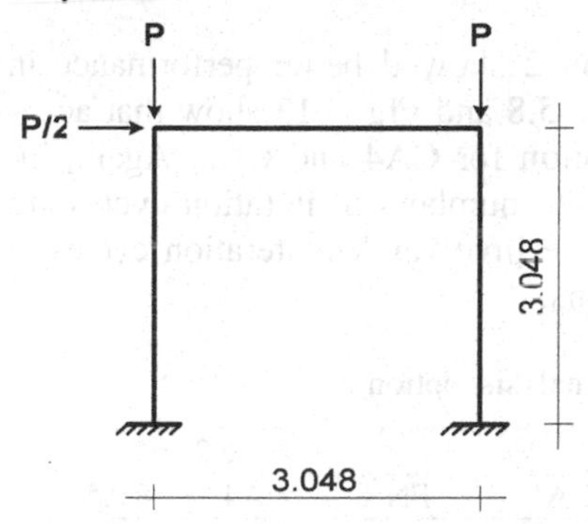

Fig. 5.11. Frame example, geometric nonlinear analysis and reanalysis

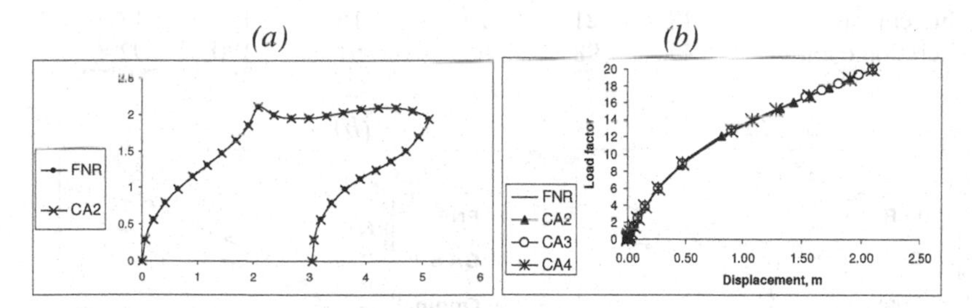

Fig. 5.12. *a* Deformed frame *b*. Displacement at top-left corner

Table 5.7. Results, geometric nonlinear analysis

Result	Strategy	CA2	CA3	CA4	CA5	CA6	MNR	FNR
Relative number	A	0.339	0.304	0.304	0.214	0.214	0.589	1.000
of factorizations	B	0.214	0.161	0.179	0.152	0.152	0.536	1.000
Increments	A	19	17	17	12	12	33	18
	B	24	18	20	17	17	60	18
Iteration cycles	A	174	150	122	57	47	383	56
	B	261	187	193	137	136	702	56
Iterations per	A	9.16	8.82	7.18	4.75	3.92	11.61	3.11
increment	B	10.88	10.39	9.65	8.06	8.00	11.70	3.11

Reanalysis of modified structures. The following two cases have been solved by the CA4, CA5 and FNR procedures:

- Case 1. The cross sectional areas and moments of inertia of the columns are increased by 100% and those of the beam are decreased by 50%. The maximum load factor is 25.
- Case 2. The cross sectional areas and moments of inertia of the columns are decreased by 50% and those of the beam are increased by 100%. The maximum load factor is 12.

It was found that, in general, option 2 showed better performance in terms of accuracy and efficiency. Table 5.8 and Fig. 5.13 show that accurate results are achieved using this option for CA4 and CA5. Again, increasing the number of basis vectors, the numbers of iteration cycles are reduced. Solution by the FNR procedure involves less iteration cycles at the expense of more matrix factorizations.

Table 5.8. Results, geometric nonlinear reanalysis, option 2

	Case 1			Case 2		
Result	FNR	CA4	CA5	FNR	CA4	CA5
Top left displacement	1.85	1.85	1.85	2.0555	2.0555	2.0555
Left base moment	45450	45450	45450	19305	19305	19305
Increments	18	21	19	16	19	17
Iteration cycles	49	98	68	51	120	92

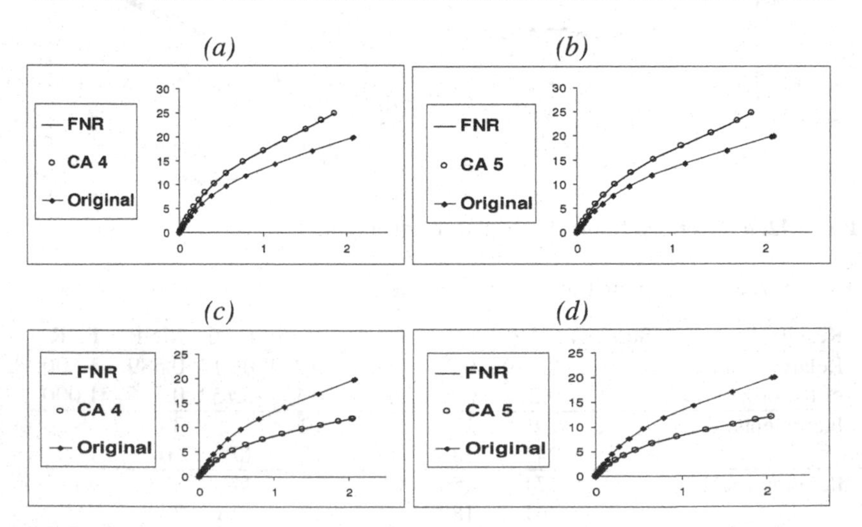

Fig. 5.13. Load factor versus top horizontal displacement, option 2:
a. case 1, CA4 ***b.*** case 1, CA5 ***c.*** case 2, CA4 ***d.*** case 2, CA5

Example 5.10

To demonstrate *snap-through of a shallow arch*, consider the classic example shown in Fig. 5.14. The geometric and material properties are taken from [20]. The half-arch is modeled with 5 co-rotational beam elements and 14 DOF. Using [*lb, inch*] units the modulus of elasticity is 10^7, the cross sectional area is 0.32 and the moment of inertia is 1.0. The reference load $P/2 = 200$ is applied by automatic load incrementation up to a maximum load factor of $\lambda = 20$. The nonlinear analysis of this example might involve accuracy difficulties due to the limit points and non-positive-definite stiffness matrix. The object is to demonstrate the accuracy achieved by the CA approach considering strategy A and only two basis vectors (CA2). The first increment has a small prescribed displacement and the initial arc length is calculated according to this displacement. The stiffness matrix is factorized only at the beginning of each increment. The apex displacement obtained by CA2 is in good agreement with the results obtained by the FNR procedure and a commercial code [20, 21], as shown in Fig. 5.15.

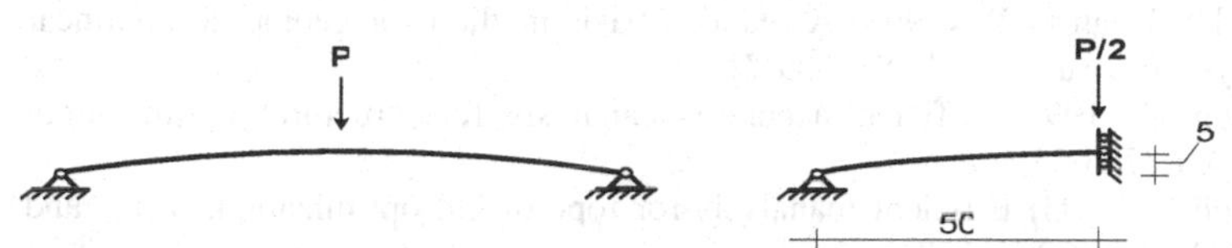

Fig. 5.14. Shallow arch example, snap-through nonlinear analysis

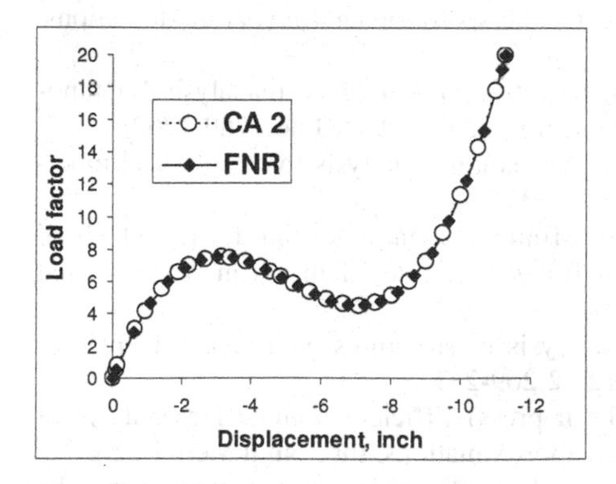

Fig. 5.15. Load factor versus displacement at apex

References

1. Kirsch U (2000) Combined approximations – a general reanalysis approach for structural optimization. Struct and Multidis Opt 20:97-106
2. Kirsch U (2002) Design-oriented analysis of structures. Kluwer Academic Publishers, Dordrecht
3. Kirsch U (2003) Design-oriented analysis of structures – a unified approach. ASCE J Engrg Mech 129:264-272
4 Kirsch U (2003) A unified rcanalysis approach for structural analysis, design and optimization. Struct and Multidis Opt 25:67-85
5. Kirsch U (1984) Approximate behavior models for optimum structural design. In: Atrek E et al (eds) New directions in optimum structural design. John Wiley &Sons, NY
6. Kirsch U (1993) Structural Optimization.Springer-Verlag, Berlin
7. Kirsch U, Toledano G (1983) Approximate reanalysis for modifications of structural geometry. Computers & Structures 16:269-279
8. Wilkinson W (1965) The algebraic eigenvalue problem. Oxford University Press, Oxford
9. Ortega JM, Rheinboldt WC (1970) Iterative solutions of nonlinear equations in several variables. Academic Press, New York
10. Leu LJ, Huang CW (1998) A reduced basis method for geometric nonlinear analysis of structures. IASS J 39:71-75
11. Kirsch U (1999) Efficient-accurate reanalysis for structural optimization. AIAA J 37:1663-1669
12. Kirsch U (1993) Efficient reanalysis for topological optimization. Struct and Multidis Opt 6:143-150
13. Kirsch U, Liu S (1995) Exact structural reanalysis by a first-order reduced basis approach. Struct and Multidis Opt 10:153-158
14. Kirsch U, Liu S (1997) Structural reanalysis for general layout modifications. AIAA J 35:382-388
15. Chen S, Huang C, Liu Z (1998) Structural approximate reanalysis for topological modifications by finite element systems. AIAA J 36:1760-1762
16. Kirsch U, Papalambros PY (2001) Structural reanalysis for topological modifications. Struct and Multidis Opt 21:333-344
17. Wu BS, Lim CW, Li ZG (2004) A finite element algorithm for reanalysis of structures with added degrees of freedom. Finite Elements in Analysis and Design 40:1791-1801
18. Wu BS, Li ZG (2006) Static reanalysis of structures with added degrees of freedom, Comm Num Meth Engrg 22:269-281
19. Amir O, Kirsch U, Sheinman I (in press) Efficient nonlinear reanalysis of skeletal structures using combined approximations. Int J Num Meth Engrg
20. Meek JL, Tan HS (1984) Geometrically nonlinear analysis of space frames by an incremental iterative technique. Comp Meth Appl Mech Engrg 47:261-282
21. ADINA R&D Inc (2004) A general propose finite element nonlinear dynamic analysis, Version 8.2

6 Vibration Reanalysis

It has been noted that in dynamic analysis by mode superposition, the main computational effort is spent in the solution of the eigenproblem. Since exact solution of the problem can be prohibitively expensive, approximate solution techniques have been developed, primarily to calculate the lowest eigenvalues and corresponding eigenvectors [1–3]. Several studies have been published on *eigenvalue reanalysis*, where the object is to calculate only the modified eigenvalues [e.g. 4–6]. Other studies deal with *eigenproblem reanalysis*, or *vibration reanalysis,* where the object is to calculate both the modified eigenvalues and eigenvectors [e.g. 7, 8]. Vibration reanalysis is needed in various problems of structural analysis, design and optimization. Different reanalysis methods, which have been used for linear static analysis, are usually not suitable for vibration reanalysis.

In this chapter effective procedures for vibration reanalysis, based on the CA approach, are developed. Using these procedures, significant improvements in the accuracy of the results and the efficiency of the calculations can be achieved [9–12].

In Sect. 6.1 vibration reanalysis by the CA approach is introduced. The approximate reduced eigenproblem is formulated and procedures for determining the basis vectors are developed. Various means intended to improve the accuracy of the results are developed in Sect. 6.2. The procedures presented, which improve the quality of the basis vectors, include Gram-Schmidt orthogonalizations of the approximate modes, shifts of the basis vectors and Gram-Schmidt orthogonalizations of the basis vectors. A general solution procedure for vibration reanalysis is presented in Sect. 6.3 and various numerical examples are demonstrated in Sect. 6.4. In Sect. 6.5 it is shown how the CA approach can be used to improve the solution efficiency when various common iterative procedures for eigenproblem analysis are considered. These iterative procedures, formulated as reanalysis problems in Sect. 4.3.2, include inverse iteration, inverse iteration with shifts and subspace iteration.

6.1 The Reduced Eigenproblem

6.1.1 Problem Formulation

Consider the given initial values $\mathbf{K}_0$, $\mathbf{M}_0$, and the initial eigenpairs $\lambda_0, \mathbf{\Phi}_0$, known from solution of the initial eigenproblem [Eq. (4.28)]

$$\mathbf{K}_0 \mathbf{\Phi}_0 = \lambda_0 \mathbf{M}_0 \mathbf{\Phi}_0 . \tag{6.1}$$

For simplicity of presentation, the subscript i denoting the ith mode is omitted. Assume a change in the design and corresponding changes $\Delta\mathbf{K}_0$ and $\Delta\mathbf{M}_0$ in the stiffness and mass matrices such that the modified matrices are [Eqs. (4.29), (4.30)]

$$\mathbf{K} = \mathbf{K}_0 + \Delta\mathbf{K}_0 \qquad \mathbf{M} = \mathbf{M}_0 + \Delta\mathbf{M}_0. \tag{6.2}$$

The modified analysis equations to be solved are

$$\mathbf{K}\mathbf{\Phi} = \lambda\, \mathbf{M}\mathbf{\Phi} . \tag{6.3}$$

Substitution of Eq. (6.2) into Eq. (6.3), the problem can be expressed as

$$(\mathbf{K}_0 + \Delta\mathbf{K}_0)\,\mathbf{\Phi} = \lambda\, \mathbf{M}\mathbf{\Phi} . \tag{6.4}$$

Thus, the reanalysis problem under consideration is to evaluate efficiently and accurately the modified eigenpairs λ, $\mathbf{\Phi}$, due to various changes in the design, such that the modified equations (6.4) are satisfied.

Similar to linear static reanalysis, we assume that the mode shape $\mathbf{\Phi}$ of a new design can be approximated by a linear combination of pre-selected s linearly independent basis vectors $\mathbf{r}_1, \mathbf{r}_2, \ldots, \mathbf{r}_s$

$$\mathbf{\Phi} = y_1\mathbf{r}_1 + y_2\mathbf{r}_2 + \ldots + y_s\mathbf{r}_s = \mathbf{r}_B\, \mathbf{y}, \tag{6.5}$$

where s is much smaller than the number of degrees of freedom n, $\mathbf{r}_B$ is the $n{\times}s$ matrix of the basis vectors and $\mathbf{y}$ is a vector of unknown coefficients

$$\mathbf{r}_B = [\mathbf{r}_1, \mathbf{r}_2, \ldots, \mathbf{r}_s], \tag{6.6}$$

$$\mathbf{y}^T = \{y_1, y_2, \ldots, y_s\}. \tag{6.7}$$

The modified equations (6.3) are now approximated by a smaller system of equations in the new unknowns $\mathbf{y}$. Substituting Eq. (6.5) into Eq. (6.3) and pre-multiplying the resultant equation by $\mathbf{r}_B^T$ gives the $s{\times}s$ system

$$\mathbf{r}_B^T \mathbf{K}\mathbf{r}_B\mathbf{y} = \lambda\, \mathbf{r}_B^T \mathbf{M}\mathbf{r}_B\mathbf{y} . \tag{6.8}$$

Introducing the notation

$$\mathbf{K}_R = \mathbf{r}_B^T \mathbf{K} \mathbf{r}_B \qquad \mathbf{M}_R = \mathbf{r}_B^T \mathbf{M} \mathbf{r}_B , \tag{6.9}$$

and substituting Eq. (6.9) into Eq. (6.8), the reduced eigenproblem becomes

$$\mathbf{K}_R \mathbf{y} = \lambda \mathbf{M}_R \mathbf{y} . \tag{6.10}$$

The $s{\times}s$ matrix $\mathbf{K}_R$ is full but is symmetric and much smaller in size than the $n{\times}n$ matrix $\mathbf{K}$. That is, rather than computing the exact solution by solving the large $n{\times}n$ system in Eq. (6.3), we first solve the smaller $s{\times}s$ system in Eq. (6.10) for the first eigenpair λ_1, $\mathbf{y}_1$. Then we evaluate the first mode shape $\mathbf{\Phi}$ for the computed $\mathbf{y}_1$ by Eq. (6.5). A similar procedure is used to evaluate higher eigenpairs, as will be shown later in this chapter

6.1.2 Determination of the Basis Vectors

The basis vectors can be determined in several different ways. One possibility is to pre-multiply the modified equations (6.4) by $\mathbf{K}_0^{-1}$ to obtain

$$(\mathbf{I} + \mathbf{B})\,\mathbf{\Phi} = \mathbf{r}_1, \tag{6.11}$$

where $\mathbf{r}_1$ and $\mathbf{B}$ are defined as

$$\mathbf{r}_1 = \mathbf{K}_0^{-1} \lambda\, \mathbf{M} \mathbf{\Phi} , \tag{6.12}$$

$$\mathbf{B} = \mathbf{K}_0^{-1} \Delta\mathbf{K}. \tag{6.13}$$

Pre-multiplying Eq. (6.11) by $(\mathbf{I} + \mathbf{B})^{-1}$ and expanding

$$(\mathbf{I} + \mathbf{B})^{-1} \cong \mathbf{I} - \mathbf{B} + \mathbf{B}^2 - \ldots, \tag{6.14}$$

we obtain the binomial series

$$\mathbf{\Phi} = (\mathbf{I} - \mathbf{B} + \mathbf{B}^2 - \ldots)\, \mathbf{r}_1. \tag{6.15}$$

Similar to static reanalysis, the terms of Eq. (6.15) could be assumed as basis vectors, with the first basis vector given by Eq. (6.12). However, since $\mathbf{r}_1$ is unknown it is convenient to consider the known expression,

$$\mathbf{r}_1 = \mathbf{K}_0^{-1} \lambda_0\, \mathbf{M}\, \mathbf{\Phi}_0 = \mathbf{K}_0^{-1} \mathbf{R}_0, \tag{6.16}$$

as the first basis vector $\mathbf{r}_1$, with $\mathbf{R}_0$ defined as

$$\mathbf{R}_0 = \lambda_0\, \mathbf{M}\, \mathbf{\Phi}_0 . \tag{6.17}$$

Since multiplication of a basis vector by a scalar does not affect the results [only the corresponding y is changed, Eq. (6.5)], we can drop λ_0 from the expression of $\mathbf{r}_1$. It is instructive to note that for $\mathbf{M} = \mathbf{M}_0$ the first basis vector is simply the initial mode shape $\mathbf{r}_1 = \mathbf{\Phi}_0$.

Similar to static reanalysis, the additional basis vectors are calculated successively by

$$\bar{\mathbf{r}}_i = -\mathbf{B}\,\mathbf{r}_{i-1} \qquad (i = 2, ..., s). \tag{6.18}$$

The basis vectors $\bar{\mathbf{r}}_i$ can then be normalized in several different ways. One possibility to do that is to divide $\bar{\mathbf{r}}_i$ by an arbitrary reference element, $ref(\bar{\mathbf{r}}_i)$, e.g. the largest element, to obtain the normalized basis vector

$$\mathbf{r}_i = \frac{\bar{\mathbf{r}}_i}{ref(\bar{\mathbf{r}}_i)}. \tag{6.19}$$

These normalizations are useful in cases of large changes in the design (large $\Delta\mathbf{K}$), when the elements of the basis vectors may become very large.

Alternatively, the basis vectors can be determined as follows. Rearranging Eq. (6.4), dropping the eigenvalue λ and writing the resulting equation as a recurrence relation, we obtain

$$\mathbf{K}_0\bar{\mathbf{r}}_i = (-\Delta\mathbf{K}_0 + \mathbf{M})\,\mathbf{r}_{i-1}. \tag{6.20}$$

Denoting

$$\mathbf{C} = \mathbf{K}_0^{-1}(\Delta\mathbf{K} - \mathbf{M}), \tag{6.21}$$

we can calculate the basis vectors successively by

$$\bar{\mathbf{r}}_i = -\mathbf{C}\,\mathbf{r}_{i-1} \qquad i = 2, ..., s, \tag{6.22}$$

where the first basis vector is chosen as the initial mode shape $\mathbf{\Phi}_0$

$$\mathbf{r}_1 = \mathbf{\Phi}_0. \tag{6.23}$$

It is observed that the expression of the recurrence relation [Eq. (6.22)] is similar to the expression of the inverse iteration [Eq. (2.46)]. One difference is that, for the expression of Eq. (6.22), once the initial stiffness matrix has been factorized by $\mathbf{K}_0 = \mathbf{U}_0^T\,\mathbf{U}_0$, it can be used for calculation of the basis vectors for any modified design. Again, calculation of each basis vector by Eq. (6.22) involves only forward and backward substitutions. On the other hand, for the expression of Eq. (2.46), the modified stiffness matrix $\mathbf{K}$ must be factorized repeatedly for each change in the design.

It can be observed that for $\mathbf{M} = \mathbf{M}_0$ both methods of Eq. (6.18) and Eq. (6.22) provide identical results. In other cases, still similar results have been obtained [10, 12].

6.2 Improved Basis Vectors

For illustrative purposes, consider again the procedure of Eqs. (6.16) – (6.18) for determining the basis vectors, and assume the normalization of Eq. (2.47). We obtain for the first basis vector

$$\bar{\mathbf{r}}_1 = \mathbf{K}_0^{-1}\mathbf{M}\boldsymbol{\Phi}_0 , \tag{6.24}$$

$$\hat{\mathbf{r}}_1 = \frac{\bar{\mathbf{r}}_1}{\left(\bar{\mathbf{r}}_1^T \mathbf{M}\,\bar{\mathbf{r}}_1\right)^{1/2}} , \tag{6.25}$$

where $\hat{\mathbf{r}}_1$ and $\bar{\mathbf{r}}_1$ are the normalized and non-normalized vectors, respectively. Additional non-normalized and normalized basis vectors are calculated by

$$\bar{\mathbf{r}}_i = -\mathbf{B}\,\hat{\mathbf{r}}_{i-1} , \tag{6.26}$$

$$\hat{\mathbf{r}}_i = \frac{\bar{\mathbf{r}}_i}{\left(\bar{\mathbf{r}}_i^T \mathbf{M}\,\bar{\mathbf{r}}_i\right)^{1/2}} \qquad i = 2, 3, \ldots, s. \tag{6.27}$$

It was found [10, 12] that high accuracy is achieved by this procedure for the first mode shape with a small number of basis vectors, even for very large changes in the design. Less accurate results might be obtained for the higher mode shapes. In this section, several procedures intended to improve the basis vectors, in order to achieve higher accuracy of the results, are introduced.

6.2.1 Gram-Schmidt Orthogonalizations of the Modes

The accuracy can be significantly improved with a small computational effort by using vector deflation, as described in Sect. 2.3.3. The basis of vector deflation is that in order to converge to a required eigenvector, the basis vectors must not be orthogonal to it. If the basis vectors are orthogonalized, with respect to $\mathbf{M}$, to the eigenvectors already calculated we eliminate the possibility that the solution converges to any one of them. To improve the accuracy of the results in calculation of the higher mode shapes, we use Gram-Schmidt orthogonalizations of the approximate mode shapes. Assume for example that we have calculated the first m eigenvectors and we want to **M**-orthogonalize $\mathbf{\Phi}_{m+1}$ to these eigenvectors. For this purpose, we **M**-orthogonalize the basis vector $\hat{\mathbf{r}}_i$ of $\mathbf{\Phi}_{m+1}$ to the lower eigenvectors. Using the expression

$$\mathbf{r}_i = \hat{\mathbf{r}}_i - \sum_{k=1}^{m} \alpha_k \mathbf{\Phi}_k , \tag{6.28}$$

the non-orthogonal basis vector $\hat{\mathbf{r}}_i$ is calculated by Eq. (6.25) or Eq. (6.27), and the coefficients α_k are obtained from the conditions

$$\mathbf{\Phi}_k^T \mathbf{M} \mathbf{r}_i = 0 \qquad k = 1, \ldots, m, \tag{6.29}$$

$$\mathbf{\Phi}_k^T \mathbf{M} \mathbf{\Phi}_j = \delta_{kj} \qquad k, j = 1, \ldots, m, \tag{6.30}$$

where δ_{kj} is the kronecker delta. Pre-multiplying both sides of Eq. (6.28) by $\mathbf{\Phi}_k^T \mathbf{M}$ and using Eqs. (6.29), (6.30), we obtain

$$\alpha_k = \mathbf{\Phi}_k^T \mathbf{M} \, \hat{\mathbf{r}}_i . \tag{6.31}$$

Substituting Eq. (6.31) into Eq. (6.28) gives

$$\mathbf{r}_i = \hat{\mathbf{r}}_i - \sum_{k=1}^{m} (\mathbf{\Phi}_k^T \mathbf{M} \hat{\mathbf{r}}_i) \mathbf{\Phi}_k \qquad i = 2, 3, \ldots, s. \tag{6.32}$$

The first basis vector for the first mode is

$$\mathbf{r}_1 = \hat{\mathbf{r}}_1 , \tag{6.33}$$

whereas for the higher modes we have from Eq. (6.32)

$$\mathbf{r}_1 = \hat{\mathbf{r}}_1 - \sum_{k=1}^{m} (\mathbf{\Phi}_k^T \mathbf{M} \hat{\mathbf{r}}_1) \mathbf{\Phi}_k . \tag{6.34}$$

6.2.2 Shifts of the Basis Vectors

To further improve the accuracy of the higher mode shapes, we can use the concept of shifts and the Rayleigh quotient iteration, discussed in Sect. 2.3.2. Introducing a shift μ, defining

$$\widehat{\lambda} = \lambda - \mu , \tag{6.35}$$

$$\widehat{\mathbf{K}} = \mathbf{K} - \mu\mathbf{M} , \tag{6.36}$$

and substituting Eqs. (6.35), (6.36) into Eq. (6.3) yields

$$\widehat{\mathbf{K}}\mathbf{\Phi} = \widehat{\lambda}\,\mathbf{M}\mathbf{\Phi} . \tag{6.37}$$

The eigenvectors of the two problems of Eqs. (6.3) and (6.37) are the same. Using inverse iteration the solution will converge to the mode having the smallest shifted eigenvalue. The object is to use the concept of shifts to improve the basis vectors. Substituting Eqs. (6.2), (6.36) into Eq. (6.37), we obtain

$$(\mathbf{K}_0 + \Delta\mathbf{K}_0 - \mu\mathbf{M})\,\mathbf{\Phi} = \widehat{\lambda}\,\mathbf{M}\mathbf{\Phi} . \tag{6.38}$$

Pre-multiplying Eq. (6.38) by $\mathbf{K}_0^{-1}$ and denoting

$$\mathbf{B}_\mu = \mathbf{K}_0^{-1}(\Delta\mathbf{K}_0 - \mu\mathbf{M}), \tag{6.39}$$

$$\mathbf{r}_0 = \mathbf{K}_0^{-1}\,\widehat{\lambda}\,\mathbf{M}\mathbf{\Phi}, \tag{6.40}$$

the resulting equation is

$$(\mathbf{I} + \mathbf{B}_\mu)\,\mathbf{\Phi} = \mathbf{r}_0. \tag{6.41}$$

Since Eqs. (6.11), (6.41) are similar, we can calculate the basis vectors by

$$\overline{\mathbf{r}}_i = -\mathbf{B}_{\mu i}\,\mathbf{r}_{i-1} \qquad i = 2, 3, \ldots, s, \tag{6.42}$$

where matrix $\mathbf{B}_{\mu i}$, which is changed for each basis vector, is given by

$$\mathbf{B}_{\mu i} = \mathbf{K}_0^{-1}(\Delta\mathbf{K}_0 - \mu_i\,\mathbf{M}). \tag{6.43}$$

The shift μ_i is calculated by the Rayleigh quotient iteration

$$\mu_i = \frac{\mathbf{r}_{i-1}^T\mathbf{K}\,\mathbf{r}_{i-1}}{\mathbf{r}_{i-1}^T\mathbf{M}\,\mathbf{r}_{i-1}} . \tag{6.44}$$

Finally, normalization of the calculated vectors $\overline{\mathbf{r}}_i$ gives

$$\mathbf{r}_i = \frac{\overline{\mathbf{r}}_i}{\left|\overline{\mathbf{r}}_i^T\mathbf{M}\,\overline{\mathbf{r}}_i\right|^{1/2}} \qquad i = 2, 3, \ldots, s. \tag{6.45}$$

6.2.3 Gram-Schmidt Orthogonalizations of the Basis Vectors

It was found that in many cases the basis vectors, determined by Eqs. (6.24) through (6.27), come close to being linearly-dependent. As a result, numerical errors might occur. To overcome this difficulty, Gram-Schmidt orthogonalizations (as described in Sect. 5.2.2 for static reanalysis) are used to generate a new set of orthogonal basis vectors $\mathbf{V}_i$ ($i = 1, 2, \ldots, s$). The advantage is that more accurate results are obtained with the new vectors that satisfy the conditions $\mathbf{V}_i^T \mathbf{M} \mathbf{V}_j = \delta_{ij}$. The first normalized basis vector $\mathbf{V}_1$ is determined by

$$\mathbf{V}_1 = \frac{\mathbf{r}_1}{|\mathbf{r}_1^T \mathbf{K} \mathbf{r}_1|^{1/2}}. \tag{6.46}$$

Additional basis vectors are generated in a way similar to the procedure presented in Sect. 5.2.2. The resulting general expression is [Eqs. (5.47)]

$$\overline{\mathbf{V}}_i = \mathbf{r}_i - \sum_{j=1}^{i-1} \left(\mathbf{r}_i^T \mathbf{K} \mathbf{V}_j\right) \mathbf{V}_j \qquad i = 2, \ldots, s. \tag{6.47}$$

The vectors $\overline{\mathbf{V}}_i$ are normalized by

$$\mathbf{V}_i = \frac{\overline{\mathbf{V}}_i}{|\overline{\mathbf{V}}_i^T \mathbf{K} \overline{\mathbf{V}}_i|^{1/2}}. \tag{6.48}$$

6.3 General Solution Procedure

Given the initial values $\mathbf{K}_0$, $\mathbf{M}_0$, $\mathbf{\Phi}_0$ and λ_0 from vibration analysis of the initial structure, we can now summarize the steps of a general solution procedure for vibration reanalysis. In the procedure presented in this section we distinguish between the two main stages of determining improved basis vectors and evaluating the eigenpairs by the reduced eigenproblem.

Determination of Improved Basis Vectors

Using the concepts presented in Sect. 6.2, the procedure for determining improved basis vectors may involve the following steps (see Table 6.1):

- *Basic calculations.* It is possible only to calculate the basic terms of the the basis veectors and to normalize them [Eqs. (6.24) through (6.27)].

- *Gram-Schmidt orthogonalizations of the modes.* The first basis vector for the first mode is unchanged [Eq. (6.33)] whereas the first basis vector of the higher modes are calculated by Eq. (6.34). The remaining basis vectors are determined by Eq. (6.32).
- *Shifts of the basis vectors.* The second and higher basis vectors are calculated by Eqs. (6.42), (6.45).
- *Gram-Schmidt orthogonalizations of the basis vectors.* The first basis vector is calculated by Eq. (6.46) and the remaining vectors by Eqs. (6.47), (6.48).

It was found that high accuracy is achieved for the higher modes if all the above steps are considered. For the lower modes it is possible to eliminate part of the steps and still obtain high accuracy.

Table 6.1. Solution procedure for determining the basis vectors

Step	Equation	Expression
Basic calculations	(6.24)	$\bar{\mathbf{r}}_1 = \mathbf{K}_0^{-1}\mathbf{M}\boldsymbol{\Phi}_0$
	(6.25)	$\hat{\mathbf{r}}_1 = \dfrac{\bar{\mathbf{r}}_1}{\left(\bar{\mathbf{r}}_1^T\mathbf{M}\bar{\mathbf{r}}_1\right)^{1/2}}$
	(6.26)	$\bar{\mathbf{r}}_i = -\mathbf{B}_i\mathbf{r}_{i-1} \qquad i = 2, 3, \ldots, s$
	(6.27)	$\hat{\mathbf{r}}_i = \dfrac{\bar{\mathbf{r}}_i}{\left(\bar{\mathbf{r}}_i^T\mathbf{M}\bar{\mathbf{r}}_i\right)^{1/2}} \qquad i = 2, 3, \ldots, s$
Gram-Schmidt orthogonalizations of the modes	(6.33)	$\mathbf{r}_1 = \hat{\mathbf{r}}_1$ (first mode)
	(6.34)	$\mathbf{r}_1 = \hat{\mathbf{r}}_1 - \sum_{k=1}^{m}(\boldsymbol{\Phi}_k^T\mathbf{M}\hat{\mathbf{r}}_1)\boldsymbol{\Phi}_k$ (higher modes)
	(6.32)	$\mathbf{r}_i = \hat{\mathbf{r}}_i - \sum_{k=1}^{m}(\boldsymbol{\Phi}_k^T\mathbf{M}\hat{\mathbf{r}}_i)\boldsymbol{\Phi}_k \qquad i = 2, 3, \ldots, s$
Shifts	(6.42)	$\bar{\mathbf{r}}_i = -\mathbf{B}_{\mu i}\mathbf{r}_{i-1} \qquad i = 2, 3, \ldots, s$
	(6.45)	$\mathbf{r}_i = \dfrac{\bar{\mathbf{r}}_i}{\left\lvert\bar{\mathbf{r}}_i^T\mathbf{M}\bar{\mathbf{r}}_i\right\rvert^{1/2}} \qquad i = 2, 3, \ldots, s$
Gram-Schmidt orthogonalizations of the basis vectors	(6.46)	$\mathbf{V}_1 = \dfrac{\mathbf{r}_1}{\lvert\mathbf{r}_1^T\mathbf{K}\mathbf{r}_1\rvert^{1/2}}$
	(6.47)	$\bar{\mathbf{V}}_i = \mathbf{r}_i - \sum_{j=1}^{i-1}\left(\mathbf{r}_i^T\mathbf{K}\mathbf{V}_j\right)\mathbf{V}_j \qquad i = 2, 3, \ldots, s$
	(6.48)	$\mathbf{V}_i = \dfrac{\bar{\mathbf{V}}_i}{\lvert\bar{\mathbf{V}}_i^T\mathbf{K}\bar{\mathbf{V}}_i\rvert^{1/2}} \qquad i = 2, 3, \ldots, s$

Evaluation of the eigenpairs

The second stage of evaluating the eigenpairs by the reduced eigenproblem involves the following steps:

- Calculation of the reduced matrices $\mathbf{K}_R$ and $\mathbf{M}_R$ [Eqs. (6.9)]

$$\mathbf{K}_R = \mathbf{r}_B^T \mathbf{K} \mathbf{r}_B \qquad \mathbf{M}_R = \mathbf{r}_B^T \mathbf{M} \mathbf{r}_B . \tag{6.49}$$

- Solution of the reduced $s{\times}s$ eigenproblem for λ_1, $\mathbf{y}_1$ [Eq. (6.10)]

$$\mathbf{K}_R \mathbf{y}_1 = \lambda_1 \mathbf{M}_R \mathbf{y}_1 , \tag{6.50}$$

  where λ_1 is the first eigenvalue and $\mathbf{y}_1$ is the corresponding eigenvector. The solution at this step is similar to that of Eq. (6.3), except that a much smaller system is considered.
- Evaluation of the mode shape [Eq. (6.5)]

$$\mathbf{\Phi} = \mathbf{r}_B \, \mathbf{y}_1 . \tag{6.51}$$

  The required eigenvalue is already given from Eq. (6.50) $\lambda = \lambda_1$.

6.4 Numerical Examples

The object of the numerical examples presented in this section is to illustrate typical results achieved by the CA approach for different design situations. The examples demonstrate the accuracy achieved by the approach compared with direct (exact) formulation of the modified equations. Small-scale examples are considered for illustrative purposes. Some of the examples describe common cases such as global changes in the design, and changes in the stiffness and in the mass for various sizes of the structure. Other examples demonstrate the performance of the approach in special cases such as switch of modes due to changes in the design, and problems with identical eigenvalues. In most examples the improved basis vectors have been determined by considering only Gram-Schmidt orthogonalizations of the approximate modes. The following notation has been used:

- **r***i*(init) = **r***i*(0) = initial *i*th mode shape.
- **r***i*(s) = modified *i*th mode shape, CA formulation with s basis vectors.
- **r***i*(ex) = modified *i*th mode shape, exact (direct) formulation.

Example 6.1

Consider the cantilever beam shown in Fig. 6.1. The beam is divided into 6 elements, the number of DOF is 12 and the initial stiffness of each element is given by $EI = 1.0$, $L = 1.0$. The mass is lumped at the joints, the initial mass at the interior joints is 1.0, and the initial mass at the end of the beam is 0.5. The change in the design is represented by additional mass of 2.0 at each of the joints. The results obtained by the CA approach with only 2 basis vectors for the first 4 mode shapes, shown in Table 6.2, demonstrate the high accuracy achieved by the approach.

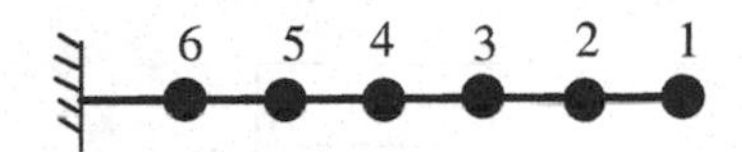

Fig. 6.1. Cantilever beam

Table 6.2. Results, mass changes in a cantilever beam

Mode	1		2		3		4	
Case	**r**1(2)	**r**1(ex)	**r**2(2)	**r**2(ex)	**r**3(2)	**r**3(ex)	**r**4(2)	**r**4(ex)
Shape	1.0000	1.0000	1.0000	1.0000	0.7413	0.7411	0.5584	0.5580
	0.7664	0.7664	0.0184	0.0185	-0.6118	-0.6115	-0.9817	-0.9809
	0.5401	0.5401	-0.7142	-0.7145	-0.7657	-0.7663	0.1348	0.1350
	0.3331	0.3331	-0.9820	-0.9822	0.2295	0.2309	1.0000	1.0000
	0.1615	0.1615	-0.7574	-0.7571	1.0000	1.0000	-0.4690	-0.4711
	0.0438	0.0438	-0.2793	-0.2789	0.6104	0.6091	-0.9876	-0.9865
λ	0.0025	0.0025	0.0990	0.0990	0.7743	0.7743	2.9352	2.9352

Example 6.2

To illustrate results obtained for switching of modes due to changes in the design, consider the simply supported beam divided into 10 elements shown in Fig. 6.2. The following parameters represent the initial design:

- Lumped masses, $M = 1.0$.
- The stiffness of each beam element is given by $EI = 1.0$, $L = 1.0$..
- The stiffness of the middle spring, $K = 0$ (there is no spring).

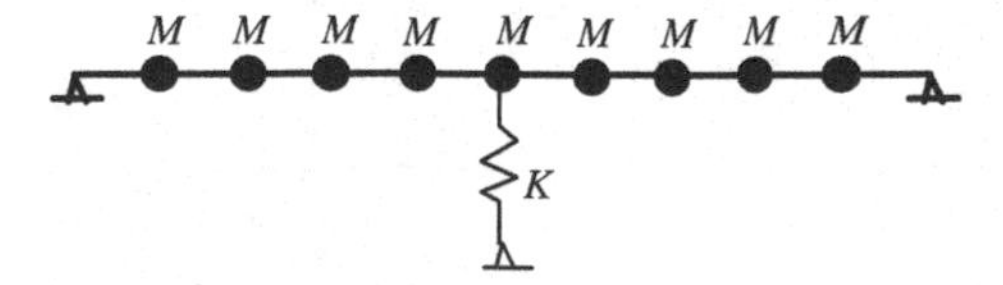

Fig. 6.2. Simply-supported beam with a spring

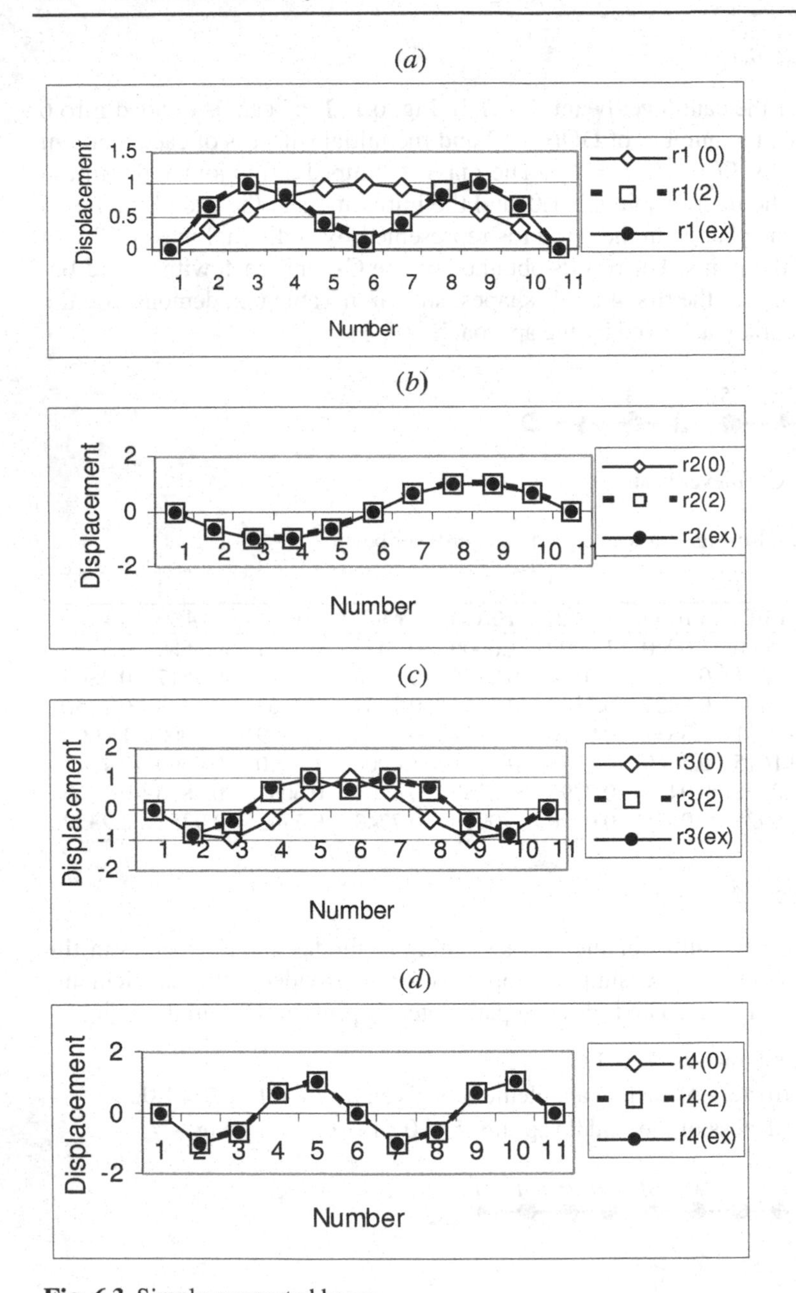

Fig. 6.3. Simply-supported beam
a. Original-mode 1, modified-mode 2 ***b***. Original-mode 2, modified-mode 1
c. Original-mode 3, modified-mode 4 ***d***. Original-mode 4, modified-mode 3

Table 6.3. Initial values, simply-supported beam

Case	r1(0)	r2(0)	r3(0)	r4(0)
Shape	0.31	0.62	0.81	-1.00
	0.59	1.00	0.95	-0.62
	0.81	1.00	0.31	0.62
	0.95	0.62	-0.59	1.00
	1.00	0.00	-1.00	0.00
	0.95	-0.62	-0.59	-1.00
	0.81	-1.00	0.31	-0.62
	0.59	-1.00	0.95	0.62
	0.31	-0.62	0.81	1.00
λ	0.01	0.16	0.79	2.48

Table 6.4. Results, simply-supported beam

Case	r1(2)*	r1(ex)*	r2(2)	r2(ex)	r3(2)	r3(ex)	r4(2)	r4(ex)
Shape	0.65	0.67	-0.62	-0.62	-0.77	-0.85	-1.00	-1.00
	1.00	1.00	-1.00	-1.00	-0.48	-0.43	-0.62	-0.62
	0.90	0.84	-1.00	-1.00	0.56	0.69	0.62	0.62
	0.44	0.39	-0.62	-0.62	1.00	1.00	1.00	1.00
	0.12	0.11	0.00	0.00	0.63	0.63	0.00	0.00
	0.44	0.39	0.62	0.62	1.00	1.00	-1.00	-1.00
	0.90	0.84	1.00	1.00	0.56	0.69	-0.62	-0.62
	1.00	1.00	1.00	1.00	-0.48	-0.43	0.62	0.62
	0.65	0.67	0.62	0.62	-0.77	-0.85	1.00	1.00
λ	0.36	0.35	0.16	0.16	2.96	2.90	2.48	2.48

* **r**1 means modified displacements corresponding to the original-mode 1

The modified design is represented by adding a spring with stiffness of $K = 12$ in the middle of the span. It is observed that there is a change in the topology of the structure. The initial eigenpairs are shown in Table 6.3. The modified eigenpairs determined by the CA approach with only 2 basis vectors, for the first 4 mode shapes, are shown in Table 6.4 and in Fig. 6.3. It is observed that high accuracy is achieved for the significant changes in the mode shapes. Moreover, there is a switch in the modes as follows: the original mode 1 becomes the modified mode 2, the original mode 2 becomes the modified mode 1, the original mode 3 becomes the modified mode 4 and the original mode 4 becomes the modified mode 3.

Example 6.3

To illustrate results for problems with identical eigenvalues, consider the system of masses and springs shown in Fig. 6.4. The initial lumped masses are given by $M = 1$ and the initial springs stiffness is given by $K = 1$, with

$K = 0$ for the middle spring (there is no middle spring). That is, the initial system consists of two independent subsystems. The modified design is represented by adding the middle spring with $K = 1$. In this case, it is necessary to consider basis vectors of the 2 subsystems.

Considering the first mode of each of the 2 subsystems as the only 2 basis vectors, the exact modified shape of mode 1 is obtained. For mode 2, an approximate solution is obtained with the usual 2 basis vectors. The initial eigenpairs and the modified eigenpairs obtained for the first 2 mode shapes, by the CA approach with only 2 basis vectors and by exact solutions, are shown in Fig. 6.5 and in Table 6.5.

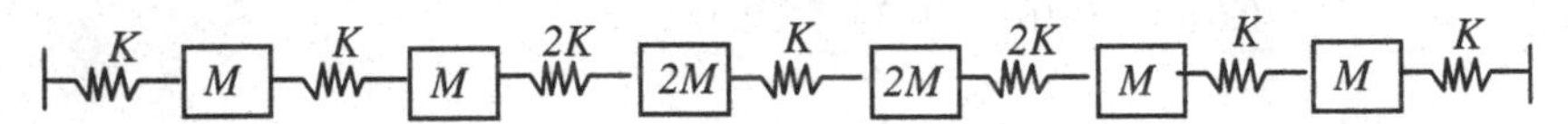

Fig. 6.4. A system of masses and springs

(*a*)

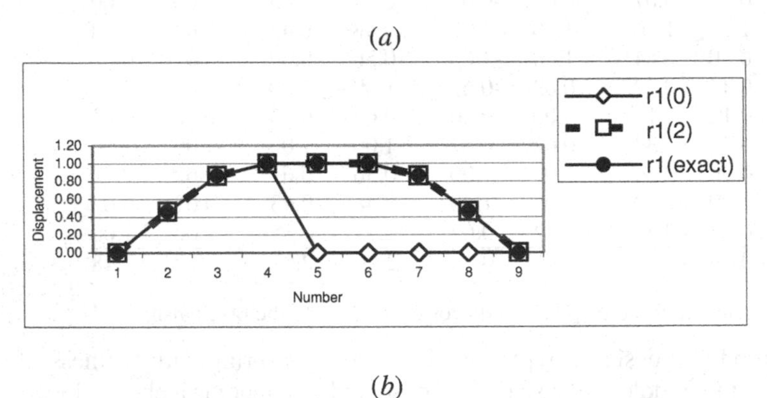

(*b*)

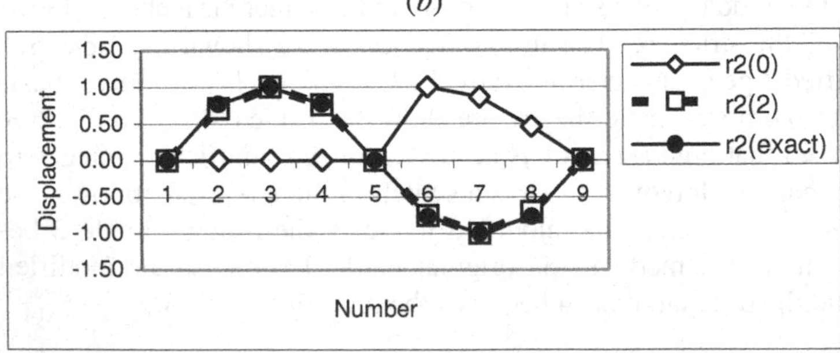

Fig. 6.5. A system of masses and springs: ***a***. Mode 1 ***b***. Mode 2

Table 6.5. Results for mode shapes 1, 2, system of masses and springs

Case	r1(0)	r1(2)	r1(ex)	r2(0)	r2(2)	r2(ex)
Shape	0.46	0.46	0.46	0.00	0.71	0.77
	0.86	0.86	0.86	0.00	1.00	1.00
	1.00	1.00	1.00	0.00	0.76	0.77
	0.00	1.00	1.00	0.00	0.00	0.00
	0.00	1.00	1.00	1.00	-0.76	-0.77
	0.00	0.86	0.86	0.86	-1.00	-1.00
	0.00	0.46	0.46	0.46	-0.71	-0.77
λ	0.139	0.139	0.139	0.139	0.699	0.697

Example 6.4

To demonstrate results for a larger structure, consider the fifty-story frame shown in Fig. 6.6. The inertia force at each joint is due to the frame self-weight and an additional concentrated mass of 25*ton* at an external joint and 50*ton* at an internal joint. The total number of DOF is 600 and the width and depth of all cross sections are 0.5*m*, 2.0*m*, respectively. The object is to evaluate the first 8 eigenpairs of a modified structure where the depth of all beams is 0.75*m* and the depths of the columns are as follows.

- Stories 1–10: 2.5*m*.
- Stories 11–20: 2.0*m*.
- Stories 21–30: 1.5*m*.
- Stories 31–40: 1.0*m*.
- Stories 41–50: 0.75*m*.

The initial eigenvalues λ(init), the modified exact eigenvalues λ(ex), and the modified approximate eigenvalues considering only Gram-Schmidt orthogonalizations of the mode shapes and 6 basis vectors, λ(CA6), are shown in Table 6.6. It is observed that small errors are obtained by the CA approach for the lower modes and larger errors for the higher modes.

To illustrate the effect of the improved basis vectors, the ratios of eigenvalues λ(CA)/ λ(ex) have been compared for the following cases, considering various numbers of basis vectors:

- Without Gram-Schmidt orthogonalizations of the basis vectors: without shifts (Fig. 6.7*a*) and with shifts (Fig. 6.7*b*).
- With Gram-Schmidt orthogonalizations of the basis vectors: without shifts (Fig. 6.8*a*) and with shifts (Fig. 6.8*b*).

Fig. 6.7 shows how the shifts improve the accuracy. However, for a large number of basis vectors errors might occur due to ill-conditioning.

To overcome this difficulty we use Gram-Schmidt orthogonalizations of the basis vectors (Fig. 6.8). The highest accuracy is achieved when Gram-Schmidt orthogonalizations of the basis vectors and shifts are considered.

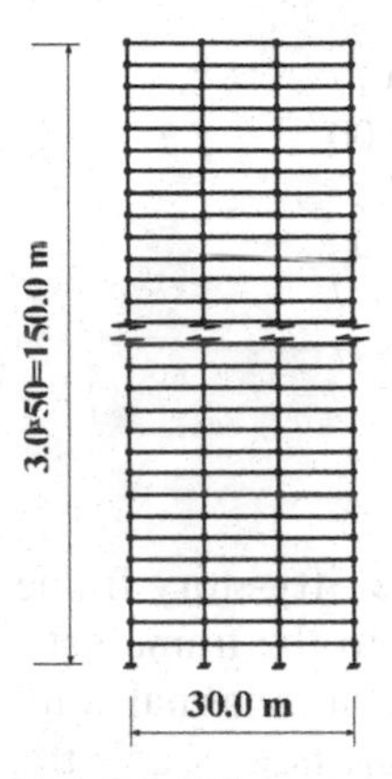

Fig. 6.6. Fifty-story frame

Table 6.6. First 8 eigenvalues with only Gram-Schmidt orthogonalizations of the approximate modes

Mode	λ(init)	λ(ex)	λ(CA6)	Error [%]
1	3.2	1.007	1.011	0.32
2	53.8	8.862	9.001	1.58
3	232.7	28.02	28.31	1.04
4	526.5	59.00	59.40	0.67
5	945.1	106.1	107.0	0.86
6	1483.6	173.6	180.0	3.68
7	2157.3	259.0	276.5	6.75
8	2971.1	371.9	408.3	9.79

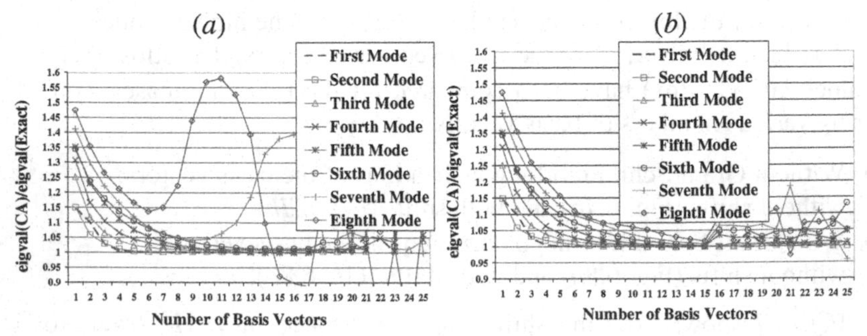

Fig. 6.7. Approximate/exact eigenvalues without Gram-Schmidt orthogonalizations of the basis vectors: ***a.*** without shifts ***b.*** with shifts

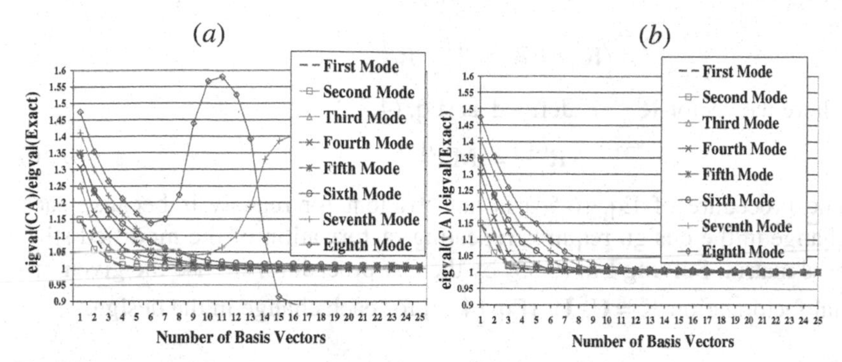

Fig. 6.8. Approximate/exact eigenvalues with Gram-Schmidt orthogonalizations of the basis vectors: ***a.*** without shifts ***b.*** with shifts

6.5 Reanalysis by Iterative Procedures

It has been noted that the general formulation of vibration reanalysis problems is given by Eq. (6.3). The modified equations solved by the CA approach are expressed in the form of Eq. (6.4)

$$(\mathbf{K}_0 + \Delta\mathbf{K}_0)\,\mathbf{\Phi} = \mathbf{R}, \tag{6.52}$$

where **R** is defined as

$$\mathbf{R} = \lambda\,\mathbf{M}\mathbf{\Phi}\,, \tag{6.53}$$

and the reduced eigenproblem to be solved is given by Eq. (6.10).

In this section, solutions of vibration reanalysis problems, using *iterative analysis* procedures presented in Sect. 4.3.2 and the CA approach, are introduced. The iterative procedures include inverse iteration (II), inverse iteration with shifts (IIS) and subspace iteration (SI). These procedures are not efficient for reanalysis of large scale systems, because each change in the design (sometimes, each iteration) requires factorization of the modified stiffness matrix, which involves much computational effort. We can reduce this effort by using the CA approach.

Inverse Iteration (II)

Considering the general inverse iteration formulation, the original reanalysis problem to be solved at the *k*th iteration cycle is [Eq. (2.46)]

$$\mathbf{K}\,\bar{\mathbf{r}}^{(k)} = \mathbf{M}\,\mathbf{r}^{(k\text{-}1)}. \tag{6.54}$$

Using the CA formulation, the problem to be solved is [Eq. (4.34)]

$$(\mathbf{K}_0 + \Delta\mathbf{K}_0)\bar{\mathbf{r}}^{(k)} = \mathbf{R}^{(k-1)}, \tag{6.55}$$

where the vector $\mathbf{R}^{(k-1)}$ is defined as [Eq. (4.35)]

$$\mathbf{R}^{(k-1)} = \mathbf{M}\,\mathbf{r}^{(k-1)}. \tag{6.56}$$

The procedure of Eq. (6.54) is not efficient for reanalysis because each change in the design requires repeating factorization of the modified stiffness matrix. Solving Eq. (6.55) by the CA procedure, we use the given initial factorization $\mathbf{K}_0 = \mathbf{U}_0^T\,\mathbf{U}_0$ [Eq. (4.2)] for each change in the design.

Inverse Iteration with Shifts (IIS)

Considering the original formulation of inverse iteration shifts, the reanalysis problem to be solved at the *k*th iteration cycle is [Eq. (2.58)]

$$(\mathbf{K} - \lambda^{(k-1)}\,\mathbf{M})\,\bar{\mathbf{r}}^{(k)} = \mathbf{M}\,\mathbf{r}^{(k-1)}. \tag{6.57}$$

The inverse iteration with shifts (IIS) procedure is not efficient for reanalysis because each change in the shift $\lambda^{(k-1)}$ (or in the design $\mathbf{K}$, $\mathbf{M}$) requires factorization of the modified matrix $(\mathbf{K} - \lambda^{(k-1)}\,\mathbf{M})$. That is, new factorization is required at each iteration cycle.

Using the CA formulation, the problem to be solved is [Eq. (4.36)]

$$(\mathbf{K}_0 + \Delta\mathbf{K}^{(k-1)})\,\bar{\mathbf{r}}^{(k)} = \mathbf{R}^{(k-1)}, \tag{6.58}$$

where $\Delta\mathbf{K}^{(k-1)}$ and $\mathbf{R}^{(k-1)}$ are defined as [Eqs. (4.35), (4.37)]

$$\Delta\mathbf{K}^{(k-1)} = \Delta\mathbf{K}_0 - \lambda^{(k-1)}\,\mathbf{M}, \tag{6.59}$$

$$\mathbf{R}^{(k-1)} = \mathbf{M}\,\mathbf{r}^{(k-1)}. \tag{6.60}$$

Again, instead of repeating the factorization of $(\mathbf{K} - \lambda^{(k-1)}\,\mathbf{M})$ at each iteration cycle, we use the given initial factorization $\mathbf{K}_0 = \mathbf{U}_0^T\,\mathbf{U}_0$ [Eq. (4.2)] to solve Eq. (6.58) by the CA procedure described in Sect. 6.3.

Subspace Iteration (SI)

Considering the original formulation of subspace iteration, the reanalysis problem to be solved at the *k*th iteration cycle is [Eq. (2.102)]

$$\mathbf{K}\,\bar{\mathbf{r}}^{(k+1)} = \mathbf{M}\,\mathbf{r}^{(k)}. \tag{6.61}$$

Similar to the previous iteration procedures, the subspace iteration (SI) is not efficient for reanalysis of multiple modified designs.

Using the CA formulation, the problem to be solved is [Eq. 4.38)]

$$(\mathbf{K}_0 + \Delta\mathbf{K}_0)\bar{\mathbf{r}}^{(k+1)} = \mathbf{R}^{(k)}, \tag{6.62}$$

where the vector $\mathbf{R}^{(k)}$ is defined as [Eq. (4.39)]

$$\mathbf{R}^{(k)} = \mathbf{M}\,\mathbf{r}^{(k)}. \tag{6.63}$$

Again, we use the given initial factorization of $\mathbf{K}_0$ to solve Eq. (6.62) by the CA procedure described in Sect. 6.3.

The various presented formulations are summarized in Table 6.7. The accuracy of the results is demonstrated by the following numerical examples. Efficiency considerations are discussed later in Sect. 10.1.

Table 6.7. Summary of modified equations

Procedure	Original formulation	CA formulation
General	(6.3) $\mathbf{K}\,\boldsymbol{\Phi} = \mathbf{R}$	(6.52) $(\mathbf{K}_0 + \Delta\mathbf{K}_0)\,\boldsymbol{\Phi} = \mathbf{R}$
II	(6.54) $\mathbf{K}\,\bar{\mathbf{r}}^{(k)} = \mathbf{M}\,\mathbf{r}^{(k\text{-}1)}$	(6.55) $(\mathbf{K}_0 + \Delta\mathbf{K}_0)\,\bar{\mathbf{r}}^{(k)} = \mathbf{R}^{(k\text{-}1)}$
IIS	(6.57) $(\mathbf{K} - \lambda^{(k\text{-}1)}\mathbf{M})\,\bar{\mathbf{r}}^{(k)} = \mathbf{M}\mathbf{r}^{(k\text{-}1)}$	(6.58) $(\mathbf{K}_0 + \Delta\mathbf{K}^{(k\text{-}1)})\,\bar{\mathbf{r}}^{(k)} = \mathbf{R}^{(k\text{-}1)}$
SI	(6.61) $\mathbf{K}\bar{\mathbf{r}}^{(k+1)} = \mathbf{M}\mathbf{r}^{(k)}$	(6.62) $(\mathbf{K}_0 + \Delta\mathbf{K}_0)\bar{\mathbf{r}}^{(k+1)} = \mathbf{R}^{(k)}$

Example 6.5

In this example the frame shown in Fig. 6.9 is solved for various numbers of stories, N. Consider 3 degrees of freedom at each story, the total number of DOF is $n = 3N$. The mass of the frame is lumped at the joints, with an initial mass $M_i = 1.0$ ($i = 1, \ldots, N$) in each story. The initial stiffness of all elements is determined by assuming $EI_i = 1.0$, $L_i = 1.0$ ($i = 1, \ldots, N$). The modified design is represented by the following parameters:

- $EI_1 = 1.5$, $EI_i = EI_{i\text{-}1} + 0.25$ ($i = 2, \ldots, N$).
- $M_1 = 5.0$, $M_N = 4.0$, $M_i = 0.2$ ($i = 2, \ldots, N - 1$).

To compare results obtained for various problem sizes, solutions of the general formulation have been obtained for the following numbers of stories: N = 14, 28, 56, 112 (336 DOF). The first 3 eigenvalues were calculated using the general formulation [λ(ex), Eq. (6.3)] and the reduced CA formulation [λ(CA), Eq. (6.10)] with various numbers of basis vectors.

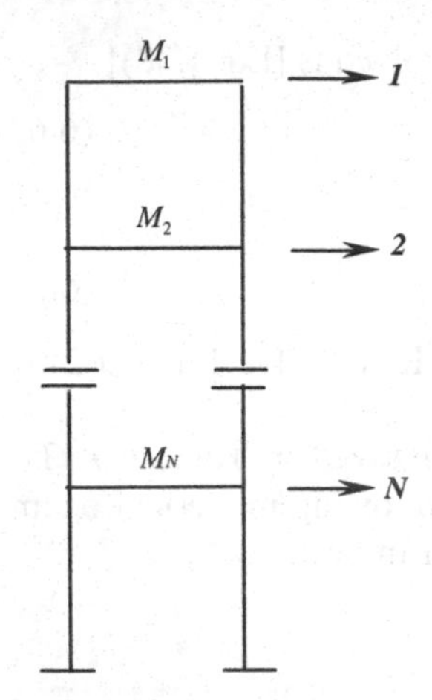

Fig. 6.9. *N*-story frame

Table 6.8. Summary of eigenvalues, general and reduced CA formulations

N (stories)	*n* (DOF)	Mode	Basis vectors	λ(CA)	λ(ex)	Error [%]
14	42	1	3	0.3042	0.3039	0.10
		2	3	6.6898	6.6814	0.13
		3	3	19.2279	19.1951	0.17
28	84	1	3	0.1913	0.1902	0.61
		2	3	2.5100	2.4888	0.85
		3	4	9.1537	9.1268	0.30
56	168	1	3	0.1255	0.1226	2.37
		2	3	1.0581	1.0189	3.85
		3	4	3.5898	3.5671	0.63
112	336	1	4	0.0799	0.0783	2.06
		2	4	0.4834	0.4662	3.68
		3	5	1.5426	1.5088	2.24

The results (Table 6.8) indicate that the number of basis vectors needed to achieve good accuracy is only slightly increased with the problem size.

To illustrate the first 3 mode shapes obtained by the above formulations, consider the 14-story frame (N = 14). The following notation is used for the results shown in Table 6.9 and Fig. 6.10:

- $\mathbf{r}$(init) = $\mathbf{r}$(0) = initial displacements.
- $\mathbf{r}$(CAs) = modified displacements, CA formulation with s basis vectors.
- $\mathbf{r}$(ex) = modified displacements, exact (or direct) formulation.

Table 6.9. Modified shapes, CA and exact formulations, 14-story frame

Mode	1		2		3	
Case	r(CA3)	r(ex)	r(CA3)	r(ex)	r(CA3)	r(ex)
Shape	1.00	1.00	-0.14	-0.15	-0.08	-0.08
	0.90	0.90	0.16	0.17	0.38	0.41
	0.80	0.79	0.45	0.48	0.79	0.81
	0.70	0.70	0.69	0.72	1.00	1.00
	0.61	0.61	0.86	0.88	0.98	0.97
	0.52	0.53	0.97	0.97	0.77	0.78
	0.45	0.45	1.00	1.00	0.44	0.47
	0.37	0.38	0.97	0.97	0.08	0.11
	0.31	0.31	0.90	0.90	-0.26	-0.23
	0.25	0.25	0.78	0.79	-0.53	-0.53
	0.19	0.19	0.64	0.65	-0.72	-0.74
	0.13	0.13	0.48	0.49	-0.84	-0.85
	0.08	0.08	0.31	0.33	-0.84	-0.85
	0.03	0.03	0.15	0.15	-0.60	-0.62
λ	0.30	0.30	6.69	6.68	19.23	19.20

Table 6.10. Solution by II and IIS procedures, second iteration cycle

Mode	1		2		3	
Case	r(CA2)	r(ex)	r(CA3)	r(ex)	r(CA4)	r(ex)
Shape	1.00	1.00	-0.27	-0.27	-0.08	-0.08
	0.91	0.90	0.03	0.05	0.34	0.41
	0.80	0.79	0.34	0.37	0.73	0.81
	0.70	0.70	0.60	0.62	0.96	1.00
	0.61	0.61	0.61	0.81	1.00	0.97
	0.52	0.53	0.93	0.93	0.86	0.78
	0.45	0.45	1.00	0.99	0.57	0.47
	0.37	0.38	1.00	1.00	0.20	0.12
	0.31	0.31	0.95	0.93	-0.18	-0.23
	0.25	0.25	0.85	0.84	-0.53	-0.52
	0.19	0.19	0.72	0.71	-0.80	-0.74
	0.13	0.13	0.57	0.55	-0.93	-0.85
	0.08	0.08	0.40	0.38	-0.91	-0.85
	0.03	0.03	0.20	0.19	-0.64	-0.62
λ	0.30	0.30	6.52	6.45	19.35	19.20

Consider now the Inverse Iteration (II) and Inverse Iteration with Shifts (IIS) procedures. Results for the second iteration cycle and various numbers of basis vectors are shown in Table 6.10 and in Fig. 6.11 for the following cases of initial values:

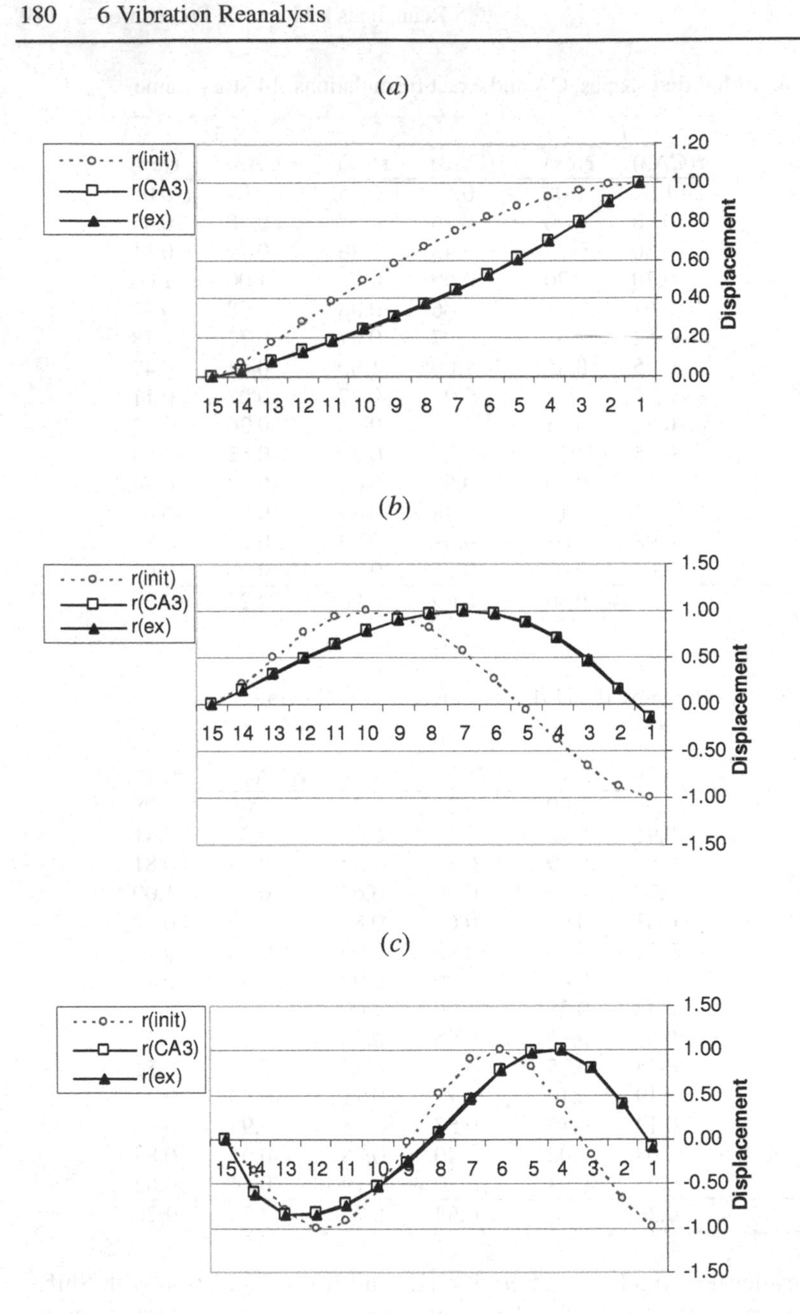

Fig. 6.10. Exact (ex) and CA3 formulations: ***a***. Mode 1 ***b***. Mode 2 ***c***. Mode 3

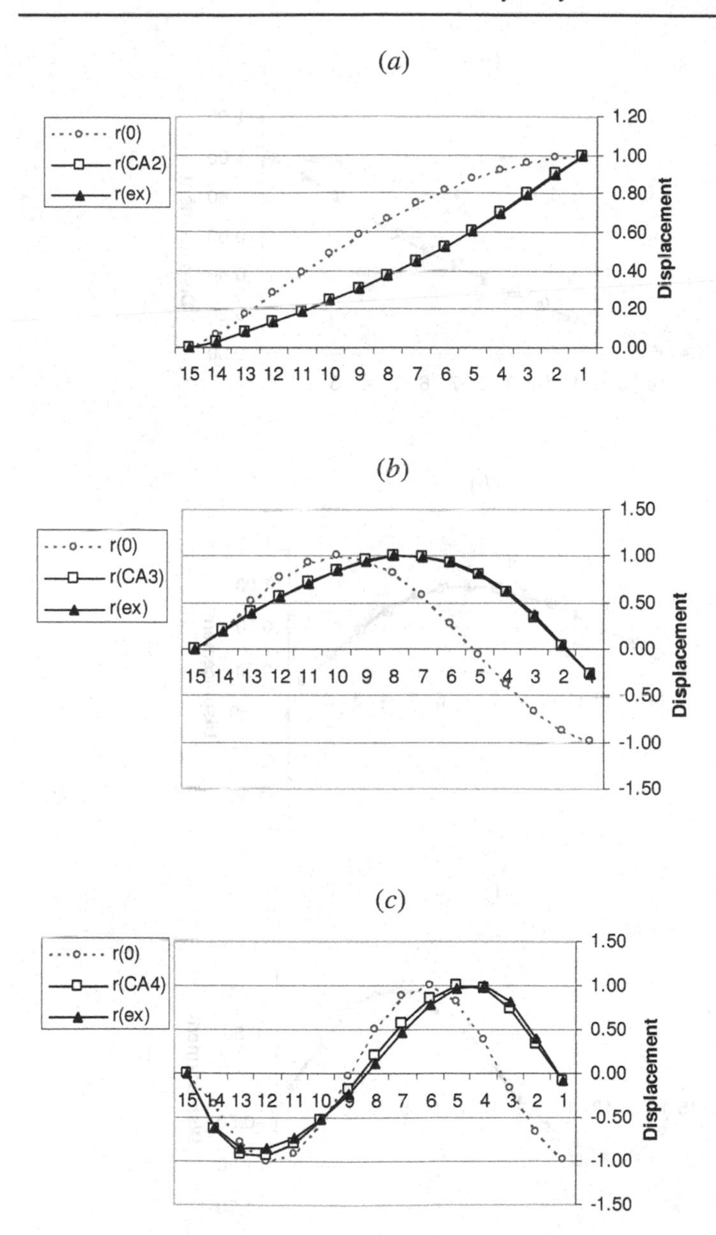

Fig. 6.11. Exact and CA formulations, II and IIS procedures, 2nd iteration cycle: ***a***. Mode 1 ***b***. Mode 2 ***c***. Mode 3

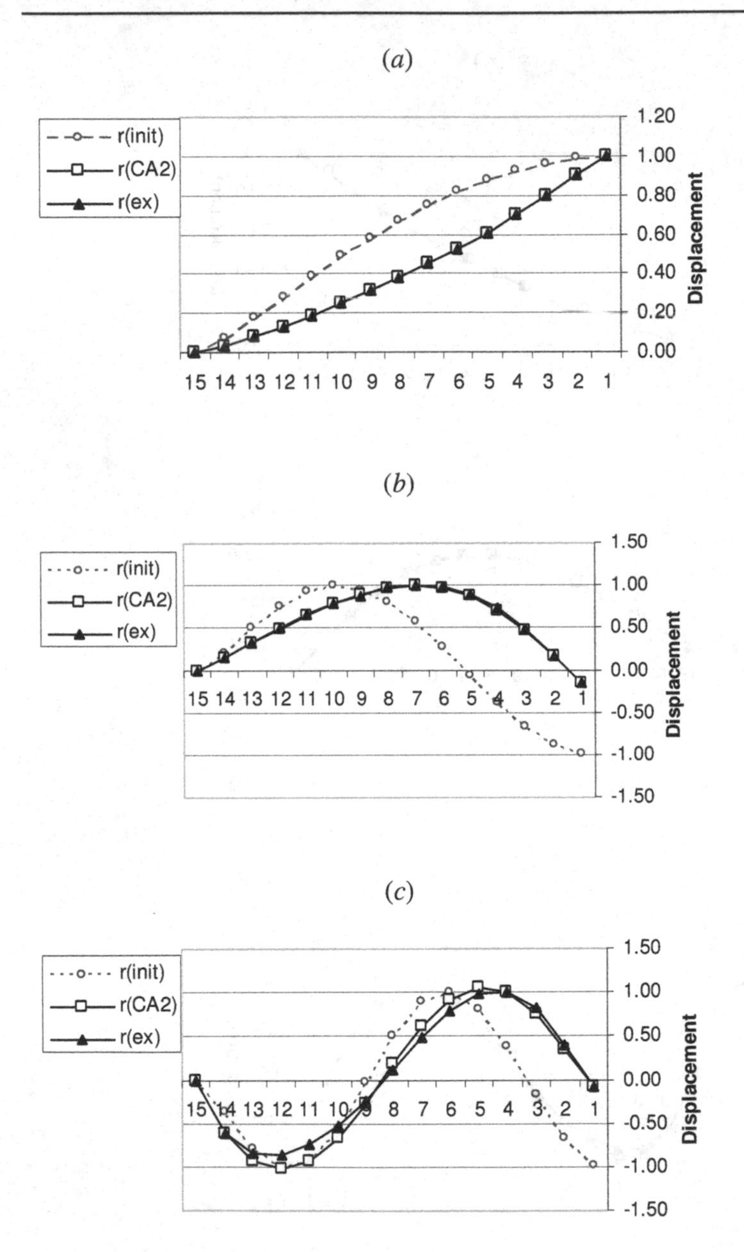

Fig. 6.12. Exact and CA formulations, SI procedure, first iteration cycle: ***a***. Mode 1 ***b***. Mode 2 ***c***. Mode 3

- Case *a*. No shift and the initial first-mode shape.
- Case *b*. Initial shift $\lambda^{(0)} = 10$ and the initial second-mode shape.
- Case *c*. Initial shift $\lambda^{(0)} = 20$ and the initial third-mode shape.

Finally, considering the Subspace Iteration (SI) procedure, results for the first iteration cycle and two basis vectors are shown in Table 6.11 and in Fig. 6.12.

It is observed that high accuracy is achieved by the various CA procedures with a small number of basis vectors.

Table 6.11. Fourteen-story frame, solution by SI, first iteration cycle

Mode	1		2		3	
Case	$\mathbf{r}$(CA2)	$\mathbf{r}$(ex)	$\mathbf{r}$(CA2)	$\mathbf{r}$(ex)	$\mathbf{r}$(CA2)	$\mathbf{r}$(ex)
Shape	1.00	1.00	-0.15	-0.15	-0.08	-0.08
	0.91	0.90	0.18	0.17	0.35	0.40
	0.80	0.79	0.49	0.49	0.75	0.82
	0.70	0.70	0.74	0.73	1.00	1.00
	0.61	0.61	0.90	0.90	1.06	0.98
	0.52	0.53	0.98	0.99	0.92	0.79
	0.45	0.45	1.00	1.00	0.61	0.48
	0.38	0.38	0.96	0.99	0.18	0.11
	0.31	0.31	0.89	0.89	-0.27	-0.23
	0.24	0.25	0.79	0.79	-0.67	-0.53
	0.19	0.19	0.66	0.66	-0.93	-0.74
	0.13	0.13	0.51	0.51	-1.01	-0.85
	0.08	0.08	0.34	0.33	-0.93	-0.85
	0.03	0.03	0.15	0.16	-0.61	-0.62
λ	0.30	0.30	6.68	6.68	19.66	19.66

References

1. Bathe KJ (1996) Finite element procedures. Prentice Hall, NJ
2. Chopra AK (2001) Dynamics of structures. Prentice Hall, NJ
3. Clough RW, Penzien JP (1993) Dynamics of structures. McGraw-Hill, NY
4. Grandhi R (1993) Structural optimization with frequency constraints - a review. AIAA J 31:2296-2303
5. Chen SH, Yang XW (2000) Extended Kirsch combined method for eigenvalue reanalysis. AIAA J 38:927-930
6. Chen SH, Yang XW, Lian HD (2000) Comparison of several eigenvalue reanalysis methods for modified structures. Struct and Multidis Opt 20:253-259
7. Wang BP, Pilkey WD, Palazzoto AB (1983) Reanalysis, modal synthesis and dynamic design. In: Noor AK, Pilkey WD (eds) State-of-the-Art Surveys on Finite Element Technology. ASME, New York

8. Chuarong Z, Yimin B (1990) Structural modification and vibration reanalysis. Comp Meth Appl Mech Engrg 83:99-108
9. Kirsch U (2002) Design-oriented analysis of structures. Kluwer Academic Publishers, Dordrecht
10. Kirsch U (2003) Approximate vibration reanalysis of structures. AIAA J 41:504-511
11. Rong F et al (2003) Structural modal reanalysis for topological modifications with extended Kirsch method. Comp Meth Appl Mech Engrg 192:697-707
12. Kirsch U, Bogomolni M (2004) Procedures for approximate eigenproblem reanalysis of structures. Int J Num Meth Engrg 60:1969-1986

7 Dynamic Reanalysis

It has been noted that, from the viewpoint of computational effort, the reduction of degrees of freedom is more important in dynamic problems than in static problems. The discretized model of a complicated system may have numerous degrees of freedom, and the solution must be performed successively at many different times to generate the time history of the response. The computational effort might become prohibitive in nonlinear dynamic reanalysis problems of large scale structures. On the other hand, the nonlinear dynamic response is usually smoother than the nonlinear static response, due to the effect of inertia forces. Therefore, convergence of the iteration is expected to be more rapid than in static analysis.

In this chapter solution of both linear and nonlinear dynamic reanalysis problems, by the CA approach, is presented. Linear dynamic reanalysis is introduced in Sect. 7.1 and nonlinear dynamic reanalysis is developed in Sect. 7.2. It will be shown that, in various problems, accurate results can be achieved with a reduced computational effort.

7.1 Linear Dynamic Reanalysis

In this section we consider the following reanalysis problem.

- Given an initial design, the corresponding stiffness matrix $\mathbf{K}_0$, mass matrix $\mathbf{M}_0$, damping matrix $\mathbf{C}_0$, and the load vector $\mathbf{R}_0(t)$; the initial displacement vector $\mathbf{r}_0(t)$, velocity vector $\dot{\mathbf{r}}_0(t)$ and acceleration vector $\ddot{\mathbf{r}}_0(t)$ are computed by the equations of motion [Eq. (3.1)]

$$\mathbf{M}_0\ddot{\mathbf{r}}_0(t) + \mathbf{C}_0\dot{\mathbf{r}}_0(t) + \mathbf{K}_0\mathbf{r}_0(t) = \mathbf{R}_0(t)\,. \tag{7.1}$$

- Assume a change in the design and corresponding changes $\Delta\mathbf{K}_0$ and $\Delta\mathbf{M}_0$ in the stiffness and mass matrices (we assume that matrix $\mathbf{C} = \mathbf{C}_0$ is unchanged) such that the modified matrices are given by

$$\mathbf{K} = \mathbf{K}_0 + \Delta\mathbf{K}_0 \qquad \mathbf{M} = \mathbf{M}_0 + \Delta\mathbf{M}_0. \tag{7.2}$$

- The objective is to evaluate the modified displacements $\mathbf{r}(t)$, velocities $\dot{\mathbf{r}}(t)$, and accelerations $\ddot{\mathbf{r}}(t)$, without solving the complete set of modified equations

$$\mathbf{M}\ddot{\mathbf{r}}(t) + \mathbf{C}\dot{\mathbf{r}}(t) + \mathbf{K}\mathbf{r}(t) = \mathbf{R}(t). \tag{7.3}$$

7.1.1 Solution by Direct Integration

Initial Analysis Formulation

As noted in Sect. 3.1.1, in implicit integration methods (e.g. the Houbolt [1], Wilson [2], and Newmark [3] methods) we use the equilibrium equations of motion at time $t+\Delta t$. The methods are unconditionally stable, and their effectiveness derives from the fact that to obtain accuracy in the integration, the time step Δt can be very large. However, a factorization of the stiffness matrix is required for the solution.

Considering the Newmark method, dynamic analysis of the initial structure involves the steps described in the following. For the initial values $\mathbf{K}_0$, $\mathbf{M}_0$, $\mathbf{C}_0$, ${}^0\mathbf{r}_0$, ${}^0\dot{\mathbf{r}}_0$, ${}^0\ddot{\mathbf{r}}_0$, we first select the time step Δt and the parameters $\delta \geq 0.5$ and $\alpha \geq 0.25(0.5+\delta)^2$. Then we calculate the integration constants $a_0, \dots, a_7$, which are functions of Δt, α, δ, by [Eq. (3.15)]

$$a_0 = \frac{1}{\alpha\Delta t^2} \quad a_1 = \frac{\delta}{\alpha\Delta t} \quad a_2 = \frac{1}{\alpha\Delta t} \quad a_3 = \frac{1}{2\alpha} - 1, \tag{7.4}$$

$$a_4 = \frac{\delta}{\alpha} - 1 \quad a_5 = \frac{\Delta t}{2}\left(\frac{\delta}{\alpha} - 2\right) \quad a_6 = \Delta t(1-\delta) \quad a_7 = \delta\Delta t,$$

and we form and triangularize the initial effective stiffness matrix [Eqs. (3.16), (3.17)]

$$\hat{\mathbf{K}}_0 = \mathbf{K}_0 + a_0\mathbf{M}_0 + a_1\mathbf{C}_0, \tag{7.5}$$

$$\hat{\mathbf{K}}_0 = \mathbf{L}_0\mathbf{D}_0\mathbf{L}_0^T. \tag{7.6}$$

For each time step we calculate the following quantities at time $t+\Delta t$:

- The effective loads [Eq. (3.18)]

$$ {}^{t+\Delta t}\hat{\mathbf{R}}_0 = {}^{t+\Delta t}\mathbf{R}_0 + \mathbf{M}_0(a_0{}^t\mathbf{r}_0 + a_2{}^t\dot{\mathbf{r}}_0 + a_3{}^t\ddot{\mathbf{r}}_0) + \mathbf{C}_0(a_1{}^t\mathbf{r}_0 + a_4{}^t\dot{\mathbf{r}}_0 + a_5{}^t\ddot{\mathbf{r}}_0)\,. \quad (7.7) $$

- The displacements [Eq. (3.19)]

$$ \mathbf{L}_0\mathbf{D}_0\mathbf{L}_0^T\ {}^{t+\Delta t}\mathbf{r}_0 = {}^{t+\Delta t}\hat{\mathbf{R}}_0\,. \quad (7.8) $$

- The accelerations and the velocities [Eqs. (3.20), (3.21)]

$$ {}^{t+\Delta t}\ddot{\mathbf{r}}_0 = a_0({}^{t+\Delta t}\mathbf{r}_0 - {}^t\mathbf{r}_0) - a_2{}^t\dot{\mathbf{r}}_0 - a_3{}^t\ddot{\mathbf{r}}_0)\,, \quad (7.9) $$

$$ {}^{t+\Delta t}\dot{\mathbf{r}}_0 = {}^t\dot{\mathbf{r}}_0 + a_6{}^t\ddot{\mathbf{r}}_0 + a_7{}^{t+\Delta t}\ddot{\mathbf{r}}_0\,. \quad (7.10) $$

Efficient Reanalysis

To evaluate the dynamic response of a modified design, represented by **K**, **M** [Eq. (7.2)], we first have to factorize the modified effective stiffness matrix $\hat{\mathbf{K}} = \mathbf{K} + a_0\mathbf{M} + a_1\mathbf{C}$ [Eq. (3.16)]. It has been noted that the major computational cost in reanalysis is involved in this factorization. Equations (7.4), (7.5) show that for the given integration constants a_0, ..., a_7, dynamic analysis of an initial design involves only a single factorization of the modified matrix $\hat{\mathbf{K}}$. However, a new factorization of the matrix is required for any change in the design, and corresponding changes $\Delta\mathbf{K}_0$, $\Delta\mathbf{M}_0$.

Using the CA approach, we express the matrix $\hat{\mathbf{K}}$ as

$$ \hat{\mathbf{K}} = \hat{\mathbf{K}}_0 + (\Delta\mathbf{K}_0 + a_0\,\Delta\mathbf{M}_0) = \hat{\mathbf{K}}_0 + \Delta\hat{\mathbf{K}}\,, \quad (7.11) $$

where $\hat{\mathbf{K}}_0$ is given by Eqs. (7.5), and $\Delta\hat{\mathbf{K}} = \Delta\mathbf{K}_0 + a_0\Delta\mathbf{M}_0$. For each time step we calculate at time $t + \Delta t$ the effective loads ${}^{t+\Delta t}\hat{\mathbf{R}}$ by Eq. (3.18) and evaluate the displacements ${}^{t+\Delta t}\mathbf{r}$ by solving the set of equations

$$ (\hat{\mathbf{K}}_0 + \Delta\hat{\mathbf{K}})\,{}^{t+\Delta t}\mathbf{r} = {}^{t+\Delta t}\hat{\mathbf{R}}\,. \quad (7.12) $$

The procedure of linear reanalysis described in Sect 5.2 can be used for this purpose. Once $\hat{\mathbf{K}}_0$ is factorized [Eq. (7.6)], it is not necessary to repeat factorizations for each change in the design. The accelerations ${}^{t+\Delta t}\ddot{\mathbf{r}}$ and the velocities ${}^{t+\Delta t}\dot{\mathbf{r}}$ are then evaluated by Eqs. (3.20), (3.21). Using the CA approach for the present formulation, the savings in the computational effort might not be significant because only a single factorization is needed for each change in the design. More significant savings are expected in other formulations, where multiple factorizations are needed for reanalysis of any specific modified design.

Example 7.1

To demonstrate reanalysis by direct integration using the CA approach, consider the eight-story plane frame shown in Fig. 7.1. The objective is to evaluate the response of the structure for the loading (ground acceleration) of the El Centro earthquake shown in Fig. 7.2, scaled to have a 10% probability of occurrence in 50 years [4].

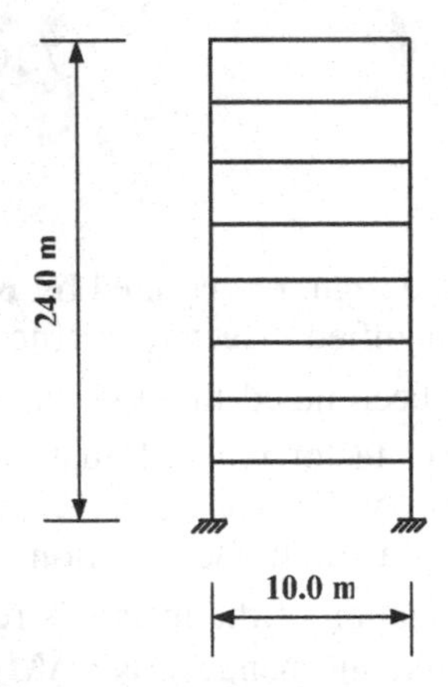

Fig. 7.1. Eight-story frame

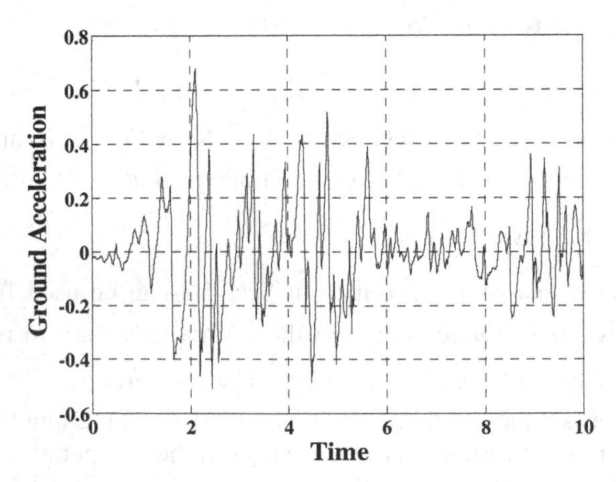

Fig. 7.2. El Centro Earthquake

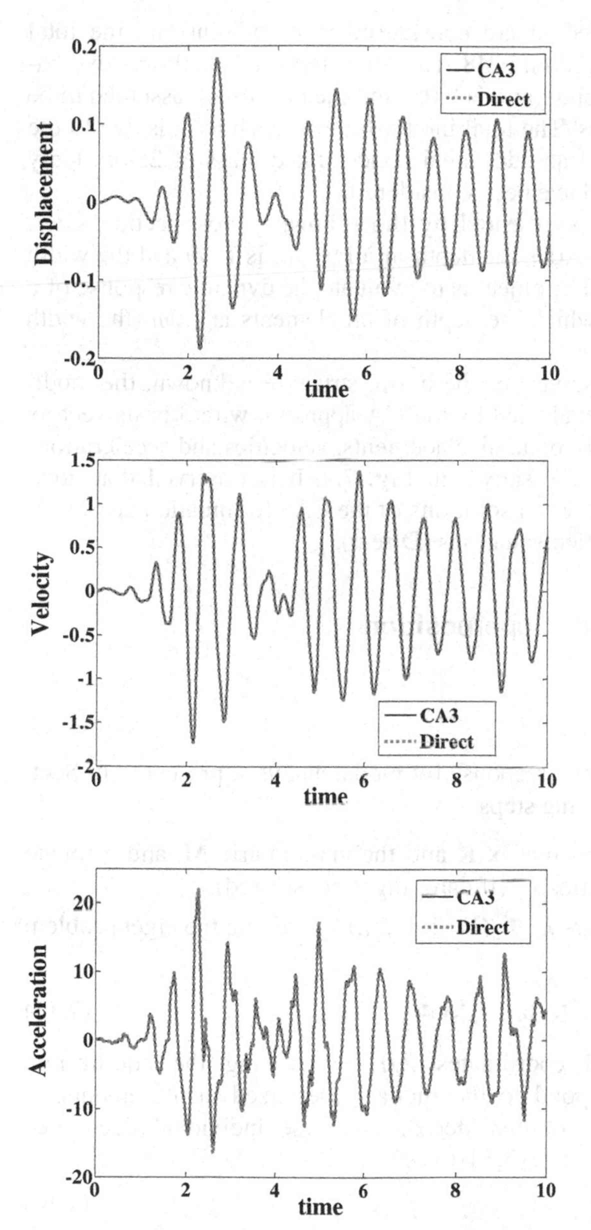

Fig. 7.3. Horizontal displacements, velocities and accelerations at the top of the frame, CA3 and direct formulation of the modified equations

Three degrees of freedom are considered at each joint and the total number of degrees of freedom is 48. The time step is $t = 0.005sec$, the material is elastic, the damping ratio is 0.05, and the masses are assumed to be concentrated at the joints. The total inertia force at each joint is due to the structure self-weight and an additional concentrated mass of 25*ton*. Only horizontal inertia forces have been considered.

The initial design is represented by the following cross-section sizes. The depth of columns is 2.0*m*, the depth of all beams is 1.5*m* and the width of all elements is 0.5*m*. The object is to evaluate the dynamic response of a modified structure, for which the depth of all elements is 1.0*m* (the width is unchanged).

Assuming that the response of the initial structure is known, the modified response has been evaluated by the CA approach with 3 basis vectors (CA3). The resulting horizontal displacements, velocities and accelerations at the top of the frame are shown in Fig. 7.3. It is observed that good agreement is obtained between solutions of the CA3 formulation and direct formulation of the modified equations (Direct).

7.1.2 Solution by Mode Superposition

Analysis Formulation

Calculation of the dynamic response by modal analysis, presented in Sect. 3.1.2, involves the following steps:

- Determine the stiffness matrix **K** and the mass matrix **M**, and estimate the modal damping ratios ζ_i (if damping is considered).
- Calculate the eigenpairs $\lambda_i, \mathbf{\Phi}_i$ $(i = 1, \ldots, n)$ by solving the eigenproblem [Eq. (3.26)]

$$\mathbf{K}\mathbf{\Phi} = \mathbf{\Lambda}\,\mathbf{M}\mathbf{\Phi}\,. \tag{7.13}$$

- Calculate the modal coordinates $Z_i(t)$ by solving the equilibrium equations that correspond to the modal generalized displacements. If damping effects are not considered, solve the individual decoupled equilibrium equations [Eq. (3.34)]

$$\ddot{Z}_i(t) + \lambda_i Z_i(t) = T_i(t)\,, \tag{7.14}$$

where $T_i(t)$ is defined as

$$T_i(t) = \mathbf{\Phi}_i^T \mathbf{R}(t)\,. \tag{7.15}$$

If damping effects are considered, solve the individual equations [Eq. (3.46)]

$$\ddot{Z}_i(t) + 2\omega_i \zeta_i \dot{Z}_i(t) + \omega_i^2 Z_i(t) = T_i(t)\,, \tag{7.16}$$

where ω_i is the ith frequency. This can be accomplished by one of several methods such as numerical finite difference method, numerical integration using Duhamel integral, or response spectra analysis.

- Calculate the nodal displacements, combining the contribution of all the modes, to determine the total response [Eq. (3.28)]

$$\mathbf{r}(t) = \sum_{i=1}^{n} \mathbf{\Phi}_i \, Z_i(t)\,. \tag{7.17}$$

- Calculate the element forces using the element stiffness properties.

Efficient Reanalysis

Solving the reanalysis problem by mode superposition, a significant part of the computational effort is involved in repeated solutions of modified eigenproblems [Eq. (7.13)]. It has been noted that some methods for solving the eigenproblem (e.g. inverse iteration with shifts) require repeated factorizations of the coefficient matrix at each iteration cycle, even when the stiffness matrix is unchanged. Solution by the CA approach, as described in Chap. 6, can significantly reduce the computational effort [5].

Example 7.2

To illustrate solution by mode superposition using the CA approach, consider again the eight-story frame of example 7.1 (Fig. 7.1). The damping ratio is 0.05 and the masses are assumed to be concentrated at the joints. The total inertia force at each joint is due to the frame self-weight and an additional concentrated mass of 25*ton*. Only horizontal inertia forces are considered and the time step is t = 0.005*sec*. The depth of all elements is 1.0*m* and the width of all elements is 0.5*m*. Again, the assumed loading is the ground acceleration of the El Centro earthquake (Fig. 7.2).

Analysis of the Initial Structure. Employing first dynamic analysis of the initial structure, the following two cases have been solved:

- Consideration of all 16 mode-shapes
- Consideration of only the first 4 mode-shapes.

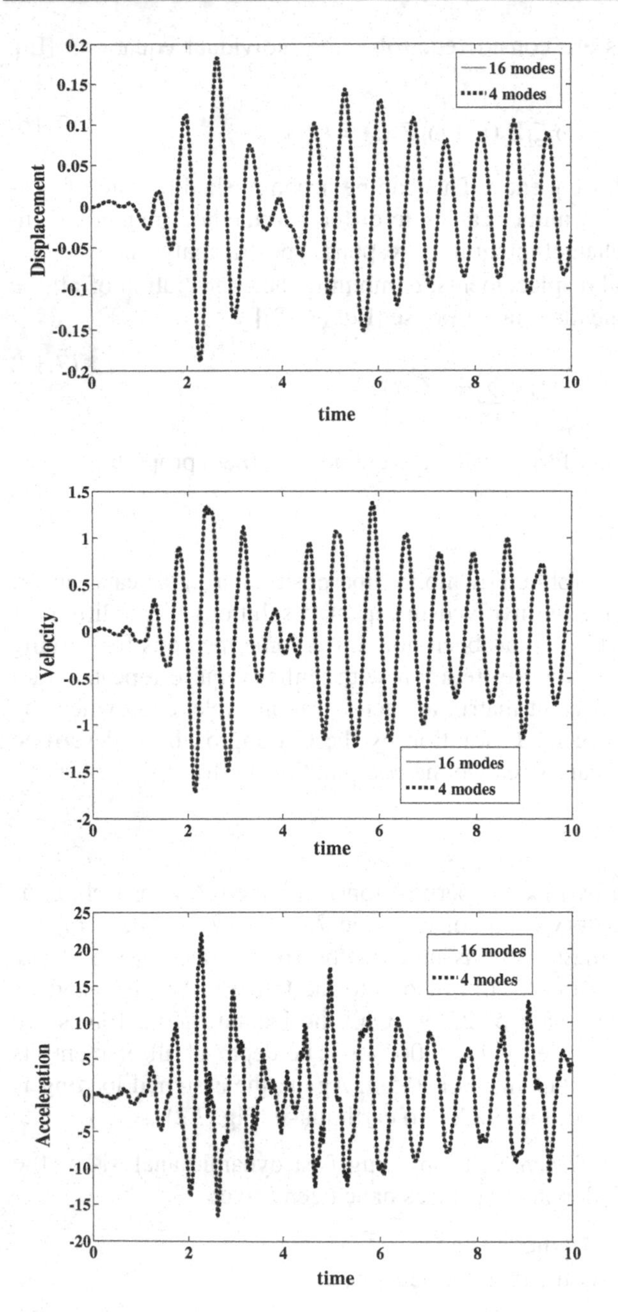

Fig. 7.4. Initial response at the top of the frame, analysis by mode superposition

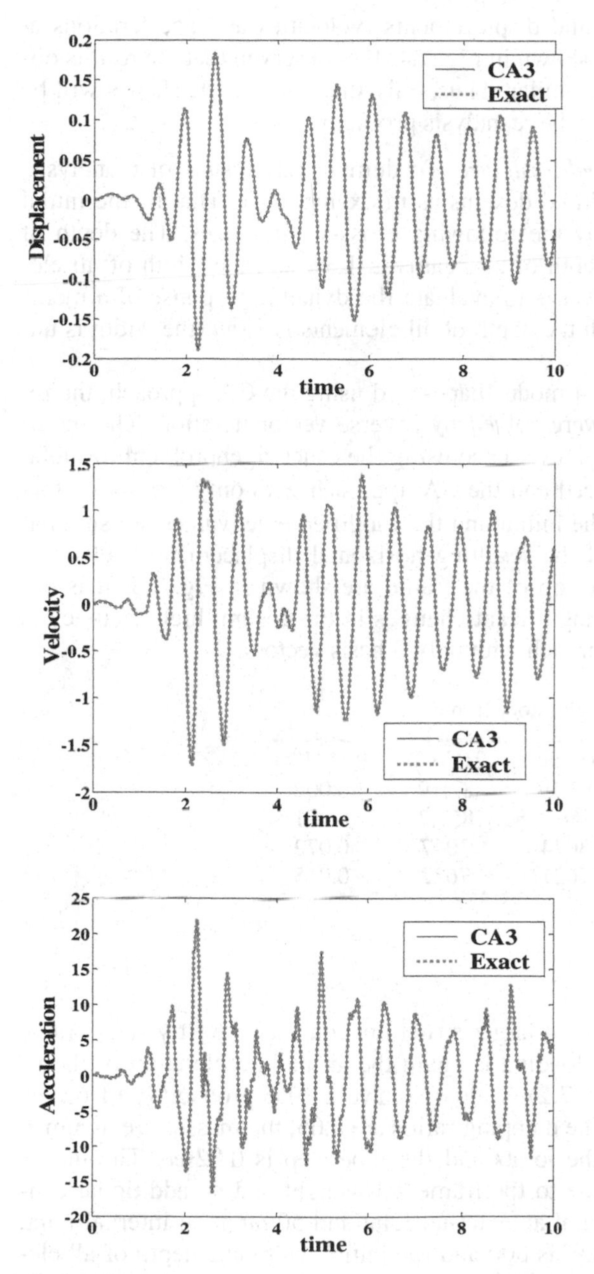

Fig. 7.5. Modified response at the top of the frame, CA3 and exact formulations

The resulting horizontal displacements, velocities and accelerations at the top of the frame are shown in Fig. 7.4. It is observed that the results obtained in both cases are similar. Thus, only the first 4 mode shapes will be considered in solution of the reanalysis problem.

Reanalysis of a Modified Structure. To demonstrate results of reanalysis, assume initial and modified designs as in example 7.1. That is, the initial design is represented by the following cross-section sizes. The depth of columns is 2.0*m*, the depth of all beams is 1.5*m* and the width of all elements is 0.5*m*. The object is to evaluate the dynamic response of a modified structure, for which the depth of all elements is 1.0*m* (the width is unchanged).

Considering the first 4 mode-shapes and using the CA approach, the reduced eigenproblems were solved by inverse vector iteration. The modified eigenpairs have been evaluated using the exact eigenproblem formulation [λ(Exact), $\mathbf{\Phi}$ (Exact)] and the CA approach with only 3 basis vectors [λ(CA3), $\mathbf{\Phi}$ (CA3)]. The initial and the modified eigenvalues are summarized in Table 7.1, and the resulting horizontal displacements, velocities and accelerations at the top of the frame are shown in Fig. 7.5. It is observed that despite the significant changes in the design, high accuracy is achieved by the CA approach with only 3 basis vectors.

Table 7.1. Eigenvalues, eight-story frame

Mode	λ(Initial)	λ(Exact)	λ(CA3)	Error [%]
1	267.33	82.19	82.19	0.002
2	3180.3	856.7	857.3	0.071
3	13370	3044	3047	0.072
4	38267	7621	7622	0.015

Example 7.3

To illustrate reanalysis of a larger structure, consider the fifty-story frame shown in Fig. 7.6. The loading is again the ground acceleration of the El Centro earthquake (Fig. 7.2), scaled to have a 10% probability of occurrence in 50 years [4]. The damping ratios are 0.05, the masses are assumed to be concentrated at the joints and the time step is 0.02*sec*. The inertia force at each joint is due to the frame self-weight and an additional concentrated mass of 25*ton* at an external joint and 50*ton* at an internal joint. The total number of DOF is 600 and the initial width and depth of all elements are 0.5*m*, 2.0*m*, respectively.

For the modified structure the depth of all beams is $0.75m$ and the depths of columns are as follows.

- Stories 1–10: $2.5m$.
- Stories 11–20: $2.0m$.
- Stories 21–30: $1.5m$.
- Stories 31–40: $1.0m$.
- Stories 41–50: $0.75m$.

The objective is to evaluate the dynamic response of the modified structure, considering the first 8 mode shapes. The effect of improved basis vectors on the accuracy of the results was demonstrated in example 6.4.

The modified response is evaluated by the CA approach with 6 basis vectors (CA6) and by exact formulation of the modified equations (Exact). The initial response (Initial) and the modified response (CA6 and Exact) at the top of the frame are shown in Fig. 7.7. It is observed that high accuracy is achieved by the CA approach with only 6 basis vectors for the very large changes in the design and in the response of the structure.

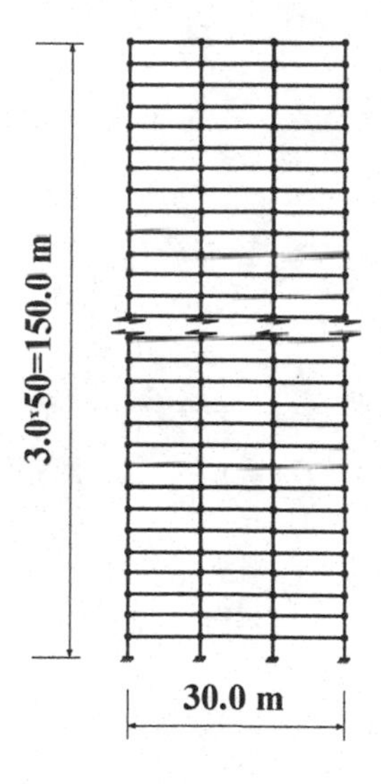

Fig. 7.6. Fifty-story frame

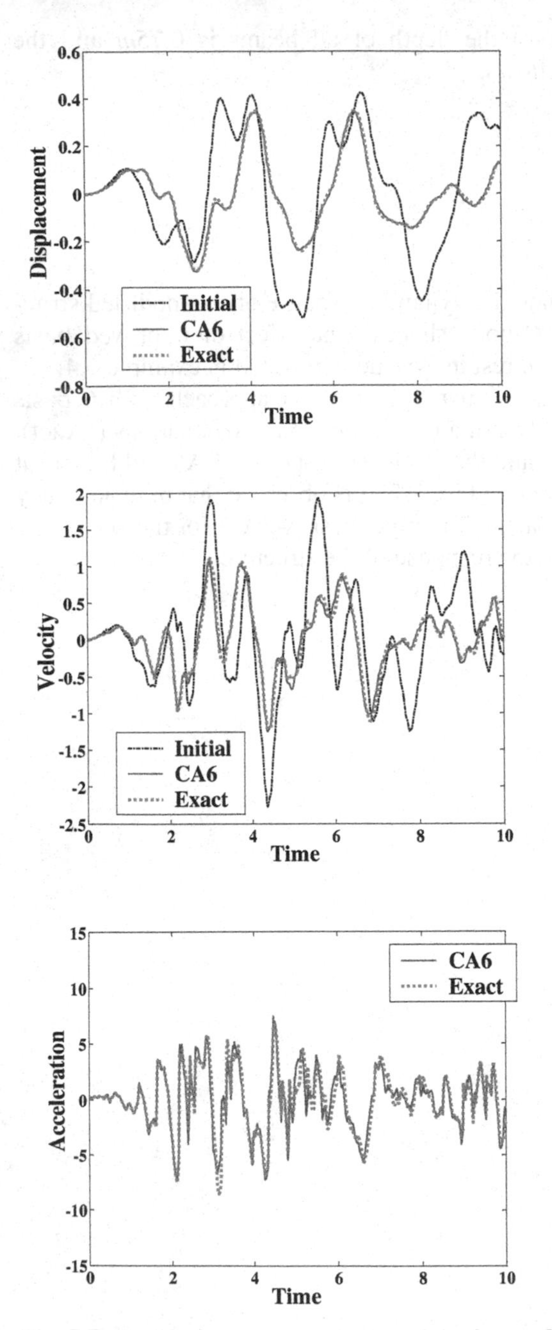

Fig. 7.7. Modified response at the top of the frame, CA6 and Exact formulations

7.2 Nonlinear Dynamic Reanalysis

7.2.1 Solution by Implicit Integration

Consider the common trapezoidal rule, which is Newmark's method with $\delta = 0.5$ and $\alpha = 0.25$. Neglecting the effect of damping and using the modified Newton-Raphson method, it has been shown in Sect. 4.4.2 that the equations to be solved at the kth iteration are [Eq. (4.56)]

$$(\hat{\mathbf{K}}_0 + {}^t\Delta\hat{\mathbf{K}})\,\delta\mathbf{r}^{(k)} = {}^{t+\Delta t}\hat{\mathbf{R}}^{(k)}\,, \tag{7.18}$$

where matrix $\hat{\mathbf{K}}_0$ is the initial effective stiffness matrix, defined as

$$\hat{\mathbf{K}}_0 = \mathbf{K}_0 + \frac{4}{\Delta t^2}\mathbf{M}_0\,. \tag{7.19)}$$

The modified effective stiffness matrix is given by [see Eq. (4.54)]

$${}^t\hat{\mathbf{K}} = {}^t\mathbf{K} + \frac{4}{\Delta t^2}\mathbf{M}\,, \tag{7.20}$$

where ${}^t\mathbf{K}$ is the stiffness matrix considered for the solution of Eq. (7.18), and ${}^t\Delta\hat{\mathbf{K}} = {}^t\hat{\mathbf{K}}({}^t\mathbf{K},\mathbf{M}) - \hat{\mathbf{K}}_0(\mathbf{K}_0,\mathbf{M}_0)$ [see Eq. (4.57)]. The effective load vector is given by [Eq. (4.55)]

$${}^{t+\Delta t}\hat{\mathbf{R}}^{(k)} = {}^{t+\Delta t}\mathbf{R}_0 - {}^{t+\Delta t}\mathbf{R}_I^{(k-1)} - \mathbf{M}\left[\frac{4}{\Delta t^2}\left({}^{t+\Delta t}\mathbf{r}^{(k-1)} - {}^t\mathbf{r}\right) - \frac{4}{\Delta t}\,{}^t\dot{\mathbf{r}} - {}^t\ddot{\mathbf{r}}\right]. \tag{7.21}$$

It is observed that the effective stiffness matrix ${}^t\hat{\mathbf{K}}$ might change at each load step whereas the effective load vector ${}^{t+\Delta t}\hat{\mathbf{R}}^{(k)}$ is changed at each iteration cycle. In addition, any change in the design results in corresponding changes in both ${}^t\hat{\mathbf{K}}$ and ${}^{t+\Delta t}\hat{\mathbf{R}}^{(k)}$. In summary, the iterative equations are similar to the nonlinear static equations, except that ${}^t\hat{\mathbf{K}}$ and ${}^{t+\Delta t}\hat{\mathbf{R}}^{(k)}$ contain contributions from the inertia of the system. Thus, the CA procedure might prove useful for the solution process.

7.2.2 Solution by Mode Superposition

Assuming proportional damping and a mass matrix which is constant in time, the solution of nonlinear dynamic reanalysis problems by mode superposition involves the following calculations at each time step:

- Calculation of the eigenpairs ${}^t\mathbf{\Phi}_i$, ${}^t\lambda_i$ at time t by solving the modified eigenproblem [Eqs. (4.63), (4.64), (4.65)]

$$(\mathbf{K}_0 + {}^t\Delta\mathbf{K})\,{}^t\mathbf{\Phi}_i = {}^t\mathbf{R}_i \qquad i = 1, \ldots, n, \tag{7.22}$$

$${}^t\mathbf{R}_i = {}^t\lambda_i\ \mathbf{M}\,{}^t\mathbf{\Phi}_i\,, \tag{7.23}$$

$${}^t\Delta\mathbf{K} = \Delta\mathbf{K}_0 + {}^t\Delta\mathbf{K}_{NL}. \tag{7.24}$$

The changes $\Delta\mathbf{K}_0$ in the elastic stiffness matrix due to changes in the design variables are constant for any given design. The changes ${}^t\Delta\mathbf{K}_{NL}$ due to the nonlinear behavior might be different for the various time steps. The tangent stiffness matrix at time t, ${}^t\mathbf{K}$, expressed as

$${}^t\mathbf{K} = \mathbf{K}_0 + {}^t\Delta\mathbf{K}, \tag{7.25}$$

is calculated by [Eq. (4.59)]

$${}^t\mathbf{K} = \frac{\partial\,{}^t\mathbf{F}_R}{\partial\mathbf{r}}. \tag{7.26}$$

- Solution of the uncoupled equations of motion [Eq. (4.60)]

$$\mathbf{I}\ {}^t\Delta\ddot{\mathbf{Z}} + {}^t\mathbf{C}_d\ {}^t\Delta\dot{\mathbf{Z}} + {}^t\mathbf{\Omega}^2\ {}^t\Delta\mathbf{Z} = {}^t\mathbf{\Phi}^T\ {}^t\Delta\mathbf{R}\,, \tag{7.27}$$

where ${}^t\Delta\mathbf{Z}$ are the generalized displacements and ${}^t\mathbf{\Phi}$ is the matrix of eigenvectors. The identity matrix $\mathbf{I} = {}^t\mathbf{\Phi}^T\mathbf{M}\,{}^t\mathbf{\Phi}$ is the mass matrix in normalized coordinates. The damping matrix $\mathbf{C}_d = {}^t\mathbf{\Phi}^T\mathbf{C}\,{}^t\mathbf{\Phi}$ and the stiffness matrix ${}^t\mathbf{\Omega}^2 = {}^t\mathbf{\Phi}^T\ {}^t\mathbf{K}\ {}^t\mathbf{\Phi}$ in these coordinates are diagonal.

- Calculation of the nodal displacements by [Eqs. (4.61), (4.62)]

$${}^t\Delta\mathbf{r} = {}^t\mathbf{\Phi}\ {}^t\Delta\mathbf{Z}\,, \tag{7.28}$$

$${}^{t+\Delta t}\mathbf{r} = {}^t\mathbf{r} + {}^t\Delta\mathbf{r}. \tag{7.29}$$

As noted earlier, a significant part of the computational effort is involved in repeated solutions of modified eigenproblems in the nonlinear region. A solution procedure based on the CA approach, similar to the procedure used for linear dynamic reanalysis, has been developed [6, 7].

Example 7.4

The object of this simple example is to demonstrate the effectiveness of the CA procedure by comparing its results with those obtained by exact reanalysis, and by an approximate procedure where the initial mode shapes are not updated during the solution process. Consider the clamped beam shown in Fig. 7.8, subjected to a uniformly distributed load of 275*kN/m* constant in time. The beam is divided into 20 elements, the number of DOF is 38 (rotation and deflection at each joint) and the time step is Δt = 0.0001*sec*. The inertia forces are due to the mass of the beam, and damping is not considered. The width and the depth of the beam are 0.5*m*, 1.0*m*, respectively. The assumed moment-curvature relation is elasto-plastic with no hardening, as shown in Fig. 7.9, the modulus of elasticity is 3 $10^7 kN/m^2$, the elastic limit stress is σ_Y =20000kN/m^2, and the plastic hinge moment is M_P = 2500*kNm*. The displacements at the middle of the beam obtained by the following methods are shown in Fig. 7.10:

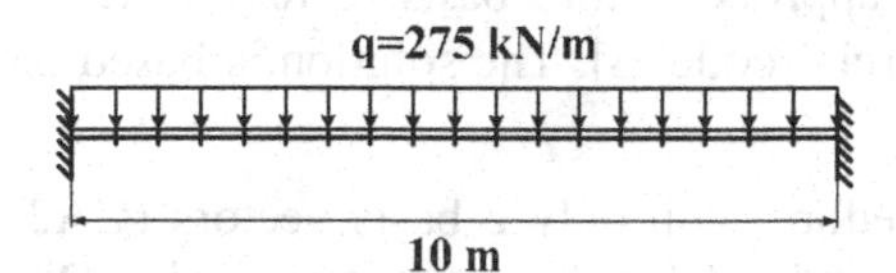

Fig. 7.8. Clamped beam

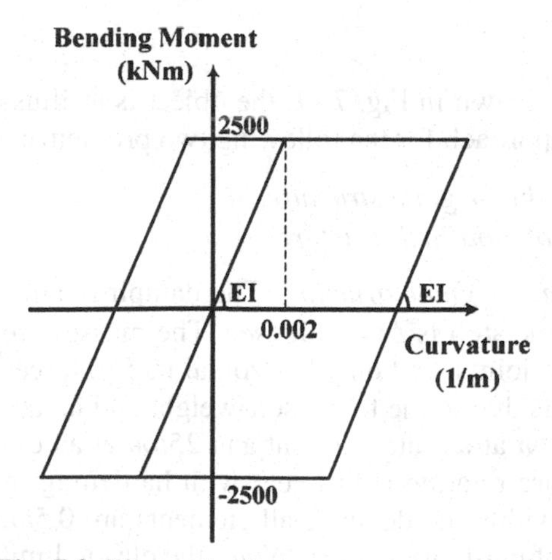

Fig. 7.9. Elasto-plastic moment-curvature relation with no hardening

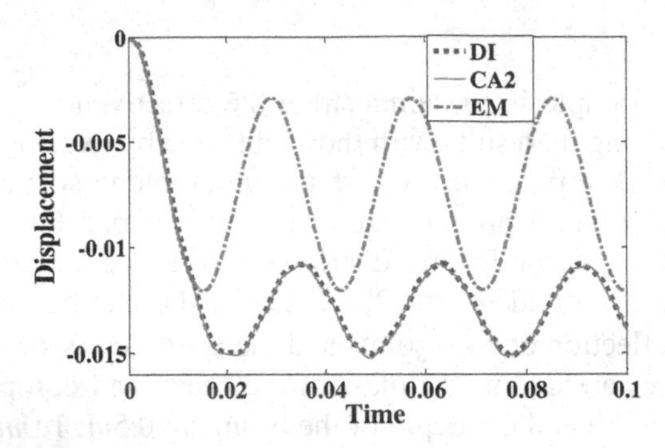

Fig. 7.10. Displacements at the middle of the beam

- Approximate mode superposition analysis considering only 3 modes (1st, 3rd and 5th mode), where the elastic modes (EM) are not updated, and are used as approximate modes during the solution process.
- Approximate solution by the CA approach with 2 basis vectors (CA2).
- Exact reanalysis using a commercial code [8]. The solution is based on direct integration (DI) scheme.

It is observed that the CA procedure with only 2 basis vectors (CA2) provides practically the solution obtained by the exact reanalysis (DI), whereas consideration of only the initial elastic modes (EM) during the mode superposition analysis provides poor results.

Example 7.5

Considering the fifty-story frame shown in Fig. 7.11, the object is to illustrate results obtained by the CA approach for the following two problems:

- *Nonlinear dynamic analysis of the original structure.*
- *Nonlinear dynamic reanalysis of modified structures.*

Nonlinear dynamic analysis of the original structure. The damping ratios for all modes are 0.05 and the time step is Δt = 0.005*sec*. The masses are assumed to be concentrated at the joints, and only horizontal inertia forces are considered. The inertia force is due to the frame self-weight and an additional concentrated mass of 50*ton* at an internal joint and 25*ton* at an external joint. The moment-curvature relation is bi-linear with hardening of 5%, as shown in Fig. 7.12. The width and depth of all elements are 0.5*m*, 1.0*m*, respectively, the modulus of elasticity is $3\ 10^7 kN/m^2$, the elastic limit stress is $\sigma_Y = 20000 kN/m^2$, and the plastic hinge moment is $M_P = 2500 kNm$.

Considering the first 8 mode shapes, the object is to evaluate the nonlinear dynamic response for the loading of the El Centro earthquake shown in Fig. 7.2. To demonstrate results for cases where the time range of the nonlinear response is larger, the loading was also multiplied by a factor of 2. The following two cases have been solved using the CA approach:

- *a*. El-Cento loading, using 4 basis vectors (CA4).
- *b*. 2x El-Cento loading, using 6 basis vectors (CA6).

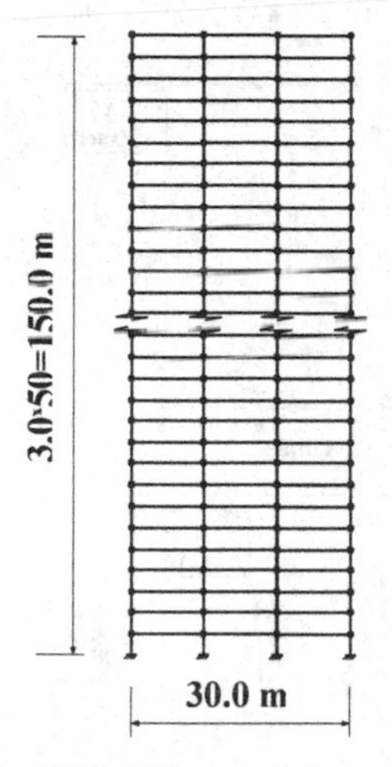

Fig. 7.11. Fifty-story frame

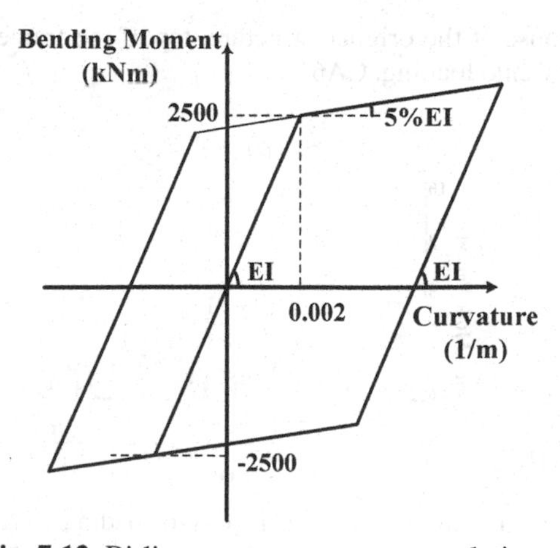

Fig. 7.12. Bi-linear moment-curvature relation with hardening

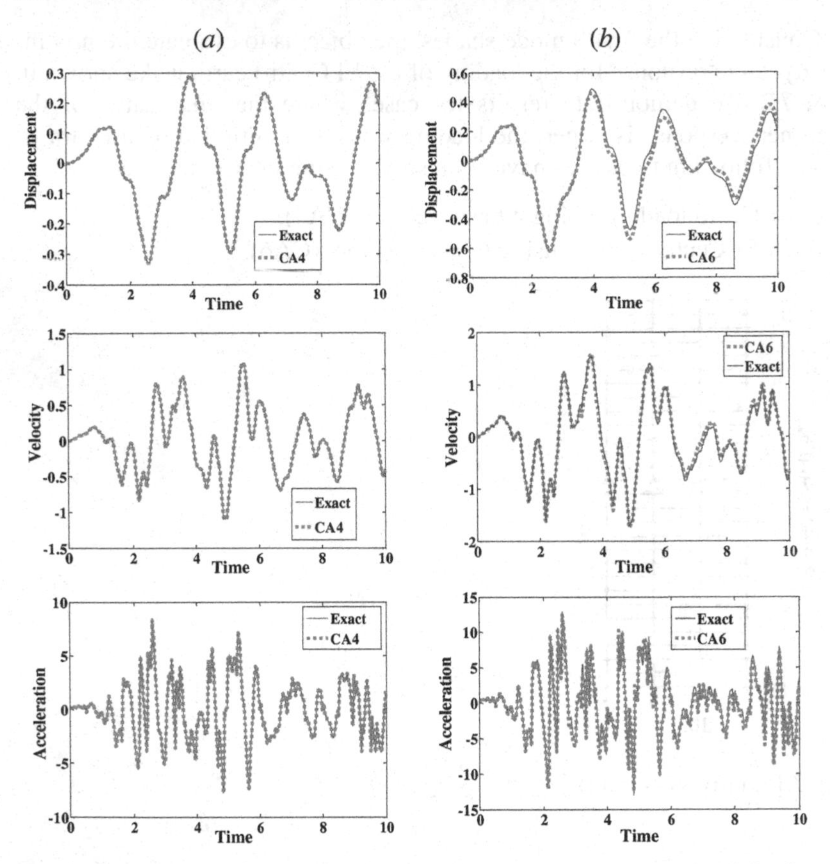

Fig. 7.13. Nonlinear dynamic response of the original structure, top of the frame: ***a***. El-Cento loading, CA4 ***b***. 2x El-Cento loading, CA6

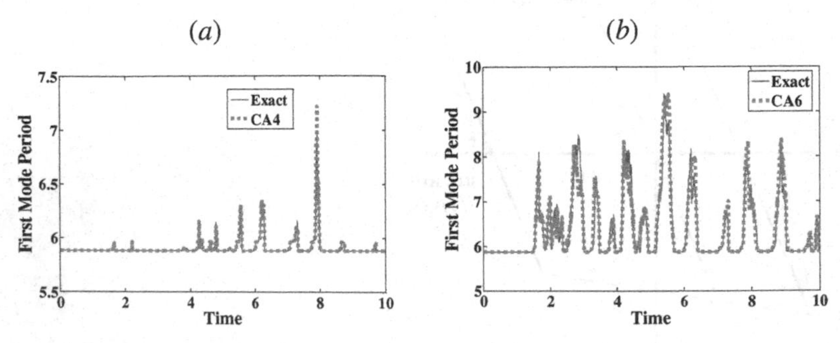

Fig. 7.14. First mode period of the original structure: ***a***. El-Cento loading, CA4 ***b***. 2x El-Cento loading, CA6

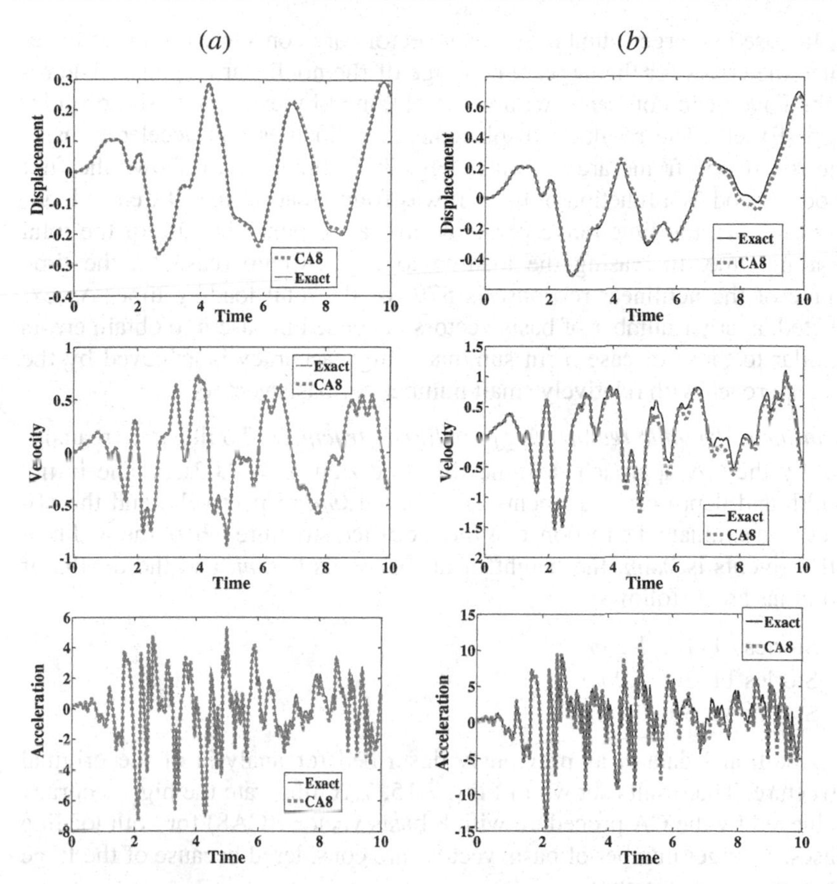

Fig. 7.15. Displacements, velocities, accelcrations, top of frame reanalysis, CA8: ***a***. El-Cento loading ***b***. 2x El-Cento loading

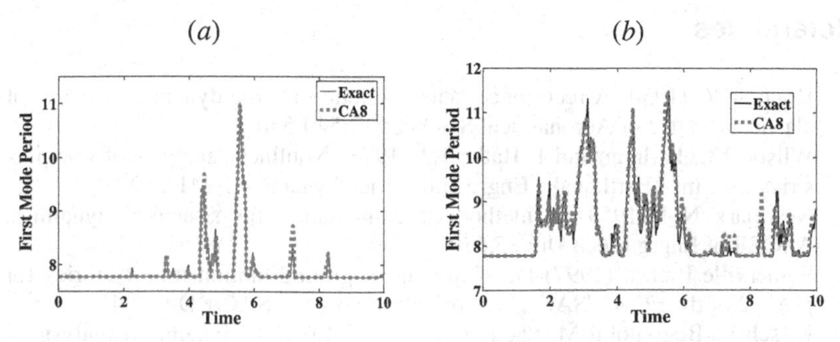

Fig. 7.16. First mode period of the modified structure, CA8: ***a***. El-Cento loading ***b***. 2x El-Cento loading

In case b a larger number of basis vectors are considered in order to reduce the errors for the larger time range of the nonlinear response. The results have been compared with those obtained by exact analysis formulation (Exact). The resulting displacements, velocities and accelerations at the top of the frame are shown in Fig. 7.13. Figure 7.14 shows the first mode period as a function of time. It was found that for the El Centro loading (case a), the time range of the nonlinear response is 20% of the total loading time. Increasing the loading to 2x El-Cento (case b), the time range of the nonlinear response is 57% of the total loading time. As expected, a larger number of basis vectors is needed in case b to obtain errors similar to those of case a. In summary, high accuracy is achieved by the CA approach with relatively small numbers of basis vectors.

Nonlinear dynamic reanalysis of modified structures. To illustrate reanalysis by the CA approach, assume the time step $\Delta t = 0.02sec$. The initial width and depth of all elements are $0.5m$, $1.0m$, respectively, and the object is to evaluate the response of the modified structure where the width of all elements is $0.4m$, the height of all beams is $0.75m$, and the depths of columns are as follows.

- Stories 1–10: $1.25m$.
- Stories 11–30: $1.00m$.
- Stories 31–50: $0.75m$.

Additional data is as previously described for analysis of the original structure. The results shown in Figs. 7.15, 7.16 illustrate the high accuracy achieved by the CA procedure with 8 basis vectors (CA8) for both loading cases. A larger number of basis vectors are considered because of the large changes in the structure.

References

1. Houbolt JC (1950) A recurrence matrix solution for the dynamic response of elastic aircraft. J of Aeronautical Sciences 17:540-550
2. Wilson EL, Farhoomand I, Bathe KJ (1973) Nonlinear analysis of complex structures. Int J Earthquake Engrg and Struct Dynamics 1:241-252
3. Newmark NM (1959) A method of computation for structural dynamics. ASCE J of Engrg Mech Div 85:67-94
4. Somerville P et al (1997) Development of ground motion time histories for phase 2 of the FEMA/SAC steel project. Report No. SAC/BD-97/04
5. Kirsch U, Bogomolni M, Sheinman I (2007) Efficient dynamic reanalysis of structures. ASCE J Struct Engrg 133:440-448

6. Kirsch U, Bogomolni M (2007) Nonlinear and dynamic structural analysis using linear reanalysis. Computers & Structures 85:566-578
7. Kirsch U, Bogomolni M, Sheinman I (2006) Nonlinear dynamic reanalysis for structural optimization. Comp Meth Appl Mech Engrg 195:4420-4432
8. ADINA R&D Inc (2004) A general propose finite element nonlinear dynamic analysis. Version 8.2

8 Direct Reanalysis

Direct (closed form) reanalysis methods are efficient for low-rank changes in the stiffness matrix. In particular, these methods are applicable to situations where a relatively small proportion of the structure is changed and the changes in the stiffness matrix can be represented by a small sub-matrix. These methods are inefficient when the sub-matrix of changes in the system stiffness matrix is of high-rank or large.

Direct methods are usually based on the Sherman-Morrison [1] and Woodbury [2] formulae for the update of the inverse of a matrix. Surveys on these methods are given elsewhere [3–5]. A comprehensive historical survey of the origin of these formulae is presented in [4]. It has been shown [5] that various reanalysis methods may be viewed as variants of these formulae. When the stiffness matrix is modified by a rank-one increment, the solution can be updated inexpensively with the Sherman-Morrison formula by solving the initial analysis equations with a different right-hand side vector, which is a factor of the matrix increment. Similarly, solution for a higher-rank change in the stiffness matrix can be carried out by superposition of rank-one changes. This is reflected in the Woodbury formula.

Direct reanalysis methods are presented in Sect. 8.1. The Sherman-Morrison formula for exact solutions for a single rank-one change is introduced in Sect. 8.1.1, the Woodbury formula for multiple rank-one changes is discussed in Sect. 8.1.2, and a procedure for general changes in the design is presented in Sect. 8.1.3. In Sect. 8.2 direct solutions, which are based on the CA approach, are developed. For multiple rank-one changes in the stiffness matrix the presented CA solution procedure and the Sherman-Morrison and Woodbury formulae are equivalent. It is shown in Sect. 8.3 that direct solutions can be obtained by the CA approach also for topological and geometrical changes.

8.1 Direct Methods

8.1.1 A Single Rank-One Change

The rank of the matrix of changes $\Delta\mathbf{K}$ (rank $\Delta\mathbf{K}$) is the dimension of the linear space spanned by its columns. Rank $\Delta\mathbf{K}$ is equal to the maximum number of linearly independent columns (or rows) of $\Delta\mathbf{K}$. Rank $\Delta\mathbf{K}$ is also equal to the order of the square sub-matrix of $\Delta\mathbf{K}$ of greatest order whose determinant does not vanish. An example of a rank-one change in the stiffness matrix is a change in the cross-sectional area of a truss member.

For a change of rank-one in the $n \times n$ stiffness matrix $\mathbf{K}_0$, the matrix of changes $\Delta\mathbf{K}$ can be expressed in terms of the vectors $\mathbf{v}$, $\mathbf{w}$ and a scalar η as

$$\Delta\mathbf{K} = \mathbf{v}\,\mathbf{w}^T = \eta\,\mathbf{v}\,\mathbf{v}^T, \tag{8.1}$$

where the scalar η is positive or negative, depending on the sign of the stiffness change, and

$$\mathbf{w} = \eta\,\mathbf{v}. \tag{8.2}$$

The Sherman-Morrison (S-M) formula, giving the change in the inverse of a matrix due to a rank-one change $\Delta\mathbf{K}$, can be expressed in terms of $\mathbf{v}$ and $\mathbf{w}$ as

$$(\mathbf{K}_0 + \Delta\mathbf{K})^{-1} = (\mathbf{K}_0 + \mathbf{v}\mathbf{w}^T)^{-1} = \mathbf{K}_0^{-1} - \mathbf{K}_0^{-1}\mathbf{v}(1 + \mathbf{w}^T\mathbf{K}_0^{-1}\mathbf{v})^{-1}\mathbf{w}^T\mathbf{K}_0^{-1}. \tag{8.3}$$

Using the relation of Eq. (8.2), we can express the S-M formula in terms of $\mathbf{v}$ and η as

$$(\mathbf{K}_0 + \eta\,\mathbf{v}\,\mathbf{v}^T)^{-1} = \mathbf{K}_0^{-1} - \mathbf{K}_0^{-1}\mathbf{v}(1 + \eta\,\mathbf{v}^T\mathbf{K}_0^{-1}\mathbf{v})^{-1}\eta\,\mathbf{v}^T\mathbf{K}_0^{-1}\cdot \tag{8.4}$$

Define the vector $\mathbf{t}$ by

$$\mathbf{t} = \mathbf{K}_0^{-1}\mathbf{v}\,. \tag{8.5}$$

Post-multiplying Eq. (8.4) by the load vector $\mathbf{R}_0$ and substituting Eq. (8.5) and $\mathbf{K}_0\,\mathbf{r}_0 = \mathbf{R}_0$, we obtain the S-M formula for the modified displacements

$$\mathbf{r} = (\mathbf{K}_0 + \Delta\mathbf{K})^{-1}\mathbf{R}_0 = (\mathbf{K}_0 + \eta\,\mathbf{v}\,\mathbf{v}^T)^{-1}\mathbf{R}_0 = \mathbf{r}_0 - \eta\,\mathbf{t}(1 + \eta\,\mathbf{v}^T\mathbf{t})^{-1}\mathbf{v}^T\mathbf{r}_0\cdot \tag{8.6}$$

Defining the scalar

$$a = \eta(1 + \eta\,\mathbf{v}^T\mathbf{t})^{-1}\mathbf{v}^T\mathbf{r}_0 \tag{8.7}$$

and substituting Eq. (8.7) into Eq. (8.6), we obtain the expression for $\mathbf{r}$

$$\mathbf{r} = \mathbf{r}_0 - a\mathbf{t} . \tag{8.8}$$

In summary, solution by the S-M formula for evaluating the modified displacements **r**, due to a rank-one change $\Delta\mathbf{K}$ in the stiffness matrix $\mathbf{K}_0$, involves the following steps:

- Calculation of the vector **t** by Eq. (8.5).
- Determination of the scalar *a* by Eq. (8.7).
- Calculation of the modified displacements by Eq. (8.8).

It can be observed that calculation of the vector **t** by Eq. (8.5) involves only forward and backward substitutions if $\mathbf{K}_0$ is given from the initial analysis in the decomposed form $\mathbf{K}_0 = \mathbf{U}_0^T \mathbf{U}_0$ [Eq. (4.2)]. In this case, calculation of **t** is equivalent to the solution of the initial analysis equations with a different right-hand side vector.

8.1.2 Multiple Rank-One Changes

For *m* rank-one changes $\Delta\mathbf{K}_i$ (i = 1, 2, ..., m) in the $n \times n$ stiffness matrix, the total change in stiffness $\Delta\mathbf{K}$ can be expressed in terms of the matrices **V**, **W** and a diagonal matrix **H** as

$$\Delta\mathbf{K} = \Delta\mathbf{K}_1 + \Delta\mathbf{K}_2 + ... + \Delta\mathbf{K}_m = \mathbf{V}\,\mathbf{W}^T = \mathbf{V}\,\mathbf{H}\,\mathbf{V}^T, \tag{8.9}$$

where matrices **V** and **W** are of order $n \times m$, matrix **H** is of order $m \times m$ and

$$\mathbf{W} = \mathbf{V}\,\mathbf{H}. \tag{8.10}$$

The Woodbury formula, giving the change in the inverse of a matrix due to a rank-m change $\Delta\mathbf{K}$, can be expressed in terms of **V** and **W** as

$$(\mathbf{K}_0 + \Delta\mathbf{K})^{-1} = (\mathbf{K}_0 + \mathbf{V}\mathbf{W}^T)^{-1} = \mathbf{K}_0^{-1} - \mathbf{K}_0^{-1}\mathbf{V}(\mathbf{I} + \mathbf{W}^T\mathbf{K}_0^{-1}\mathbf{V})^{-1}\mathbf{W}^T\mathbf{K}_0^{-1}. \tag{8.11}$$

Using Eq. (8.10), we can express the Woodbury formula in terms of **V** and **H** as

$$(\mathbf{K}_0 + \mathbf{V}\,\mathbf{H}\,\mathbf{V}^T)^{-1} = \mathbf{K}_0^{-1} - \mathbf{K}_0^{-1}\mathbf{V}(\mathbf{I} + \mathbf{H}\,\mathbf{V}^T\mathbf{K}_0^{-1}\mathbf{V})^{-1}\mathbf{H}\,\mathbf{V}^T\mathbf{K}_0^{-1}. \tag{8.12}$$

Define the matrix **T** by

$$\mathbf{T} = \mathbf{K}_0^{-1}\mathbf{V} . \tag{8.13}$$

Post-multiplying Eq. (8.12) by $\mathbf{R}_0$ and substituting $\mathbf{K}_0\mathbf{r}_0 = \mathbf{R}_0$ and Eq. (8.13), we obtain the Woodbury formula for the modified displacements **r**

$$\mathbf{r} = \left(\mathbf{K}_0 + \mathbf{V}\,\mathbf{H}\,\mathbf{V}^T\right)^{-1}\mathbf{R}_0 = \mathbf{r}_0 - \mathbf{T}\left(\mathbf{I} + \mathbf{H}\,\mathbf{V}^T\mathbf{T}\right)^{-1}\mathbf{H}\mathbf{V}^T\mathbf{r}_0\,. \tag{8.14}$$

Denoting the right-hand term in parentheses, called the capacitance matrix, by **C**

$$\mathbf{C} = \mathbf{I} + \mathbf{H}\mathbf{V}^T\mathbf{T}\,, \tag{8.15}$$

defining the vector **A** as

$$\mathbf{A} = \mathbf{C}^{-1}\mathbf{H}\mathbf{V}^T\mathbf{r}_0\,, \tag{8.16}$$

and substituting into Eq. (8.14), we obtain the expression for the modified displacements

$$\mathbf{r} = \mathbf{r}_0 - \mathbf{T}\mathbf{A}\,. \tag{8.17}$$

In summary, solution by the Woodbury formula for evaluating the modified displacements, due to m rank-one changes in the stiffness matrix, involves the following steps:

- Calculation of matrix **T** by Eq. (8.13).
- Determination of the vector **A** by Eq. (8.16).
- Calculation of the modified displacements by Eq. (8.17).

Again, it can be observed that calculation of matrix **T** by Eq. (8.13) is equivalent to the solution of the initial analysis equations with m different right-hand side vectors. In addition, calculation of the vector **A** involves solution of the $m \times m$ set of Eq. (8.16). When the rank of $\Delta\mathbf{K}$ is small compared to the order of **K**, the main computational cost is involved in the solution of Eq. (8.13). For a banded matrix **K** of order n and band-width m_k, this requires about $m\,n\,m_k$ multiplications. The factorization of the matrix requires about $n\,m_k^2$ multiplications. Therefore, the above procedure is effective only when the ratio m/m_k is small.

Example 8.1

To illustrate calculation of the exact displacements by the S-M Formula, consider the ten-bar truss shown in Fig. 8.1. The truss is subjected to two concentrated loads, the modulus of elasticity is E =30000 and the eight analysis unknowns are the horizontal (to the right) and the vertical (downward) displacements at joints 1, 2, 3 and 4, respectively.

Assuming the initial cross-sectional areas **X** = **1.0**, solutions are presented for the following two separate cases of changes:

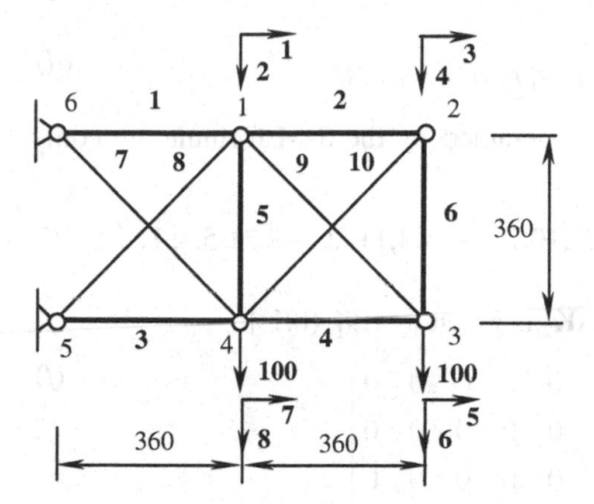

Fig. 8.1. Ten-bar truss

- *Case 1*. A change $\Delta X_1 = 1.0$ in member 1.
- *Case 2*. A change $\Delta X_2 = 1.0$ in member 2.

The terms related to case 1 are denoted by subscript 1 and those related to case 2 by subscript 2.

Case 1. The matrix of changes, $\Delta\mathbf{K}_1$ is given by [Eq. (8.1)]

$$\Delta\mathbf{K}_1 = \frac{E\Delta X_1}{L_1}\begin{bmatrix} 1 & 0 & 0 & 0 & 0 & 0 & 0 & 0 \\ 0 & 0 & 0 & 0 & 0 & 0 & 0 & 0 \\ 0 & 0 & 0 & 0 & 0 & 0 & 0 & 0 \\ 0 & 0 & 0 & 0 & 0 & 0 & 0 & 0 \\ 0 & 0 & 0 & 0 & 0 & 0 & 0 & 0 \\ 0 & 0 & 0 & 0 & 0 & 0 & 0 & 0 \\ 0 & 0 & 0 & 0 & 0 & 0 & 0 & 0 \\ 0 & 0 & 0 & 0 & 0 & 0 & 0 & 0 \end{bmatrix} = \eta_1 \mathbf{v}_1 \mathbf{v}_1^T, \tag{a}$$

where L_1 is the length of the member and

$$\eta_1 = 83.333 \qquad \mathbf{v}_1^T = \{1, 0, 0, 0, 0, 0, 0, 0\}. \tag{b}$$

The vector $\mathbf{t}_1$ is computed by $\mathbf{t}_1 = \mathbf{K}_0^{-1}\mathbf{v}_1$ [Eq. (8.5)],

$$\mathbf{t}_1^T = \{0.0106, 0.0066, 0.0107, 0.0179, -0.0013, 0.0181, -0.0014, 0.0054\}, \tag{c}$$

and the scalar a_1 is computed by [Eq. (8.7)]

$$a_1 = \eta_1\left(1 + \eta_1 \mathbf{v}_1^T \mathbf{t}_1\right)^{-1} \mathbf{v}_1^T \mathbf{r}_0 = 103.7468\,. \qquad (d)$$

The final modified displacements obtained by the S-M formula are computed by [Eq. (8.8)]

$$\mathbf{r}^T = (\mathbf{r}_0 - a_1 \mathbf{t}_1)^T = \{1.24, 4.89, 1.71, 10.79, -3.04, 11.26, -2.31, 5.45\}\,. \qquad (e)$$

Case 2. The matrix of changes, $\Delta\mathbf{K}_2$ is given by [Eq. (8.1)]

$$\Delta\mathbf{K}_2 = \frac{E\Delta X_2}{L_2}\begin{bmatrix} 1 & 0 & -1 & 0 & 0 & 0 & 0 & 0 \\ 0 & 0 & 0 & 0 & 0 & 0 & 0 & 0 \\ -1 & 0 & 1 & 0 & 0 & 0 & 0 & 0 \\ 0 & 0 & 0 & 0 & 0 & 0 & 0 & 0 \\ 0 & 0 & 0 & 0 & 0 & 0 & 0 & 0 \\ 0 & 0 & 0 & 0 & 0 & 0 & 0 & 0 \\ 0 & 0 & 0 & 0 & 0 & 0 & 0 & 0 \\ 0 & 0 & 0 & 0 & 0 & 0 & 0 & 0 \end{bmatrix} = \eta_2 \mathbf{v}_2 \mathbf{v}_2^T, \qquad (f)$$

where L_2 is the length of the member and

$$\eta_2 = 83.333 \qquad \mathbf{v}_2^T = \{1,\ 0,\ -1,\ 0,\ 0,\ 0,\ 0,\ 0\}\,. \qquad (g)$$

The vector $\mathbf{t}_2$ is computed by $\mathbf{t}_2 = \mathbf{K}_0^{-1}\mathbf{v}_2$ [Eq. (8.5)],

$$\mathbf{t}_2^T = \left\{ \begin{matrix} -0.0001, -0.0006, -0.0109, -0.0066, & \\ & 0.0011, -0.0054, -0.0001, 0.0006 \end{matrix} \right\}, \qquad (h)$$

and the scalar a_2 is computed by [Eq. (8.7)]

$$a_2 = \eta_2\left(1 + \eta_2 \mathbf{v}_2^T \mathbf{t}_2\right)^{-1} \mathbf{v}_2^T \mathbf{r}_0 = -21.1721\,. \qquad (i)$$

The final modified displacements obtained by the S-M formula are computed by [Eq. (8.8)]

$$\mathbf{r}^T = (\mathbf{r}_0 - a_2 \mathbf{t}_2)^T = \{2.34, 5.57, 2.60, 12.51, -3.15, 13.02, -2.46, 6.02\}. \qquad (j)$$

8.1.3 General Procedure

Assume the general case where the incremental stiffness matrix $\Delta\mathbf{K}$ can be compressed, by eliminating zero columns and rows, to form a reduced incremental matrix $\Delta\mathbf{K}_R$ of size equal to the number of changed columns (or rows) in matrix $\mathbf{K}$. The relation between $\Delta\mathbf{K}$ and $\Delta\mathbf{K}_R$ is given by

$$\Delta\mathbf{K} = \mathbf{b}^T \Delta\mathbf{K}_R\, \mathbf{b}, \tag{8.18}$$

where $\mathbf{b}$ is a Boolean matrix with linearly independent rows, each of which contains all zeroes except for one unit value, located at the column number where a change in $\mathbf{K}$ occurs.

It has been shown [6] that the following formula for computing $\mathbf{K}^{-1}$ can be derived from the Sherman-Morrison identity

$$\mathbf{K}^{-1} = (\mathbf{K}_0 + \mathbf{b}^T \Delta\mathbf{K}_R\, \mathbf{b})^{-1} = \mathbf{K}_0^{-1} - \mathbf{K}_0^{-1}\mathbf{b}^T(\mathbf{I} + \Delta\mathbf{K}_R\mathbf{b}\mathbf{K}_0^{-1}\mathbf{b}^T)^{-1}\Delta\mathbf{K}_R\mathbf{b}\mathbf{K}_0^{-1}, \tag{8.19}$$

where $\mathbf{I}$ is the identity matrix. Note that both matrices $\mathbf{I}$ and $\Delta\mathbf{K}_R\,\mathbf{b}\mathbf{K}_0^{-1}\mathbf{b}^T$ are of size equal to that of $\Delta\mathbf{K}_R$. In addition, for $\mathbf{V} = \mathbf{b}^T$ and $\mathbf{W}^T = \Delta\mathbf{K}_R\,\mathbf{b}$ Eqs. (8.11) and (8.19) are equivalent.

The following procedure [6] utilizes symmetry and positive definiteness properties to compute the modified displacements directly. Assume that the initial stiffness matrix $\mathbf{K}_0$ is available in the decomposed form

$$\mathbf{K}_0 = \mathbf{U}_0^T\, \mathbf{U}_0, \tag{8.20}$$

where $\mathbf{U}_0$ is an upper triangular matrix. Denote

$$\mathbf{r} = \mathbf{r}_0 + \Delta\mathbf{r}, \tag{8.21}$$

where $\Delta\mathbf{r}$ is the change in the displacements. Post multiplying Eq. (8.19) by $\mathbf{R}_0$ and substituting $\mathbf{K}_0\, \mathbf{r}_0 = \mathbf{R}_0$ and Eq. (8.21) into the resulting equation, we obtain

$$\Delta\mathbf{r} = -\mathbf{K}_0^{-1}\mathbf{b}^T(\mathbf{I} + \Delta\mathbf{K}_R\mathbf{b}\,\mathbf{K}_0^{-1}\mathbf{b}^T)^{-1}\Delta\mathbf{K}_R\,\mathbf{b}\,\mathbf{r}_0\,. \tag{8.22}$$

The reduced unsymmetrical matrix $(\mathbf{I} + \Delta\mathbf{K}_R\mathbf{b}\,\mathbf{K}_0^{-1}\mathbf{b}^T)$ can readily be shown to be nonsingular, even when $\Delta\mathbf{K}_R$ is singular. Using $\mathbf{U}_0$ from Eq. (8.20) we define the symmetric influence matrix of unit changes, $\mathbf{Q}$, by

$$\mathbf{Q} = \mathbf{b}\,\mathbf{K}_0^{-1}\mathbf{b}^T = \mathbf{b}\,\mathbf{U}_0^{-1}(\mathbf{U}_0^T)^{-1}\mathbf{b}^T = \mathbf{Z}^T\mathbf{Z}\,, \tag{8.23}$$

where the rectangular matrix $\mathbf{Z}$ is defined as

$$\mathbf{Z} = (\mathbf{U}_0^T)^{-1}\,\mathbf{b}^T. \tag{8.24}$$

The matrix $\mathbf{Z}$ can be produced by a forward-substitution process performed on $\mathbf{b}^T$.

Using Eqs. (8.23) and (8.24), we can write Eq. (8.22) as

$$\Delta\mathbf{r} = -\mathbf{U}_0^{-1}\mathbf{Z}(\mathbf{I} + \Delta\mathbf{K}_R\,\mathbf{Q})^{-1}\Delta\mathbf{K}_R\,\mathbf{b}\,\mathbf{r}_0\,. \tag{8.25}$$

This equation can be written in convenient positive-definite form by extracting the matrix **Q** as a common factor from the matrix to be inverted, as follows

$$\Delta\mathbf{r} = -\mathbf{U}_0^{-1}\mathbf{Z}\mathbf{Q}^{-1}(\mathbf{Q}^{-1} + \Delta\mathbf{K}_R)^{-1}\Delta\mathbf{K}_R\mathbf{b}\,\mathbf{r}_0\,. \tag{8.26}$$

Matrices **Q** and $(\mathbf{Q}^{-1} + \Delta\mathbf{K}_R)$ can be shown to be positive definite, thus pivoting is unnecessary in the triangularization. An optimal order of calculation has been proposed elsewhere [6, 7] and is illustrated in the following numerical example.

Example 8.2

To illustrate calculation of exact displacements by the procedure presented in this section, consider a structure with the following initial values

$$\mathbf{K}_0 = \begin{bmatrix} 9 & -6 & 0 \\ -6 & 8 & -2 \\ 0 & -2 & 2 \end{bmatrix} \qquad \mathbf{U}_0 = \begin{bmatrix} 3 & -2 & 0 \\ 0 & 2 & -1 \\ 0 & 0 & 1 \end{bmatrix}, \tag{a}$$

$$\mathbf{R}_0 = \begin{Bmatrix} 0 \\ 24 \\ 84 \end{Bmatrix} \qquad \mathbf{r}_0 = \begin{Bmatrix} 36 \\ 54 \\ 96 \end{Bmatrix}.$$

Changes are made to the structure as follows

$$\Delta\mathbf{K} = \begin{bmatrix} 0 & 0 & 0 \\ 0 & 5 & -5 \\ 0 & -5 & 5 \end{bmatrix} \tag{b}$$

$$\Delta\mathbf{K}_R = \begin{bmatrix} 5 & -5 \\ -5 & 5 \end{bmatrix} \qquad \mathbf{b} = \begin{bmatrix} 0 & 1 & 0 \\ 0 & 0 & 1 \end{bmatrix}.$$

To determine $\Delta\mathbf{r}$ by Eq. (8.26), we calculate the following intermediate values

$$\mathbf{r}_R = \mathbf{b}\,\mathbf{r}_0 = \begin{Bmatrix} 54 \\ 96 \end{Bmatrix} \qquad \mathbf{r}_1 = \Delta\mathbf{K}_R\,\mathbf{r}_R = \begin{Bmatrix} -210 \\ 210 \end{Bmatrix}, \tag{c}$$

$$\mathbf{Z} = \begin{bmatrix} 0 & 0 \\ 0.5 & 0 \\ 0.5 & 1 \end{bmatrix} \qquad \mathbf{Q} = \begin{bmatrix} 0.5 & 0.5 \\ 0.5 & 1 \end{bmatrix} \qquad \mathbf{Q}^{-1} = \begin{bmatrix} 4 & -2 \\ -2 & 2 \end{bmatrix}. \tag{d}$$

The matrix $\mathbf{P} = (\mathbf{Q}^{-1} + \Delta\mathbf{K}_R)$ is factorized into $\mathbf{P} = \mathbf{U}_P^T\mathbf{U}_P$ where

$$\mathbf{U}_P = \begin{bmatrix} 3 & -2.33 \\ 0 & 0.33(14)^{1/2} \end{bmatrix}, \tag{e}$$

and the vectors, $\mathbf{r}_{11}$, $\mathbf{r}_{111}$ are calculated by

$$\mathbf{r}_{11} = \mathbf{Q}^{-1}\mathbf{P}^{-1}\mathbf{r}_1 = \begin{Bmatrix} -60 \\ 60 \end{Bmatrix}, \tag{f}$$

$$\mathbf{r}_{111} = \mathbf{Z}\,\mathbf{r}_{11} = \begin{Bmatrix} 0 \\ -30 \\ 30 \end{Bmatrix}.$$

The final results for Eqs. (8.26) and (8.21) are

$$\Delta\mathbf{r}^T = -\{0,\ 0,\ 30\}, \tag{g}$$

$$\mathbf{r}^T = (\mathbf{r}_0 + \Delta\mathbf{r})^T = \{36,\ 54,\ 66\}.$$

8.2 Direct Solutions by Combined Approximations

In general, the CA approach provides approximate solutions for high-rank changes in the stiffness matrix. It is shown in this section that exact solutions are efficiently obtained by the approach for a small number of simultaneous rank-one changes in the stiffness matrix. In such cases, solutions obtained by the CA approach and the Sherman-Morrison-Woodbury formulae are equivalent. In Sect. 8.2.1 we present the expressions for calculating the exact modified displacements. Using these expressions, a general solution procedure is introduced in Sect. 8.2.2.

It is instructive to note that exact solutions can be obtained by the CA approach also in other particular cases discussed in Sect. 10.2. These include cases when a newly created basis vector becomes a linear combination of the previous vectors, or if a modified design is a scaled design.

8.2.1 Multiple Rank-One Changes

In this section we present the expressions for calculating the exact modified displacements by the CA approach for simultaneous rank-one changes in the stiffness matrix. Such exact solutions are efficient in cases where the number of changes is much smaller than the number of DOF.

Consider the common case where the first basis vector is $\mathbf{r}_1 = \mathbf{r}_0$ and the second basis vector is calculated by $\mathbf{r}_2 = -\mathbf{B}\,\mathbf{r}_0$. These two vectors are linearly dependent if

$$\mathbf{B}\,\mathbf{r}_0 = y\,\mathbf{r}_0, \tag{8.27}$$

where y is a scalar different from zero and $\mathbf{r}_0$ is an arbitrary displacement vector. Substituting $\mathbf{B} = \mathbf{K}_0^{-1}\Delta\mathbf{K}$ into Eq. (8.27) and pre-multiplying the resulting equation by $\mathbf{K}_0$ yields

$$\Delta\mathbf{K}\,\mathbf{r}_0 = y\,\mathbf{K}_0\,\mathbf{r}_0 = y\,\mathbf{R}_0. \tag{8.28}$$

It is observed that the condition of Eq. (8.28) is equivalent to the case of uniform scaling where $\Delta\mathbf{K} = y\,\mathbf{K}_0$.

Two successive basis vectors $\mathbf{B}^{j-1}\mathbf{r}_0$ and $\mathbf{B}^{j}\,\mathbf{r}_0$ are linearly dependent if

$$\mathbf{B}^{j}\,\mathbf{r}_0 = y\,\mathbf{B}^{j-1}\,\mathbf{r}_0, \tag{8.29}$$

for some scalar y different from zero. The condition of Eq. (8.29) is satisfied for arbitrary displacements $\mathbf{r}_0$ if

$$\mathbf{B}^{j} = y\,\mathbf{B}^{j-1}. \tag{8.30}$$

For the second and the third basis vectors this condition becomes

$$\mathbf{B}^{2} = y\,\mathbf{B}. \tag{8.31}$$

The condition of Eq. (8.31) is satisfied in cases of rank-one changes in $\mathbf{K}$.

Consider for example the typical case of a change in the cross-sectional area of a single truss member. It has been shown [8] that for a change in the ith member, $\Delta\mathbf{K}_i$, an exact solution is obtained by the CA approach with only two basis vectors by

$$\mathbf{r} = \mathbf{r}_0 + y_i\,\mathbf{r}_i, \tag{8.32}$$

where the basis vector $\mathbf{r}_i$ is defined as

$$\mathbf{r}_i = -\mathbf{K}_0^{-1}\,\Delta\mathbf{K}_i\,\mathbf{r}_0 = -\mathbf{B}_i\,\mathbf{r}_0. \tag{8.33}$$

That is, the matrix $\mathbf{B}_i$ corresponds to the change $\Delta\mathbf{K}_i$ in the stiffness matrix of member i

$$\mathbf{B}_i = \mathbf{K}_0^{-1}\,\Delta\mathbf{K}_i. \tag{8.34}$$

The term $\Delta\mathbf{K}_i\,\mathbf{r}_0$ in Eq. (8.33) represents a pair of collinear forces on the ends of the modified truss member. Thus, the basis vector $\mathbf{r}_i$ may be viewed as an influence coefficient vector measuring the effect of the change $\Delta\mathbf{K}_i$ in element i on the displacement vector [Eq. (8.32)].

The expression of Eq. (8.32) for a change in a single member can be extended to the general case of simultaneous changes in m members, for which the exact solution is given by [8]

$$\mathbf{r} = \mathbf{r}_0 + \sum_{i=1}^{m} y_i \mathbf{r}_i\,, \tag{8.35}$$

where $\mathbf{r}_i$ ($i = 1, \ldots, m$) are defined by Eq. (8.33) for all changed members. Exact solutions by Eq. (8.35) are efficient in cases where the number of changes is much smaller than the number of DOF for the structure, that is $m << n$.

It has been shown [5] that for a change in the cross-sectional area of a single truss member, the exact solutions obtained by the CA approach [Eq. (8.32)] and the S-M formula [Eq. (8.8)] are equivalent. Moreover, for simultaneous changes in m truss members, the exact solutions obtained by the CA approach [Eq. (8.35)] and the Woodbury formula [Eq. (8.17)] involve exactly the same calculations.

8.2.2 Solution Procedure

Using the expressions presented in Sect. 8.2.1, we introduce in this section a general solution procedure for the case of m rank-one simultaneous changes in the stiffness matrix (for example, simultaneous changes in m truss members). It has been noted that this procedure is particularly efficient when the number of changed members is much smaller than the number of DOF. If some of the basis vectors are linearly dependent, the exact solution is obtained with a smaller number of basis vectors.

Consider a common problem where it is necessary to repeat the analysis many times, due to various combinations of changes in m members. It will be shown that, using the procedure presented, it is necessary to calculate m basis vectors only once. These vectors can then be used to achieve exact solution for any modified design. The procedure presented is most effective in various applications, such as response surface calculations.

For the given initial values $\mathbf{K}_0$ and $\mathbf{r}_0$, the solution procedure involves the following two main stages.

Stage a, calculation of the constant vectors $\mathbf{r}_{0i}$. It has been noted in Sect. 5.2.1 that multiplying a basis vector by any scalar does not change the solution but only the corresponding scalar y_i [Eq. (5.33)]. Therefore, a basis vector $\mathbf{r}_{0i}$ related to a change $\Delta\mathbf{K}_{0i}$ in the ith member and defined as

$$\mathbf{r}_{0i} = -\mathbf{K}_0^{-1}\Delta\mathbf{K}_{0i}\,\mathbf{r}_0 \qquad i = 1, \ldots, m, \tag{8.36}$$

can be used for any magnitude of change $\Delta\mathbf{K}_i$ obtained by

$$\Delta\mathbf{K}_i = \Delta X_i\,\Delta\mathbf{K}_{0i} \qquad i = 1, \ldots, m, \tag{8.37}$$

where ΔX_i is a scalar multiplier. As noted earlier, calculation of the basis vectors involves only forward and backward substitutions. Thus, the matrix of basis vectors $\mathbf{r}_{B0}$,

$$\mathbf{r}_{B0} = [\mathbf{r}_{01}, \mathbf{r}_{02}, \ldots, \mathbf{r}_{0m}], \tag{8.38}$$

can be used for any modified design

Stage b, repeated calculations. This stage consists of the following simple repeated calculations for any modified design:

- *Calculation of the modified stiffness matrix* **K**. For any assumed set of of changes ΔX_i, the modified stiffness matrix **K** is calculated by

$$\mathbf{K} = \mathbf{K}_0 + \sum_{i=1}^{m}\Delta\mathbf{K}_i = \mathbf{K}_0 + \sum_{i=1}^{m}\Delta X_i\,\Delta\mathbf{K}_{0i}\,. \tag{8.39}$$

  Note that the matrices $\Delta\mathbf{K}_{0i}$ are already given from calculation of the constant basis vectors.

- *Calculation of the coefficients* y_i. For the given matrix of basis vectors $\mathbf{r}_{B0}$, the reduced stifffness matrix $\mathbf{K}_R$ and the reduced load vector $\mathbf{R}_R$ are calculated by

$$\mathbf{K}_R = \mathbf{r}_{B0}^T\mathbf{K}\mathbf{r}_{B0} \qquad \mathbf{R}_R = \mathbf{r}_{B0}^T(\mathbf{R} - \mathbf{K}\mathbf{r}_0)\,, \tag{8.40}$$

  and the vector of unknown coefficients,

$$\mathbf{y}^T = \{y_1, y_2, \ldots, y_m\}, \tag{8.41}$$

  is calculated by solving the $m\times m$ set of equations

$$\mathbf{K}_R\,\mathbf{y} = \mathbf{R}_R. \tag{8.42}$$

• *Calculation of the final exact displacements* $\mathbf{r}$. These are calculated by

$$\mathbf{r} = \mathbf{r}_0 + \sum_{i=1}^{m} y_i \, \mathbf{r}_i \, . \tag{8.43}$$

Example 8.3

To illustrate calculation of the exact displacements by the CA approach, consider again the ten-bar truss shown in Fig. 8.2, solved in example 8.1 by the S-M Formula. The truss is subjected to a single loading condition of two concentrated loads, the initial cross-sectional areas equal unity, the modulus of elasticity is 30000 and the eight analysis unknowns are the horizontal and the vertical displacements at joints 1, 2, 3 and 4.

The given initial displacement vector $\mathbf{r}_0$ and the constant basis vectors $\mathbf{r}_{0i}$ [Eq. (8.36)], calculated for changes $\Delta X_i = +1.0$ in each of the 10 cross sections, are shown in Table 8.1. As noted earlier, these vectors can be used in reanalysis of any modified design. The final exact displacements can readily be determined for any assumed set of changes in the cross sections by the procedure described in this section.

Assume, for example, the following two separate changes:

- Case 1. A change $\Delta X_1 = 1.0$ in member 1.
- Case 2. A change $\Delta X_2 = 1.0$ in member 2.

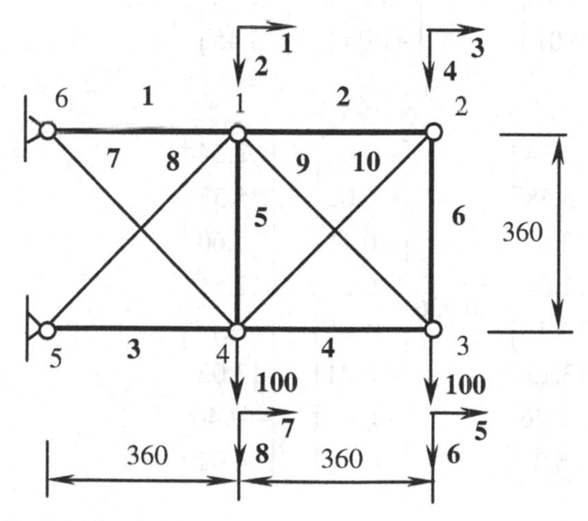

Fig. 8.2. Ten-bar truss

Table 8.1. Vectors $\mathbf{r}_0$ and $\mathbf{r}_{0i}$

$\mathbf{r}_0$	$\mathbf{r}_{01}$	$\mathbf{r}_{02}$	$\mathbf{r}_{03}$	$\mathbf{r}_{04}$	$\mathbf{r}_{05}$	$\mathbf{r}_{06}$	$\mathbf{r}_{07}$	$\mathbf{r}_{08}$	$\mathbf{r}_{09}$	$\mathbf{r}_{010}$
2.34	-2.07	-0.01	-0.29	0.01	0.04	-0.01	-0.41	0.38	0.02	-0.02
5.58	-1.30	-0.02	-1.10	0.03	0.17	-0.02	-1.59	-1.79	0.09	-0.06
2.83	-2.10	-0.44	-0.26	-0.07	0.08	0.04	-0.37	0.34	-0.19	0.13
12.65	-3.50	-0.27	-3.70	-0.32	-0.02	0.21	-1.80	-1.60	-0.91	-0.75
-3.17	0.25	0.04	2.20	0.65	0.08	0.04	-0.37	0.34	-0.19	0.13
13.13	-3.53	-0.21	-3.67	-0.40	0.02	-0.22	-1.75	-1.64	-1.12	-0.61
-2.46	0.27	-0.01	2.17	0.01	0.04	-0.01	-0.41	0.38	0.02	-0.02
6.01	-1.05	0.02	-1.36	-0.03	-0.17	0.02	-1.96	-1.45	-0.09	0.06

The scalars $y_1 = 0.531$ and $y_2 = 0.526$ are calculated separately by solving a single equation [Eq. (8.42)]. The final exact displacements are as follows [Eq. (8.43)]:

Case 1

$$\mathbf{r} = \mathbf{r}_0 + y_1\mathbf{r}_1 = \begin{Bmatrix} 2.34 \\ 5.58 \\ 2.83 \\ 12.65 \\ -3.17 \\ 13.13 \\ -2.46 \\ 6.01 \end{Bmatrix} + 0.531 \begin{Bmatrix} -2.07 \\ -1.30 \\ -2.10 \\ -3.50 \\ 0.25 \\ -3.53 \\ 0.27 \\ -1.05 \end{Bmatrix} = \begin{Bmatrix} 1.24 \\ 4.89 \\ 1.71 \\ 10.79 \\ -3.04 \\ 11.26 \\ -2.31 \\ 5.45 \end{Bmatrix}.$$

Case 2

$$\mathbf{r} = \mathbf{r}_0 + y_2\mathbf{r}_2 = \begin{Bmatrix} 2.34 \\ 5.58 \\ 2.83 \\ 12.65 \\ -3.17 \\ 13.13 \\ -2.46 \\ 6.01 \end{Bmatrix} + 0.526 \begin{Bmatrix} -0.01 \\ -0.02 \\ -0.44 \\ -0.27 \\ 0.04 \\ -0.21 \\ -0.01 \\ 0.02 \end{Bmatrix} = \begin{Bmatrix} 2.34 \\ 5.57 \\ 2.60 \\ 12.51 \\ -3.15 \\ 13.02 \\ -2.46 \\ 6.02 \end{Bmatrix}.$$

As expected, these results are identical to those obtained by the S-M formula in example 8.1 [Eqs. (*e*), (*j*)].

8.3 Topological and Geometrical Changes

In this section exact solutions for topological and geometrical changes in the structure are presented. It is shown that the solution procedure developed in Sect. 8.2.2 is suitable also for these types of changes.

8.3.1 Topological Changes

Considering topological changes, it has been shown [8–11] that exact solutions can be obtained by the CA approach also for various cases where members are deleted or added. Moreover, in cases where the number of DOF is decreased or increased, the formulations presented in Sect. 5.3 can be used together with the procedure developed in Sect. 8.2.2 to obtain exact solutions. Exact solutions for changes affecting only a small number of members are demonstrated by the numerical example that follows. For simplicity of presentation a small-scale truss structure is considered. It has been noted earlier that the number of basis vectors needed to achieve the exact solution is equal to the number of changed members. However, it was found that a smaller number of vectors is often sufficient.

Example 8.4

To illustrate exact solutions for conditionally-unstable modified structures obtained by elimination of members, consider again the initial ten-bar truss shown in Fig. 8.3. The cross-sectional areas equal unity, the modulus of elasticity is 30000 and the eight analysis unknowns are the horizontal and the vertical displacements at joints 1, 2, 3 and 4, respectively. The following cases of eliminated members have been solved:

Case a: members 2+6+10 (Fig. 8.4*a*).
Case b: members 2+5+6+10 (Fig. 8.4*b*).
Case c: members 2+6+7+10 (Fig. 8.4*c*).
Case d: members 2+6+8+10 (Fig. 8.4*d*).
Case e: members 2+6 (Fig. 8.4*e*).
Case f: members 4+9 (Fig. 8.4*f*).
Case g: members 5+8+9 (Fig. 8.4*g*).
Case h: members 4+5+8+9 (Fig. 8.4*h*).

The resulting exact displacements obtained the CA approach with only 3 basis vectors are summarized in Table 8.2. It is observed that cases *a, b, c, d* represent stable structures, whereas cases *e, f, g, h* represent conditionally unstable structures. As a result, some of the displacements and joints are irrelevant and can be eliminated in order to obtain a stable structure.

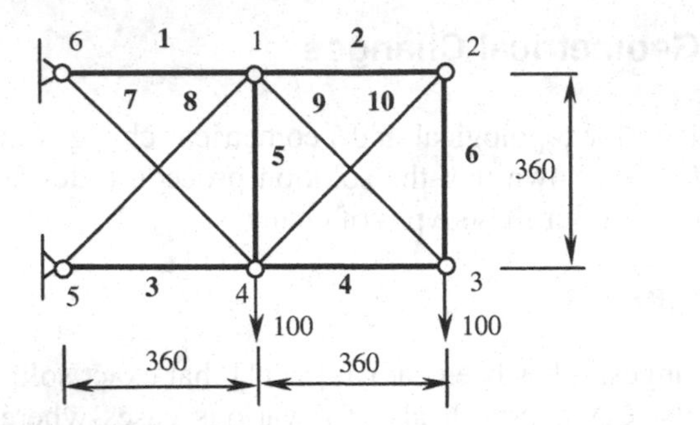

Fig. 8.3. Initial structure

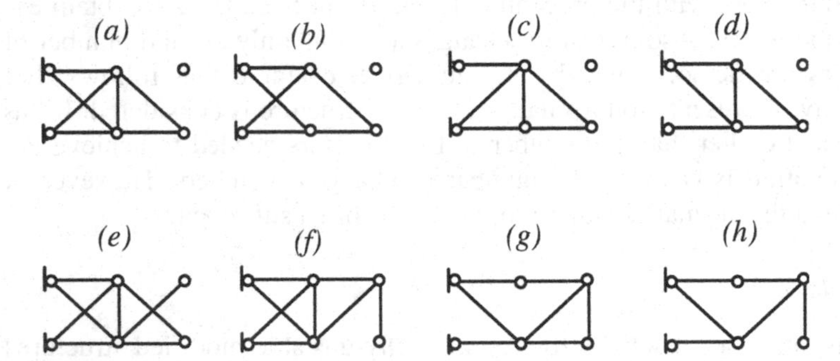

Fig. 8.4. Modified structures obtained by elimination of members

Table 8.2. Exact solutions obtained by the CA approach with 3 basis vectors

Case	Eliminated members	Displacements							
		1	2	3	4	5	6	7	8
a	2+6+10	2.40	5.80	*	*	-3.60	15.18	-2.40	5.80
b	2+5+6+10	2.40	5.80	*	*	-3.60	15.18	-2.40	5.80
c	2+6+7+10	3.60	10.37	*	*	-2.40	19.77	-1.20	11.57
d	2+6+8+10	1.20	11.57	*	*	-4.80	20.96	-3.60	10.37
e	2+6	2.40	5.80	*	*	-3.60	15.18	-2.40	5.80
f	4+9	2.11	4.67	3.30	13.62	*	14.81	-1.35	5.57
g	5+8+9	*	*	2.40	19.76	-3.60	20.96	-3.60	10.38
h	4+5+8+9	*	*	2.40	19.76	*	20.96	-3.60	10.38

* Irrelevant displacements that can be eliminated due to conditional instability

8.3.2 Geometrical Changes

Geometrical changes are conceptually similar to cross-sectional changes in the sense that the number of DOF is usually unchanged. However, since the displacements are highly nonlinear functions of the design variables, it might be difficult to achieve accurate approximations. In addition, changes in the geometry often significantly affect the response of the structure.

It is shown in this section that exact solutions can be achieved efficiently by the CA approach for geometrical changes that affect a small number of elements. This can be done by viewing these changes as corresponding topological changes [12]. Changing, for example, coordinates of a single joint we obtain the exact modified solution by viewing the change in the geometry as the following two successive changes in the topology:

- All members connected to that joint are deleted.
- New members are added at the modified location of the joint.

The solution procedure is demonstrated in the following example.

Example 8.5

Consider the nine-bar truss with the initial geometry shown in Fig. 8.5*a*. To calculate displacements for the modified geometry shown in Fig. 8.5*b*, we first assume that members 8 and 9 connected to joint 4 are deleted from the structure. Then we assume that the new members 10 and 11 are added at the new location of joint 4. The resulting matrix of changes in the stiffness matrix is given by

$$\Delta\mathbf{K} = \Delta\mathbf{K}\,(8, 9) + \Delta\mathbf{K}\,(10, 11) = \begin{bmatrix} \Delta\mathbf{K}_{11} & \Delta\mathbf{K}_{12} \\ \Delta\mathbf{K}_{21} & \Delta\mathbf{K}_{22} \end{bmatrix}, \tag{a}$$

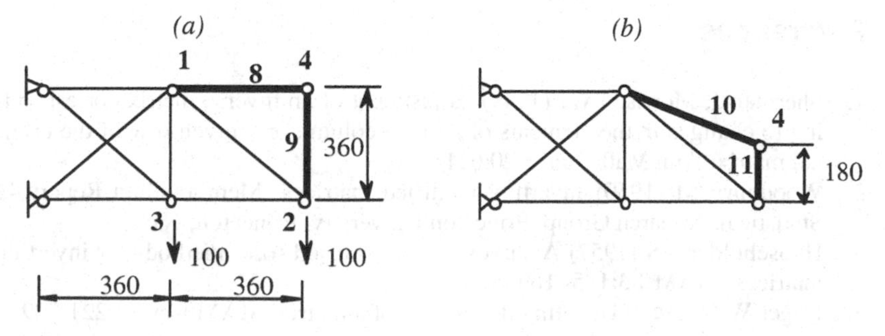

Fig. 8.5. ***a***. Initial truss geometry ***b***. Modified truss geometry

where

$$\Delta\mathbf{K}_{11} = 83.333\begin{bmatrix} -0.2844 & 0.3578 & 0 & 0 \\ 0.3578 & 0.1789 & 0 & 0 \\ 0 & 0 & 0 & 0 \\ 0 & 0 & 0 & 0 \end{bmatrix}, \tag{b}$$

$$\Delta\mathbf{K}_{22} = 83.333\begin{bmatrix} 0 & 0 & 0 & 0 \\ 0 & 0 & 0 & 0 \\ 0 & 0 & -0.2844 & 0.3578 \\ 0 & 0 & 0.3578 & 1.1789 \end{bmatrix}, \tag{c}$$

$$\Delta\mathbf{K}_{21} = \Delta\mathbf{K}_{12}^T = 83.333\begin{bmatrix} 0 & 0 & 0 & 0 \\ 0 & 0 & 0 & 0 \\ 0.2844 & -0.3578 & 0 & 0 \\ -0.3578 & -0.1789 & 0 & -1 \end{bmatrix}. \tag{d}$$

Since four members have been changed, four basis vectors are needed to achieve the exact solution. In the present case, the matrices corresponding to the two vertical members 9 and 11 are linearly dependent and, therefore, only three basis vectors are required to obtain the exact modified displacements

$$\mathbf{r}^T = \{2.40, 5.80, -3.60, 15.19, -2.40, 5.80, -2.30, 15.19\}. \tag{e}$$

References

1. Sherman J, Morrison WJ (1949) Adjustment of an inverse matrix corresponding to changes in the elements of a given column or a given row of the original matrix. Ann Math Statist 20:621
2. Woodbury M (1950) Inverting modified matrices. Memorandum Report 42 Statistical Research Group, Princeton University, Princeton, NJ
3. Householder AS (1957) A survey of some closed form methods for inverting matrices. SIAM J 3:155 -169
4. Hager WW (1989) Updating the inverse of a matrix. SIAM Rev 31:221-239

5. Akgun MA, Garcelon JH, Haftka RT (2001) Fast exact linear and nonlinear structural reanalysis and the Sherman-Morrison-Woodbury formulas. Int J Num Meth Engrg 50:1587-1606
6. Argyris JH Roy JR (1972) General treatment of structural modifications. J Struct Div ASCE 98:462-492
7. Kirsch U (1981) Optimum structural design. McGraw Hill, New York
8. Kirsch U, Liu S (1995) Exact structural reanalysis by a first-order reduced basis approach. Struct and Multidis Opt 10:153-158
9. Kirsch U (1993) Efficient reanalysis for topological optimization. Struct and Multidis Opt 6:143-150
10. Kirsch U, Liu S (1997) Structural reanalysis for general layout modifications. AIAA J 35:382-388
11. Kirsch U, Papalambros PY (2001) Structural reanalysis for topological modifications. Struct and Multidis Opt 21:333-344
12. Kirsch U, Papalambros PY (2001) Exact and accurate reanalysis of structures for geometrical changes. Engrg with Comp 17:363-372

9 Repeated Sensitivity Analysis

Design sensitivity analysis of structures deals with the calculation of changes in the response resulting from changes in the parameters describing the structure. The derivatives of the response with respect to the system parameters, called the sensitivity coefficients, are used in the solution of various problems. In design optimization, the sensitivity coefficients are used to select a search direction. Their calculation often involves much computational effort, particularly in the optimization of large scale systems. Moreover, calculation of the derivatives for a given design involves structural analysis of the design. As a result, there has been much interest in efficient procedures for calculating the sensitivity coefficients [1].

Design sensitivities are often used in generating approximations for the response of a modified system, including approximate reanalysis models and explicit approximations of the constraint functions in terms of the structural parameters (e.g. first-order Taylor series approximations). In addition, the sensitivities are required for assessing the effects of uncertainties in the structural properties (e.g. material or geometric properties) on the system response.

Similar to reanalysis, the following factors are considered in choosing a suitable method for repeated sensitivity analysis (or sensitivity reanalysis):

- The accuracy of the calculations.
- The computational effort involved.
- The ease-of-implementation.

The present chapter deals with repeated sensitivity analysis for discrete systems. The two general approaches used for calculating the sensitivity coefficients are:

- *The direct approach*, considered in this chapter, which is based on the implicit differentiation of the analysis equations that describe the system response with respect to the desired parameters, and the solution of the resulting equations.
- *The adjoint-variable approach*, not mentioned in this book, where an adjoint physical system is introduced whose solution permits the rapid evaluation of the desired sensitivity coefficients.

Methods of design sensitivity analysis for discretized systems can be divided into the following classes:

- *Finite-difference methods*, which are easy to implement but might involve numerous repeated analyses and involve high computational cost, particularly in problems with many design variables or response variables. The efficiency can be improved by using fast reanalysis techniques. In addition, the solution might involve accuracy problems.
- *Analytical methods*, which provide exact solutions but might not be easy to implement in some problems such as shape optimization.
- *Semi-analytical methods*, which are based on a compromise between finite-difference methods and analytical methods. These methods use finite-difference evaluation of the right-hand-side vector. They are easy to implement but may provide inaccurate results [2, 3]. The accuracy can be improved by using one of several procedures.
- Computational or automated derivatives that rely on differentiation of the software itself.

The outline of this chapter is as follows. Finite-difference derivatives are described in Sect. 9.1. Repeated sensitivity reanalysis by the CA approach, using analytical and finite-difference derivatives, is developed in Sects. 9.2 – 9.5. Static problems are presented in Sect. 9.2 and vibration problems are introduced in Sect. 9.3. Solutions of linear and nonlinear dynamic problems are discussed in Sects. 9.4 and 9.5, respectively.

9.1 Finite-Difference Derivatives

For simplicity of presentation assume a single design variable. Consider the problem of calculating the derivatives $\partial \mathbf{r}_0 / \partial X$ of the displacement vector $\mathbf{r}$ with respect to a design variable X at the point X_0. In the *forward-difference* method, the derivatives are approximated from the exact displacements at the original point X_0 and at the perturbed point $X_0+\delta X$ by

$$\frac{\partial \mathbf{r}_0}{\partial X} = \frac{\mathbf{r}(X_0 + \delta X) - \mathbf{r}(X_0)}{\delta X}, \tag{9.1}$$

where δX is a predetermined step-size. The accuracy can be improved by the *central-difference* approximation, where the derivatives are computed from the exact displacements at the points $X_0 - \delta X$ and $X_0 + \delta X$ by

$$\frac{\partial \mathbf{r}_0}{\partial X} = \frac{\mathbf{r}(X_0 + \delta X) - \mathbf{r}(X_0 - \delta X)}{2\delta X}. \tag{9.2}$$

Finite-difference methods are the easiest to implement and therefore they are attractive in many applications. When $\mathbf{r}(X_0)$ is known, application of Eq. (9.1) involves only one additional calculation of the displacements at $X_0 + \delta X$ while Eq. (9.2) requires calculation at the two points $X_0 - \delta X$ and $X_0 + \delta X$. For a problem with n design variables, finite difference derivative calculations require repetition of the analysis for n+1 [Eq. (9.1)] or $2n$+1 [Eq. (9.2)] different design points. This procedure is usually not efficient compared to analytical and semi-analytical methods.

As noted earlier, finite-difference approximations might involve accuracy problems. The following two sources of errors should be considered whenever these approximations are used:

- The *truncation error*, which is a result of neglecting terms in the Taylor series expansion of the perturbed response. Considering the forward-difference method [Eq. (9.1)] the Taylor series expansion of $\mathbf{r}(X_0 + \delta X)$ about X_0 yields

$$\mathbf{r}(X_0 + \delta X) = \mathbf{r}(X_0) + \delta X \frac{\partial \mathbf{r}_0}{\partial X} + \frac{(\delta X)^2}{2!}\frac{\partial^2 \mathbf{r}_0}{\partial X^2} + \frac{(\delta X)^3}{3!}\frac{\partial^3 \mathbf{r}_0}{\partial X^3} - \ldots \tag{9.3}$$

or, after rearranging

$$\frac{\partial \mathbf{r}_0}{\partial X} = \frac{\mathbf{r}(X_0 + \delta X) - \mathbf{r}(X_0)}{\delta X} - \delta X \frac{\partial^2 \mathbf{r}_0}{2\partial X^2} + (\delta X)^2 \frac{\partial^3 \mathbf{r}_0}{3!(\partial X)^3} - \ldots \tag{9.4}$$

It is observed that the truncation error in this case is

$$-\delta X \frac{\partial^2 \mathbf{r}_0}{2\partial X^2} + (\delta X)^2 \frac{\partial^3 \mathbf{r}_0}{3!(\partial X)^3} - \ldots \tag{9.5}$$

and the largest term in this expression is proportional to δX. Similarly, for the central-difference method [Eq. (9.2)], the Taylor series expansion of the displacements $\mathbf{r}(X_0 - \delta X)$ about X_0 yields

$$\mathbf{r}(X_0 - \delta X) = \mathbf{r}(X_0) - \delta X \frac{\partial \mathbf{r}_0}{\partial X} + \frac{(\delta X)^2}{2}\frac{\partial^2 \mathbf{r}_0}{\partial X^2} - \frac{(\delta X)^3}{3!}\frac{\partial^3 \mathbf{r}_0}{\partial X^3} + \ldots. \tag{9.6}$$

From Eqs. (9.3), (9.6) we obtain

$$\frac{\partial \mathbf{r}_0}{\partial X} = \frac{\mathbf{r}(X_0 + \delta X) - \mathbf{r}(X_0 - \delta X)}{2\delta X} - \frac{(\delta X)^2}{3!}\frac{\partial^3 \mathbf{r}_0}{\partial X^3} + \ldots \tag{9.7}$$

It is observed that the largest truncation error in this case is proportional to $(\delta X)^2$.

- The *condition error*, which is the difference between the numerical evaluation of the function and its exact value. Examples for this type of error include round-off error in calculating $\partial \mathbf{r}_0 / \partial X$ from the original and perturbed values of $\mathbf{r}$, and calculation of the response by approximate analysis. Approximate analysis can also be the result of an insufficient number of iterations being used during the solution process.

Considerations related to the truncation error and the condition error are conflicting. A small step size δX will reduce the truncation error, but may increase the condition error. In some cases there may not be any step size which yields an acceptable error. In certain applications, truncation errors are not of major importance since it might be sufficient to find the average rate of change in the structural response and not necessarily the accurate local rate of change at a given point. Therefore, to eliminate round-off errors due to approximations it is recommended to increase the step-size.

It is known that small response values are not calculated as accurately as large response values. The same applies to derivatives [4]. Thus, it would be difficult to evaluate accurately small response derivatives by finite difference or other approximations. However, it is usually less important to evaluate accurately such derivative values. The relative magnitude of the derivatives can be estimated from the ratio $(\partial \mathbf{r} / \mathbf{r})/(\partial X / X)$.

9.2 Static Problems

9.2.1 Analytical Derivatives

Given a design X_0 and the corresponding stiffness matrix $\mathbf{K}_0$, the resulting displacements $\mathbf{r}_0$ are computed by the linear static equilibrium equations

$$\mathbf{K}_0 \, \mathbf{r}_0 = \mathbf{R}_0, \tag{9.8}$$

where $\mathbf{R}_0$ is the load vector. The stiffness matrix is first factorized by

$$\mathbf{K}_0 = \mathbf{U}_0^T \, \mathbf{U}_0, \tag{9.9}$$

where $\mathbf{U}_0$ is an upper triangular matrix. Differentiating Eq. (9.8) with respect to the design variable X and rearranging yields

$$\mathbf{K}_0 \frac{\partial \mathbf{r}_0}{\partial X} = \frac{\partial \mathbf{R}_0}{\partial X} - \frac{\partial \mathbf{K}_0}{\partial X} \mathbf{r}_0 \, . \tag{9.10}$$

The right-hand side of this equation is often referred to as the *pseudo-load vector*. Equations (9.8) and (9.10) have the same coefficient matrix $\mathbf{K}_0$. If the decomposed form of Eq. (9.9) is available, then only forward and backward substitutions are needed to solve for $\partial \mathbf{r}_0 / \partial X$ [Eq. (9.10)].

The load vector **R** is often assumed to be independent of the design variables. In such cases Eq. (9.10) is reduced to the form

$$\mathbf{K}_0 \frac{\partial \mathbf{r}_0}{\partial X} = -\frac{\partial \mathbf{K}_0}{\partial X} \mathbf{r}_0 . \tag{9.11}$$

Analytical methods are widely used and often demonstrate good performance in terms of both accuracy and efficiency. However, implementation of these methods is difficult in some problems such as shape optimization, where the derivatives in the right-hand-side vector of Eq. (9.10) are not easy to obtain. Employing finite-difference methods can improve significantly the ease of implementation at the expense of less accurate results. The effect of the finite difference approximations on the computational efficiency strongly depends on the problem under consideration.

9.2.2 Semi-Analytical Derivatives

Semi-analytical methods may improve the ease-of-implementation of analytical methods. These methods are based on finite-difference evaluation of the right-hand-side vector of Eq. (9.10). Using the forward-difference method, the displacement derivatives $\partial \mathbf{r}_0 / \partial X$ are computed from the exact values of **R** and **K** at the two points X_0 and $X_0 + \delta X$ by

$$\mathbf{K}_0 \frac{\partial \mathbf{r}_0}{\partial X} = \frac{\mathbf{R}(X_0 + \delta X) - \mathbf{R}(X_0)}{\delta X} - \frac{\mathbf{K}(X_0 + \delta X) - \mathbf{K}(X_0)}{\delta X} \mathbf{r}_0 . \tag{9.12}$$

Using the central-difference method, the requested derivatives are computed from the exact values of **R** and **K** at $X_0 - \delta X$ and $X_0 + \delta X$ by

$$\mathbf{K}_0 \frac{\partial \mathbf{r}_0}{\partial X} = \frac{\mathbf{R}(X_0 + \delta X) - \mathbf{R}(X_0 - \delta X)}{2\delta X} - \frac{\mathbf{K}(X_0 + \delta X) - \mathbf{K}(X_0 - \delta X)}{2\delta X} \mathbf{r}_0 . \tag{9.13}$$

Semi-analytical methods combine ease-of-implementation and computational efficiency, and they have been implemented in several finite element programs. However, the errors associated with the finite-difference approximations of the right-hand-side vector can be substantial [2–3]. Various methods have been proposed to improve the accuracy [5–9].

In this section the *refined semi-analytical method* [9], which is useful for shape design variables, is described. It was found that inaccuracies arise when large rigid body motions are identified for the individual elements. Using the refined semi-analytical method, the contribution to the design sensitivities corresponding to the rigid body motion are evaluated by exact differentiation of the rigid body modes. The non-self equilibrating contributions to the pseudo-load vector are also corrected using exact differentiation of rigid body modes. Consider the forward-difference method and assume that the load vector **R** is independent of the design variables. The pseudo-load vector **q** is the right-hand side of Eq. (9.11)

$$\mathbf{q} = -\frac{\partial \mathbf{K}}{\partial X}\mathbf{r}, \tag{9.14}$$

which is evaluated by summing the individual element contributions $\mathbf{q}_e$

$$\mathbf{q}_e = -\frac{\partial \mathbf{K}_e}{\partial X}\mathbf{r}_e, \tag{9.15}$$

where $\mathbf{K}_e$ is the element stiffness matrix and $\mathbf{r}_e$ is the element displacement vector. The latter vector can be decomposed into a part of pure deformations $\mathbf{r}_e^{\varepsilon}$ and a part of pure rigid body motions $\boldsymbol{\varphi}_k$. The latter are orthogonal vectors that can be easily found. Obviously, it holds true that

$$\mathbf{K}_e\boldsymbol{\varphi}_k = \mathbf{0}. \tag{9.16}$$

Differentiating Eq. (9.16) and rearranging gives

$$\frac{\partial \mathbf{K}_e}{\partial X}\boldsymbol{\varphi}_k = -\mathbf{K}_e\frac{\partial \boldsymbol{\varphi}_k}{\partial X}. \tag{9.17}$$

The displacement vector $\mathbf{r}_e$ can be expressed in terms of the M rigid body motions $\boldsymbol{\varphi}_k$ and corresponding scalars α_k as

$$\mathbf{r}_e = \mathbf{r}_e^{\varepsilon} + \sum_{k=1}^{M}\alpha_k\boldsymbol{\varphi}_k. \tag{9.18}$$

Using some additional algebraic operations we obtain the following expression for the individual element contributions $\mathbf{q}_e$ [9]

$$\mathbf{q}_e = -\frac{\partial \mathbf{K}_e}{\partial X}\mathbf{r}_e^{\varepsilon} + \left(\frac{\boldsymbol{\varphi}_k^T\dfrac{\partial \mathbf{K}_e}{\partial X}\mathbf{r}_e^{\varepsilon}}{\boldsymbol{\varphi}_k^T\boldsymbol{\varphi}_k}\right)\boldsymbol{\varphi}_k + \left(\frac{\dfrac{\partial \boldsymbol{\varphi}_k^T}{\partial X}\mathbf{K}_e\mathbf{r}_e^{\varepsilon}}{\boldsymbol{\varphi}_k^T\boldsymbol{\varphi}_k}\right)\boldsymbol{\varphi}_k + \sum_{k=1}^{M}\alpha_k\mathbf{K}_e\frac{\partial \boldsymbol{\varphi}_k}{\partial X}. \tag{9.19}$$

The third and fourth terms in this equation can be calculated exactly by an analytical differentiation. The first and the second terms are also corrected using exact differentiation of rigid body modes. As a result, the method reduces the errors involved in the semi-analytical method.

9.2.3 Repeated Analytical Derivatives

The problem considered in this section is to evaluate analytical derivatives $\partial \mathbf{r}/\partial X$ for various modified designs where exact calculations of $\mathbf{r}$ are not available. It has been shown [10–12] that the CA procedure presented in Chap. 5 can be used to effectively calculate the required derivatives. When results of exact analysis for modified designs are available, calculation of analytical displacement derivatives might be straightforward, using the formulation presented in Sect. 9.2.1.

Assume that the elements of $\mathbf{K}$ are some functions of X, expressed as

$$\mathbf{K} = \mathbf{K}_0 + \Delta\mathbf{K}(X), \tag{9.20}$$

such that the derivatives $\partial \mathbf{K}/\partial X$ are readily available.

Differentiation of Eq. (5.24) yields

$$\frac{\partial \mathbf{r}}{\partial X} = \frac{\partial \mathbf{r}_B}{\partial X}\mathbf{y} + \mathbf{r}_B \frac{\partial \mathbf{y}}{\partial X}. \tag{9.21}$$

Differentiating Eq. (5.29) and rearranging, we obtain for $\partial \mathbf{y}/\partial X$

$$\mathbf{K}_R \frac{\partial \mathbf{y}}{\partial X} = \frac{\partial \mathbf{R}_R}{\partial X} - \frac{\partial \mathbf{K}_R}{\partial X}\mathbf{y}. \tag{9.22}$$

Assuming, for simplicity of presentation,

$$\mathbf{R} = \mathbf{R}_0, \tag{9.23}$$

then from Eqs. (5.28) we have

$$\frac{\partial \mathbf{K}_R}{\partial X} = \frac{\partial \mathbf{r}_B^T}{\partial X}\mathbf{K}\mathbf{r}_B + \mathbf{r}_B^T \frac{\partial \mathbf{K}}{\partial X}\mathbf{r}_B + \mathbf{r}_B^T \mathbf{K}\frac{\partial \mathbf{r}_B}{\partial X}, \tag{9.24}$$

$$\frac{\partial \mathbf{R}_R}{\partial X} = \frac{\partial \mathbf{r}_B^T}{\partial X}\mathbf{R}_0. \tag{9.25}$$

In summary, evaluation of the derivatives $\partial \mathbf{r}/\partial X$ by the CA approach for any assumed X involves the following steps:

- Determine the matrices $\partial \mathbf{r}_B/\partial X$ and $\partial \mathbf{K}/\partial X$. Derivatives of the basis vectors, $\partial \mathbf{r}_B/\partial X$, can be calculated in several ways, of which two simple and efficient possibilities are described in the next sub-section.
- Calculate the $\partial \mathbf{K}_R/\partial X$ and $\partial \mathbf{R}_R/\partial X$ [Eqs. (9.24), (9.25)].
- Calculate the $\partial \mathbf{y}/\partial X$ by solving a reduced set $s \times s$ set [Eq. (9.22)].
- Evaluate the derivatives $\partial \mathbf{r}/\partial X$ [Eq. (9.21)].

Table 9.1 shows that the computations of **r** and $\partial \mathbf{r}/\partial X$ by the CA procedure for any modified design involve similar calculations.

Table 9.1. Summary of calculations

Calculation	Displacements	Displacement derivatives
Coefficient matrices	$\mathbf{K}$	$\partial \mathbf{K}/\partial X$
Basis vectors	$\mathbf{r}_B$	$\partial \mathbf{r}_B/\partial X$
Reduced matrices and vectors	$\mathbf{K}_R$	$\partial \mathbf{K}_R/\partial X$
	$\mathbf{R}_R$	$\partial \mathbf{R}_R/\partial X$
Reduced unknowns	$\mathbf{y}$	$\partial \mathbf{y}/\partial X$
Final unknowns	$\mathbf{r}$	$\partial \mathbf{r}/\partial X$

Derivatives of the Basis Vectors

The basis vectors are implicit functions of X. In such cases it is possible to consider explicit approximations [12]. Assume for example the quadratic fitting

$$\mathbf{r}_i(X) = \mathbf{a}_i + \mathbf{b}_i X + \mathbf{c}_i X^2, \tag{9.26}$$

where $\mathbf{a}_i$, $\mathbf{b}_i$, $\mathbf{c}_i$ are constant vectors. The latter vectors can be determined from the basis vectors calculated at three design points, which are in between the bounds on X. Differentiating Eq. (9.26), we obtain

$$\partial \mathbf{r}_i/\partial X = \mathbf{b}_i + 2\mathbf{c}_i X. \tag{9.27}$$

The expressions of Eqs. (9.26), (9.27) can be used for any modified design.

The basis vectors are explicit functions of X. In such cases the solution is often straightforward. Consider for example rank-one changes in the stiffness matrix (Sect. 8.2.1), where Eq. (9.20) is expressed as

$$\mathbf{K} = \mathbf{K}_0 + X\,\Delta\mathbf{K}_0. \tag{9.28}$$

The elements of matrices $\Delta\mathbf{K}_0$ and $\partial \mathbf{K}/\partial X = \Delta\mathbf{K}_0$ are constant. In this case the exact displacements obtained by the CA approach are [Eq. (8.32)]

$$\mathbf{r} = \mathbf{r}_0 + y_1\,\mathbf{r}_1, \tag{9.29}$$

where the basis vector $\mathbf{r}_1$ is given by [see Eq. (8.33)]

$$\mathbf{r}_1 = -\mathbf{K}_0^{-1}\,\Delta\mathbf{K}\,\mathbf{r}_0 = -\,X\,\mathbf{B}_0\,\mathbf{r}_0, \tag{9.30}$$

and $\mathbf{B}_0$ is a matrix of constant elements defined as

$$\mathbf{B}_0 = \mathbf{K}_0^{-1}\,\Delta\mathbf{K}_0. \tag{9.31}$$

Since multiplication of a basis vector by a scalar does not change the results we can use, instead of $\mathbf{r}_1$ [Eq. (9.30)], the constant basis vector

$$\mathbf{r}_{01} = -\,\mathbf{B}_0\,\mathbf{r}_0, \tag{9.32}$$

for any modified design. The constant elements of matrix $\partial\mathbf{r}_B/\partial X$ are obtained by differentiation of Eqs. (9.30)

$$\partial\mathbf{r}_1/\partial X = -\,\mathbf{B}_0\,\mathbf{r}_0 \tag{9.33}$$

Example 9.1

To illustrate the accuracy of the results achieved by the CA procedure, consider the ten-bar truss shown in Fig. 9.1. The modulus of elasticity is 30000 and the eight analysis unknowns are the horizontal and the vertical displacements at joints 1, 2, 3 and 4, respectively. The initial cross-sectional areas are $\mathbf{X}_0$ = **1.0**, the stress constraints are $-\mathbf{25.0} \le \sigma \le \mathbf{25.0}$, and the minimum size constraints are $\mathbf{0.001} \le \mathbf{X}$. Assuming minimum weight design, the optimum is

$$\mathbf{X}^T_{opt} = \{8.0,0.001,8.0,4.0,0.001,0.001,5.667,5.667,5.667,0.001\}. \tag{a}$$

Consider the line from the initial design to the optimal design

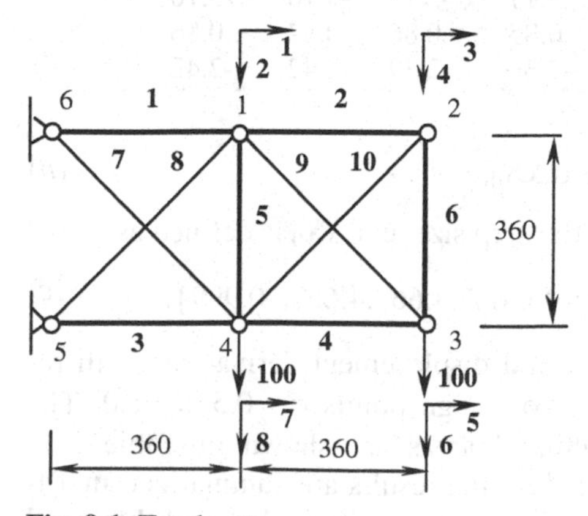

Fig. 9.1. Ten-bar truss

Table 9.2. Cross section areas of elements for $\alpha = 0.5$ and $\alpha = 1.0$

Element	$\alpha = 0.5$	$\alpha = 1.0$
1	4.500	8.000
2	0.501	0.001
3	4.500	8.000
4	2.500	4.000
5	0.501	0.001
6	0.501	0.001
7	3.334	5.667
8	3.334	5.667
9	3.334	5.667
10	0.501	0.001

Table 9.3. Approximate displacements and displacement derivatives for $\alpha = 0.5$

Number of basis vectors	2	3	4	5	Exact
Displacements	0.50	0.52	0.52	0.52	0.52
	1.53	1.46	1.48	1.49	1.49
	0.71	0.76	0.77	0.77	0.77
	3.56	3.63	3.64	3.65	3.65
	-0.89	-0.98	-0.99	-0.98	-0.98
	3.77	3.87	3.89	3.90	3.90
	-0.54	-0.55	-0.55	-0.55	-0.55
	1.71	1.64	1.62	1.62	1.62
Displacement derivatives	-0.80	-0.81	-0.80	-0.79	-0.79
	-2.16	-2.16	-2.11	-2.09	-2.09
	-1.05	-1.04	-1.02	-1.01	-1.01
	-4.96	-4.93	-4.91	-4.89	-4.89
	1.24	1.23	1.21	1.24	1.24
	-5.21	-5.17	-5.13	-5.10	-5.10
	0.85	0.85	0.86	0.87	0.86
	-2.37	-2.36	-2.39	-2.42	-2.42

$$\mathbf{X} = \mathbf{X}_0 + \alpha\Delta\mathbf{X}_0, \qquad (b)$$

where α is a variable representing the step size and $\Delta\mathbf{X}_0$ is defined as

$$\Delta\mathbf{X}_0^T = \{7,-0.999,7,3,-0.999,-0.999,4.667,4.667,4.667,-0.999\}. \qquad (c)$$

Evaluation of the displacements and displacement derivatives with respect to α are illustrated for the two design points $\alpha = 0.5$, $\alpha = 1.0$. The corresponding modified cross-sectional areas are shown in Table 9.2. Solving by the presented CA procedure the results are summarized in Tables 9.3, 9.4 for various numbers of basis vectors. It is observed that good accuracy is achieved with 2–3 basis vectors for these very large changes in the design, for both displacements and displacement derivatives. The accuracy is further improved by considering additional basis vectors.

Table 9.4. Approximate displacements and displacement derivatives for $\alpha = 1.0$

Number of basis vectors	2	3	4	5	Exact
Displacements	0.28	0.29	0.29	0.30	0.30
	0.90	0.84	0.88	0.90	0.90
	0.41	0.45	0.47	0.49	0.49
	2.10	2.17	2.19	2.21	2.21
	-0.53	-0.61	-0.62	-0.60	-0.60
	2.24	2.34	2.37	2.40	2.40
	-0.30	-0.31	-0.31	-0.30	-0.30
	1.01	0.95	0.93	0.90	0.90
Displacement derivatives	0.25	0.25	0.25	0.24	0.24
	0.74	0.72	0.69	0.64	0.64
	*	*	*	*	*
	*	*	*	*	*
	0.43	0.46	0.44	0.47	0.47
	1.81	1.84	1.81	1.74	1.74
	0.26	0.27	0.27	0.29	0.29
	0.82	0.80	0.81	0.87	0.88

* Irrelevant results, joint 2 is practically eliminated

9.2.4 Repeated Finite-Difference Derivatives

It has been noted that the evaluation of derivatives by finite-difference approximations involves multiple repeated analyses. In cases where the derivatives must be calculated for many design variables or design points, the resulting computational effort might be prohibitive. The CA approach is most suitable to overcome this difficulty. It can be used to evaluate efficiently the response for numerous modified designs.

Given an initial design, we assume that the corresponding stiffness matrix $\mathbf{K}_0$ is given in the decomposed form of Eq. (9.9), and the displacements $\mathbf{r}_0$ are computed by the initial equilibrium equations [Eq. (9.8)]. For any change in the design, calculation of the modified displacements by the CA approach can be carried out using the procedure developed in Sect. 5.2. It was found that accurate results can be achieved efficiently [13]. In this section, some numerical examples are demonstrated. Accuracy and efficiency considerations are discussed later in Chap. 10.

We may distinguish between the following two cases of calculating the sensitivity coefficients by the CA approach:

- Calculation of the sensitivity coefficients at the initial design point X_0. In this case sensitivity analysis by the forward-difference method involves an additional analysis for $X_0 + \delta X$. The central-difference method requires two additional analyses (for $X_0 - \delta X$ and $X_0 + \delta X$). Since the change in the design is small, high accuracy of the approximations is expected with a small number of basis vectors.
- Calculation of the sensitivity coefficients at a design point X where the factorized stiffness matrix $\mathbf{U}$ and the exact displacements $\mathbf{r}$ are unknown. Using the CA approach in this case, the basis vectors are calculated by forward and backward substitutions for those design points where results of analysis are required. It is expected that the accuracy of the results will not be as good as for the previous case.

Example 9.2

Consider again the ten-bar truss of example 9.1 shown in Fig. 9.1. The design variables are the cross-sectional areas $\mathbf{X}$, and the initial cross sections $\mathbf{X}_0$ are all unity. Assuming that results of exact displacements are given only at the initial point $\mathbf{X}_0$, displacement derivatives calculated by the following methods have been compared:

- $\partial \mathbf{r}/\partial X$(Ex) = exact analytical derivatives.
- $\partial \mathbf{r}/\partial X$(FD) = finite-difference derivatives using exact analysis.
- $\partial \mathbf{r}/\partial X$(CA2) = finite-difference derivatives using CA analysis with 2 basis vectors.

The following cases have been solved:

- Calculation of $\partial \mathbf{r}/\partial X_1$ and $\partial \mathbf{r}/\partial X_2$ at $\mathbf{X}_0$. Assuming the forward-difference method and $\delta X_1 = \delta X_2 = 0.001$, the results in Tables 9.5, 9.6 show that both finite-difference solutions, $\partial \mathbf{r}/\partial X$(FD), $\partial \mathbf{r}/\partial X$(CA2) are identical, with very small errors compared with the exact solution.
- Assume the line from the initial design given by

$$\mathbf{X} = \mathbf{X}_0 + \alpha \Delta \mathbf{X}_0, \tag{a}$$

where α is a variable representing the step size and $\Delta \mathbf{X}_0$ is defined as

$$\Delta \mathbf{X}_0^T = \{7, -0.999, 7, 3, -0.999, -0.999, 4.667, 4.667, 4.667, -0.999\}. \tag{b}$$

Assuming the forward-difference method and $\delta\alpha = 0.0001$, the derivatives $\partial \mathbf{r}/\partial \alpha$ computed by the various methods at $\alpha = 0$ are shown in Table 9.7. It is observed that $\partial \mathbf{r}/\partial \alpha$(FD) and $\partial \mathbf{r}/\partial \alpha$(CA2) are practically identical and very close to $\partial \mathbf{r}/\partial \alpha$(Ex).

- Assuming forward-differences and $\delta\alpha$ =0.0001, the derivatives $\partial\mathbf{r}/\partial\alpha$ computed by the various methods at $\alpha = 0.5$ are shown in Table 9.8. In this case calculation of $\partial\mathbf{r}/\partial\alpha$(FD) requires exact analysis at $\alpha = 0.5$ and $\alpha = 0.5001$, whereas $\partial\mathbf{r}/\partial\alpha$(CA2) requires only approximate analysis at these points. It is observed that the accuracy obtained by $\partial\mathbf{r}/\partial\alpha$(CA2), even for the very large change in the design, is relatively good.

Table 9.5. Sensitivities $\partial\mathbf{r}/\partial X_1$ at $\mathbf{X}_0 = \mathbf{1.0}$, $\delta X_1 = 0.001$, ten-bar truss

Number	$\partial\mathbf{r}/\partial X_1$(Ex)	$\partial\mathbf{r}/\partial X_1$(FD) = $\partial\mathbf{r}/\partial X_1$(CA2)
1	-2.0703	-2.0685
2	-1.2950	-1.2939
3	-2.0987	-2.0968
4	-3.5024	-3.4993
5	0.2457	0.2455
6	-3.5308	-3.5276
7	0.2741	0.2738
8	-1.0493	-1.0484

Table 9.6. Sensitivities $\partial\mathbf{r}/\partial X_2$ at $\mathbf{X}_0 = \mathbf{1.0}$, $\delta X_2 = 0.001$, ten-bar truss

Number	$\partial\mathbf{r}/\partial X_2$(Ex)	$\partial\mathbf{r}/\partial X_2$(FD) = $\partial\mathbf{r}/\partial X_2$(CA2)
1	-0.0058	-0.0058
2	-0.0223	-0.0223
3	-0.4369	-0.4365
4	-0.2660	-0.2657
5	0.0446	0.0446
6	-0.2155	-0.2153
7	-0.0058	-0.0058
8	0.0223	0.0223

Table 9.7. Sensitivities $\partial\mathbf{r}/\partial\alpha$ at $\alpha = 0$, $\delta\alpha = 0.0001$, ten-bar truss

Number	$\partial\mathbf{r}/\partial\alpha$(Ex)	$\partial\mathbf{r}/\partial\alpha$(FD)	$\partial\mathbf{r}/\partial\alpha$(CA2)
1	-16.548	-16.536	-16.536
2	-32.028	-32.009	-32.008
3	-17.541	-17.53	-17.53
4	-70.619	-70.575	-70.576
5	17.733	17.721	17.722
6	-71.612	-71.568	-71.569
7	17.052	17.040	17.041
8	-33.214	-33.194	-33.194

Table 9.8. Sensitivities $\partial\mathbf{r}/\partial\alpha$ at $\alpha = 0.5$, $\delta\alpha = 0.0001$, ten-bar truss

Number	$\partial\mathbf{r}/\partial\alpha$(Ex)	$\partial\mathbf{r}/\partial\alpha$(FD)	$\partial\mathbf{r}/\partial\alpha$(CA2)
1	-0.79	-0.79	-0.80
2	-2.09	-2.09	-2.16
3	-1.01	-1.01	-1.05
4	-4.89	-4.89	-4.96
5	1.24	1.24	1.24
6	-5.10	-5.10	-5.20
7	0.86	0.86	0.85
8	-2.42	-2.42	-2.37

9.2.5 Errors Due to Rigid Body Motions

It has been noted in Sect. 9.2.2 that, using semi-analytical derivatives, the errors associated with the finite-difference approximations of the right-hand-side vector can be substantial. It was found that such errors arise when relatively large rigid body motions can be identified for the individual elements. It is shown in this section that the expressions of the semi-analytical derivatives are similar to the expressions of the basis vectors in finite-difference calculations by the CA approach. Therefore, for relatively large rigid body motions, the errors obtained by this approach might also be substantial. It is possible to reduce these errors by improving the choice of the basis vectors, using central-difference basis vectors, as will be shown later in Sect. 9.3.3. These basis vectors can be used in various sensitivity analysis problems, including static, vibration and dynamic problems.

Assuming a perturbation δX, and corresponding perturbation $\delta\mathbf{K}$ in the stiffness matrix, the basis vectors at the perturbed design are usually calculated by the recurrence relation

$$\mathbf{r}_i = -\mathbf{K}_0^{-1}\,\delta\mathbf{K}\,\mathbf{r}_{i\text{-}1} \quad (i = 2, ..., s). \tag{9.34}$$

For simplicity of presentation assume a constant load vector

$$\mathbf{R}_0 = \mathbf{R}(X_0) = \mathbf{R}(X_0 + \delta X). \tag{9.35}$$

That is, the first basis vector $\mathbf{r}_1$ is simply

$$\mathbf{r}_1 = \mathbf{r}_0, \tag{9.36}$$

and the second basis vector $\mathbf{r}_2$ is given by

$$\mathbf{r}_2 = -\mathbf{K}_0^{-1}\,\delta\mathbf{K}\,\mathbf{r}_0. \tag{9.37}$$

Considering the forward-difference method, matrix $\Delta\mathbf{K}$ is given by

$$\delta\mathbf{K} = \mathbf{K}(X_0 + \delta X) - \mathbf{K}(X_0). \tag{9.38}$$

Substituting Eq. (9.38) into Eq. (9.37) and rearranging we obtain for $\mathbf{r}_2$

$$\mathbf{K}_0\,\mathbf{r}_2 = -\,[\mathbf{K}(X_0{+}\delta X) - \mathbf{K}(X_0)]\,\mathbf{r}_0. \tag{9.39}$$

Considering the semi-analytical method and a constant load vector [Eq. (9.35)], the forward-difference expression [Eq. (9.12)] becomes

$$\mathbf{K}_0\frac{\partial \mathbf{r}_0}{\partial X} = -\frac{\mathbf{K}(X_0 + \delta X) - \mathbf{K}(X_0)}{\delta X}\mathbf{r}_0. \tag{9.40}$$

Comparing Eq. (9.39) with Eq. (9.40), and noting that multiplying a basis vector by a scalar does not change the result, it is observed that the expression for calculating $\mathbf{r}_2$ is identical to the expression of $\partial \mathbf{r}_0 / \partial X$. Therefore, the errors involved in using Eq. (9.39) might be substantial. To overcome this difficulty we could calculate the second basis vector $\mathbf{r}_2$ by [14]

$$\mathbf{K}_0\,\mathbf{r}_2 = -\frac{\partial \mathbf{K}_0}{\partial X}\mathbf{r}_0, \tag{9.41}$$

instead of Eq. (9.39), which may significantly reduce the errors.

In summary, the *Refined Basis Vectors* (RBV) are calculated by

$$\mathbf{K}_0\,\mathbf{r}_i = -\frac{\partial \mathbf{K}_0}{\partial X}\mathbf{r}_{i-1} \quad (i = 2, \ldots, s), \tag{9.42}$$

instead of the usual CA procedure, where the basis vectors are calculated by

$$\mathbf{K}_0\,\mathbf{r}_i = -\delta\mathbf{K}\,\mathbf{r}_{i-1} \quad (i = 2, \ldots, s). \tag{9.43}$$

It is observed that in both cases of Eq. (9.42) and Eq. (9.43), calculation of the basis vectors involves only forward and backward substitutions. One possible difficulty is that $\partial \mathbf{K}_0 / \partial X$ must be calculated. To overcome this problem, an alternative expression is introduced in Sect. 9.3.3.

Example 9.3

Consider the cantilever beam shown in Fig. 9.2, consisting of n equally sized beam elements. The uniform bending stiffness is $EI = 1.0$, the length of the elements is the design variable X and the initial length is $X_0 = 1.0$. Denote the perturbation by δX and the relative perturbation by $\eta = \delta X/X_0$. Assuming first $n = 6$, $\delta X = 0.01$, the resulting sensitivities $\partial\mathbf{r}/\partial X$, calculated by the following methods are shown in Table 9.9:

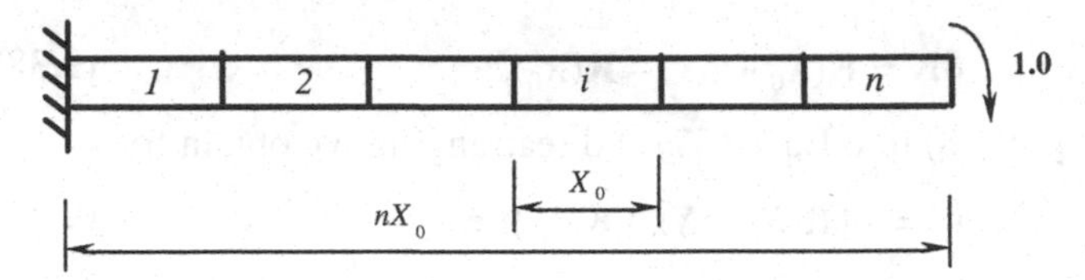

Fig. 9.2. Cantilever beam

- Exact analytical derivatives (EX).
- Finite-difference derivatives using exact analysis (FD).
- Finite-difference derivatives using CA with s basis vectors (CAs).

It is observed that the results obtained by CA3 and EX are very close.

It has been noted [3] that the errors semi-analytical derivatives are proportional to $\eta\, n^2$. Thus, similar errors might be obtained for the CA solutions. It is shown in this example how the refined basis vectors [RBV, Eq. (9.42)] improve the accuracy of the results for large n values.

Assume the following values for n and δX: n = 200, 300, 400, 500, 600 and δX = 0.01, 0.001, 0.0001. The percentage errors E(CAs) for CA with s basis vectors are calculated by

$$E(\text{CA}s) = 100\frac{\|\mathbf{r}(\text{CA}s) - \mathbf{r}(\text{EX})\|}{\|\mathbf{r}(\text{EX})\|}. \tag{a}$$

Similar errors E(RBVs) are calculated for s refined basis vectors. The results in Table 9.10 show that very large errors are obtained by CA2 and CA3. Additional basis vectors do not improve the results, whereas very small errors are obtained by RBV2.

Table 9.9. Derivatives $\partial\mathbf{r}/\partial X$ for n = 6, δX = 0.01, cantilever beam

Number	FD	CA3	EX
1	1.01	1.00	1.00
2	1.00	1.00	1.00
3	4.02	4.01	4.00
4	2.00	2.00	2.00
5	9.05	9.03	9.00
6	3.00	3.00	3.00
7	16.08	16.05	16.00
8	4.00	4.00	4.00
9	25.13	25.09	25.00
10	5.00	5.00	5.00
11	36.18	36.15	36.00
12	6.00	6.00	6.00

Table 9.10. True percentage errors of displacements, cantilever beam

n	δX	E(CA2)	E(CA3)	E(RBV2)
200	0.01	100.2	98.7	$2.2\ 10^{-7}$
	0.001	21.4	2.0	$3.1\ 10^{-8}$
	0.0001	0.2	$2.1\ 10^{-4}$	$1.6\ 10^{-7}$
300	0.01	100.1	99.9	$1.9\ 10^{-5}$
	0.001	70.7	32.4	$2.9\ 10^{-6}$
	0.0001	1.1	$5.0\ 10^{-3}$	$3.0\ 10^{-6}$
400	0.01	100.0	100.0	$3.7\ 10^{-4}$
	0.001	94.1	78.9	$3.1\ 10^{-6}$
	0.0001	3.4	$5.1\ 10^{-2}$	$1.3\ 10^{-6}$
500	0.01	100.0	100.0	$6.9\ 10^{-3}$
	0.001	99.2	94.4	$6.7\ 10^{-5}$
	0.0001	8.4	0.3	$3.9\ 10^{-5}$
600	0.01	100.0	100.0	$6.7\ 10^{-2}$
	0.001	100.1	98.3	$7.3\ 10^{-4}$
	0.0001	17.5	1.3	$7.6\ 10^{-6}$

9.3 Vibration Problems

9.3.1 Analytical Derivatives

Consider the initial eigenproblem [Eq. (6.1)]

$$\mathbf{K}_0 \mathbf{\Phi}_0 = \lambda_0 \mathbf{M}_0 \mathbf{\Phi}_0 , \tag{9.44}$$

where the eigenvector $\mathbf{\Phi}_0$ is normalized such that

$$\mathbf{\Phi}_0^T \mathbf{M}_0 \mathbf{\Phi}_0 = 1 . \tag{9.45}$$

Differentiating Eqs. (9.44), (9.45) with respect to X and rearranging gives

$$(\mathbf{K}_0 - \lambda_0 \mathbf{M}_0) \frac{\partial \mathbf{\Phi}_0}{\partial X} - \frac{\partial \lambda_0}{\partial X} \mathbf{M}_0 \mathbf{\Phi}_0 = -\left(\frac{\partial \mathbf{K}_0}{\partial X} - \lambda_0 \frac{\partial \mathbf{M}_0}{\partial X} \right) \mathbf{\Phi}_0 , \tag{9.46}$$

$$\mathbf{\Phi}_0^T \mathbf{M}_0 \frac{\partial \mathbf{\Phi}_0}{\partial X} = -\frac{1}{2} \mathbf{\Phi}_0^T \frac{\partial \mathbf{M}_0}{\partial X} \mathbf{\Phi}_0 ,$$

or, in matrix form

$$\begin{bmatrix} \mathbf{K}_0 - \lambda_0 \mathbf{M}_0 & -\mathbf{M}_0 \mathbf{\Phi}_0 \\ \mathbf{\Phi}_0^T \mathbf{M}_0 & 0 \end{bmatrix} \begin{Bmatrix} \dfrac{\partial \mathbf{\Phi}_0}{\partial X} \\ \dfrac{\partial \lambda_0}{\partial X} \end{Bmatrix} = - \begin{Bmatrix} \left(\dfrac{\partial \mathbf{K}_0}{\partial X} - \lambda_0 \dfrac{\partial \mathbf{M}_0}{\partial X} \right) \mathbf{\Phi}_0 \\ \dfrac{1}{2} \mathbf{\Phi}_0^T \dfrac{\partial \mathbf{M}_0}{\partial X} \mathbf{\Phi}_0 \end{Bmatrix}. \tag{9.47}$$

In the solution of Eq. (9.47) care must be taken because the principal minor ($\mathbf{K}_0 - \lambda_0 \mathbf{M}_0$) is singular. In many cases we are interested only in the derivatives of the eigenvalues. These derivatives may be obtained by premultiplying the first equation in (9.46) by $\mathbf{\Phi}_0^T$ and rearranging to obtain

$$\frac{\partial \lambda_0}{\partial X} \mathbf{\Phi}_0^T \mathbf{M}_0 \mathbf{\Phi}_0 = \mathbf{\Phi}_0^T \left(\frac{\partial \mathbf{K}_0}{\partial X} - \lambda_0 \frac{\partial \mathbf{M}_0}{\partial X} \right) \mathbf{\Phi}_0 + \mathbf{\Phi}_0^T (\mathbf{K}_0 - \lambda_0 \mathbf{M}_0) \frac{\partial \mathbf{\Phi}_0}{\partial X}. \tag{9.48}$$

Equation (9.44) can be written as $\mathbf{\Phi}_0^T (\mathbf{K}_0 - \lambda_0 \mathbf{M}_0) = \mathbf{0}$, i.e., the last term in Eq. (9.48) equals zero. Rearranging the resulting equation we obtain

$$\frac{\partial \lambda_0}{\partial X} = \frac{\mathbf{\Phi}_0^T \left(\dfrac{\partial \mathbf{K}_0}{\partial X} - \lambda_0 \dfrac{\partial \mathbf{M}_0}{\partial X} \right) \mathbf{\Phi}_0}{\mathbf{\Phi}_0^T \mathbf{M}_0 \mathbf{\Phi}_0}. \tag{9.49}$$

Note that this is correct only if λ_0 is distinct.

9.3.2 Repeated Finite-Difference Derivatives

Vibration reanalysis by the CA approach has been discussed in Chap. 6. Given the initial values $\mathbf{K}_0$, $\mathbf{M}_0$, $\mathbf{\Phi}_0$, λ_0 from the initial vibration analysis, the CA reanalysis procedure described in Sect. 6.3 can be used to evaluate the modified eigenpairs $\mathbf{\Phi}$, λ, due to changes in the design. It was found that accurate results can be achieved efficiently in calculating finite-difference design-sensitivities [13], as illustrated in the following example. Various means, discussed earlier in Sect. 6.2 and later in Sect. 9.3.3, can be used to improve the accuracy of the results.

Example 9.4

Consider the eight-story frame shown in Fig. 9.3. The mass of the frame is lumped in the girders, with initial values $M_1 = 1.0$, $M_2 = 1.5$, $M_3 = 2.0$. The girders are assumed to be non-deformable and the initial lateral stiffness of each of the stories is given by $EI/L^3 = 5.0$. Consider a single design variable X, representing the lateral stiffness of the bottom story, and assume the perturbation $\delta X = 0.01$. The eigenpair derivatives $\partial \mathbf{r}/\partial X$, $\partial \lambda/\partial X$ have been calculated for the first 3 mode shapes by the following procedures:

- Forward-difference using exact reanalysis (FD).
- Forward-difference using CA with 2 basis vectors (CA2).
- Exact analytical derivatives (Exact).

The derivatives $\partial \lambda/\partial X$ and $100\,\partial \mathbf{r}/\partial X$, shown in Tables 9.11, 9.12 and in Fig. 9.4, demonstrate the accuracy obtained with only 2 basis vectors.

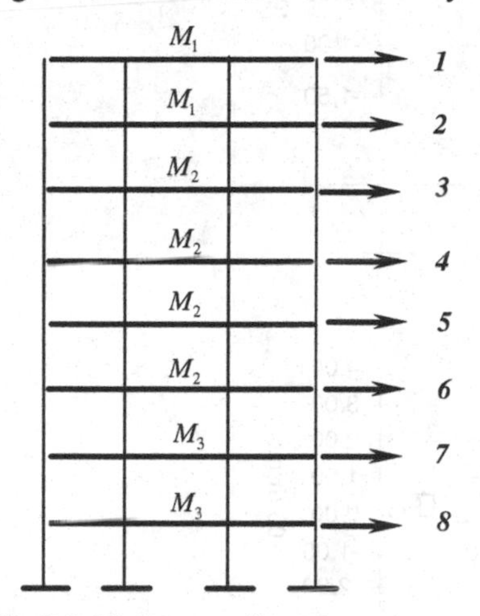

Fig. 9.3. Eight-story frame

Table 9.11. Eigenvalue derivatives, first 3 mode shapes

Mode	1			2			3		
Method	FD	CA2	Exact	FD	CA2	Exact	FD	CA2	Exact
$\partial\lambda/\partial X$	0.0830	0.0830	0.0831	0.6703	0.6703	0.6711	1.1339	1.1338	1.1346

Table 9.12. Eigenvector derivatives, first 3 mode shapes

Mode	1			2			3		
Method	FD	CA2	Exact	FD	CA2	Exact	FD	CA2	Exact
$100\partial r/\partial X$	0.61	0.55	0.61	0.58	0.20	0.58	-0.06	-0.46	-0.06
	0.54	0.49	0.54	0.00	-0.19	0.00	-0.87	-0.89	-0.87
	0.39	0.37	0.39	-0.96	-0.85	-0.96	-1.64	-1.24	-1.64
	0.16	0.16	0.16	-1.91	-1.57	-1.92	-0.76	-0.52	-0.76
	-0.16	-0.12	-0.16	-2.22	-1.92	-2.22	1.77	1.45	1.77
	-0.53	-0.47	-0.53	-1.45	-1.47	-1.45	3.83	3.43	3.83
	-0.92	-0.87	-0.92	0.37	0.00	0.37	2.98	3.25	2.98
	-1.32	-1.31	-1.32	2.88	2.81	2.88	-2.09	-1.99	-2.09

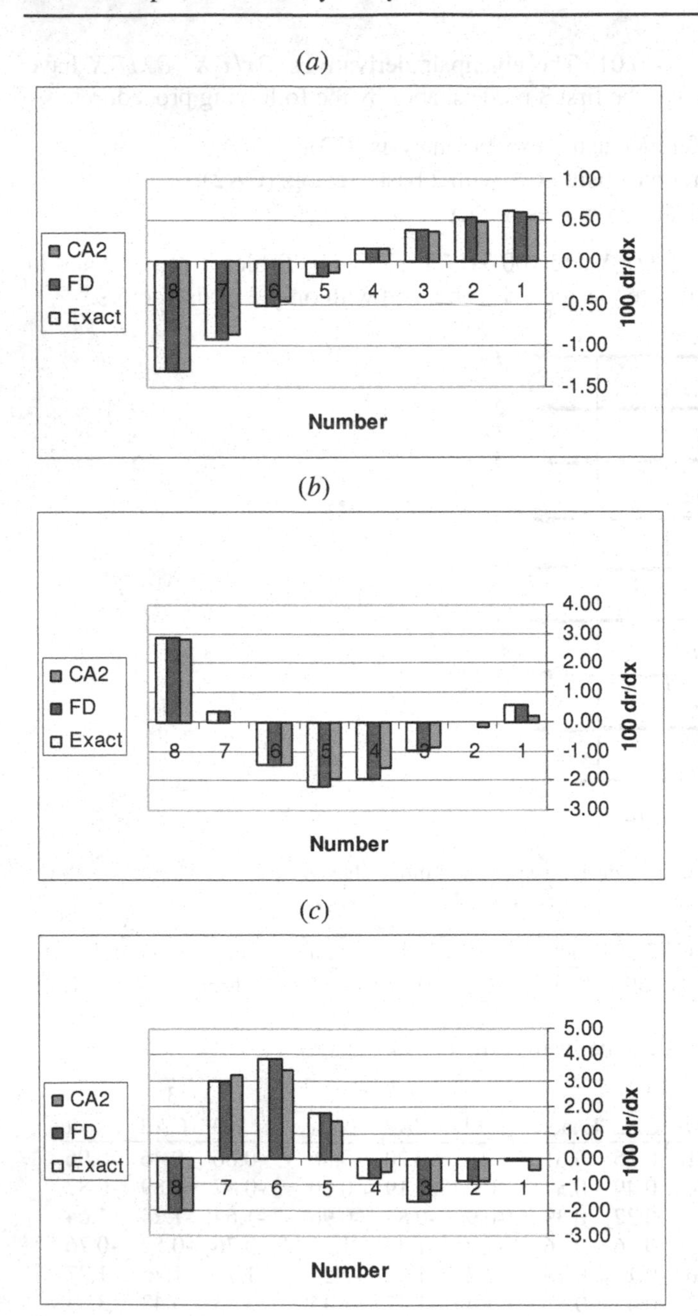

Fig. 9.4. Eigenvector derivatives $100\partial \mathbf{r}/\partial X$: ***a***. Mode 1 ***b***. Mode 2 ***c***. Mode 3

9.3.3 Improved Basis Vectors using Central-Differences

Various means intended to improve the basis vectors have been presented in Sect. 6.2. These include Gram-Schmidt orthogonalizations of the approximate modes, shifts of the basis vectors, and Gram-Schmidt orthogonalizations of these vectors. It has been shown in Sect. 9.2.5 that in cases of large rigid body motions, the errors obtained in sensitivity analysis by the CA approach might be substantial. In this section a central-difference expression, intended to improve the accuracy of the results, is introduced. The presented improved basis vectors can be used in various sensitivity analysis problems, including static, vibration and dynamic problems.

Consider the usual basis vectors determined by the CA approach

$$\mathbf{r}_1 = \mathbf{K}_0^{-1}\,\mathbf{M}\,\mathbf{\Phi}_0\,, \tag{9.50}$$

$$\mathbf{r}_i = -\mathbf{K}_0^{-1}\,\delta\mathbf{K}\,\mathbf{r}_{i\text{-}1} \qquad (i = 2, \ldots, s), \tag{9.51}$$

where matrix $\delta\mathbf{K}$ represents changes in the stiffness matrix due to δX

$$\delta\mathbf{K} = \mathbf{K}(X + \delta X) - \mathbf{K}_0. \tag{9.52}$$

It has been noted in Sect. 9.2.5 that Eq. (9.51) might cause inaccuracies in problems of shape design variables. To improve the accuracy, it is possible to use the central-difference expression

$$\delta\overline{\mathbf{K}} = \mathbf{K}(X + \delta X) - \mathbf{K}(X - \delta X), \tag{9.53}$$

in Eq. (9.51), instead of the forward-difference expression $\delta\mathbf{K}$ [Eq. (9.52)]. This modification may reduce significantly the number of basis vectors required to achieve sufficiently accurate results. It should be emphasized that, using Eq. (9.53), it is only necessary to evaluate $\mathbf{K}(X + \delta X)$, $\mathbf{K}(X - \delta X)$ and not $\mathbf{r}(X + \delta X)$, $\mathbf{r}(X - \delta X)$. The procedure presented can be used for both forward-difference derivatives and central-difference derivatives.

In summary, once $\delta\overline{\mathbf{K}}$ is calculated, we use the following expressions for calculating the basis vectors [instead of Eqs. (9.50) – (9.52)]

$$\overline{\mathbf{r}}_1 = \mathbf{r}_1 = \mathbf{K}_0^{-1}\,\mathbf{M}\,\mathbf{\Phi}_0\,, \tag{9.54}$$

$$\overline{\mathbf{r}}_i = -\overline{\mathbf{B}}\;\overline{\mathbf{r}}_{i-1} \qquad (i = 2, \ldots, s), \tag{9.55}$$

where

$$\overline{\mathbf{B}} = \mathbf{K}_0^{-1}\,\delta\overline{\mathbf{K}}\,. \tag{9.56}$$

Example 9.5

Consider the column shown in Fig. 9.5 consisting of n equally sized elements. The uniform bending stiffness is $EI = 10^9$, the distributed mass is 10^3, the length of the elements is the design variable $X = L/n$ and the initial design is $X_0 = 1.0$. Denote the relative perturbation by $\eta = \delta X/X$, where δX is the perturbation. Assume the following values for n and η:
n = 200, 300 and η = 0.01, 0.001, 0.0001. The eigenvalue derivatives $\partial\lambda_i/\partial X$ (i = 1, 2, ..., 5) have been calculated by the following methods:

- $\partial\lambda_i/\partial X$(FD) = forward-difference derivatives using exact reanalysis.
- $\partial\lambda_i/\partial X$(CA$s$) = forward-difference derivatives using CA reanalysis with s basis vectors, assuming either the usual $\delta\mathbf{K}$ [Eq. (9.52)] or the improved central-difference expression $\delta\overline{\mathbf{K}}$ [Eq. (9.53)].

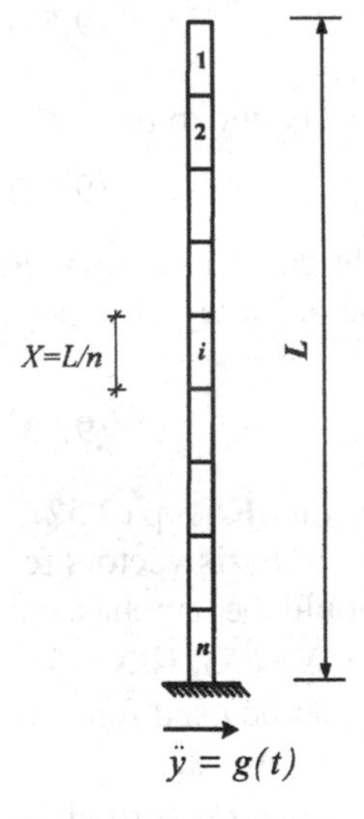

Fig. 9.5. Column example

Consider the following percentage errors in Eigenvalue Sensitivities

$$ES(\lambda_i) = 100\,\frac{\partial\lambda_i/\partial X\,(\mathrm{FD}) - \partial\lambda_i/\partial X\,(\mathrm{CA}_s)}{\partial\lambda_i/\partial X\,(\mathrm{FD})}.$$

The results in Table 9.13 show that significantly smaller numbers of basis vectors are required by the improved $\delta\overline{\mathbf{K}}$ to obtain errors similar to those obtained by $\delta\mathbf{K}$, particularly for large n and η values.

Table 9.13. Percentage errors in eigenvalue sensitivities, using $\delta\mathbf{K}$ and $\delta\overline{\mathbf{K}}$

n	η	Method	CAs	$ES(\lambda_1)$	$ES(\lambda_2)$	$ES(\lambda_3)$	$ES(\lambda_4)$	$ES(\lambda_5)$
200	0.01	$\delta\mathbf{K}$	CA10	0.03720	0.00010	0.00007	0.00004	0.00010
		$\delta\overline{\mathbf{K}}$	CA4	0.00182	0.00016	0.00038	0.00010	0.00020
	0.001	$\delta\mathbf{K}$	CA4	0.01553	0.00014	0.00020	0.00011	0.00010
		$\delta\overline{\mathbf{K}}$	CA2	0.01507	0.00015	0.00020	0.00011	0.00010
	0.0001	$\delta\mathbf{K}$	CA3	0.05068	0.04500	0.00729	0.00086	0.00065
		$\delta\overline{\mathbf{K}}$	CA2	0.05069	0.04494	0.00730	0.00086	0.00064
300	0.01	$\delta\mathbf{K}$	CA13	0.06554	0.00164	0.00003	0.00531	0.00308
		$\delta\overline{\mathbf{K}}$	CA4	0.06166	0.07698	0.20689	0.15456	0.01501
	0.001	$\delta\mathbf{K}$	CA5	0.14591	0.00400	0.00179	0.00120	0.00007
		$\delta\overline{\mathbf{K}}$	CA2	0.14835	0.00465	0.00184	0.00118	0.00008
	0.0001	$\delta\mathbf{K}$	CA3	1.64950	0.11213	0.01491	0.00421	0.00509
		$\delta\overline{\mathbf{K}}$	CA2	1.64930	0.11247	0.01491	0.00422	0.00509

9.4 Linear Dynamic Problems

9.4.1 Modal Analysis Equations

Calculation of linear dynamic response by modal analysis, presented in Sect. 3.1.2 and summarized in Sect. 7.1.2, involves the following steps:

- Determine the matrices **K**, **M**, and **C**.
- Determine the eigenpairs λ_i, $\mathbf{\Phi}_i$ by solving the eigenproblem [Eq. (7.13)]

$$\mathbf{K}\mathbf{\Phi} = \mathbf{\Lambda}\,\mathbf{M}\mathbf{\Phi}\,. \tag{9.57}$$

- Calculate the modal coordinates $Z_i(t)$ by solving [Eq. (7.16)]

$$\ddot{Z}_i(t) + 2\omega_i\zeta_i\dot{Z}_i(t) + \omega_i^2 Z_i(t) = T_i(t) \qquad i = 1, \ldots, p, \tag{9.58}$$

where p is the number of modes considered and

$$T_i(t) = \mathbf{\Phi}_i^T\mathbf{R}(t)\,. \tag{9.59}$$

In various problems (e.g. earthquake loading) the load vector **R** [and therefore $T_i(t)$] are not given analytically but as discrete values.

- Calculate the nodal displacements **r** by [Eq. (7.17)]

$$\mathbf{r}(t) = \sum_{i=1}^{p}\mathbf{\Phi}_i\,Z_i(t)\,. \tag{9.60}$$

- Calculate the element forces using the element stiffness properties.

9.4.2 Analytical Derivatives

The derivative expressions, $\partial \mathbf{r} / \partial X_j$, of the displacements $\mathbf{r}$ with respect to a design variable X_j are obtained by differentiating Eq. (9.60)

$$\frac{\partial \mathbf{r}}{\partial X_j} = \sum_{i=1}^{p} \left(\frac{\partial \mathbf{\Phi}_i}{\partial X_j} Z_i + \mathbf{\Phi}_i \frac{\partial Z_i}{\partial X_j} \right). \tag{9.61}$$

Assuming that the damping ratios ζ_i are independent of design variables, we calculate $\partial Z_i / \partial X_j$ by differentiating Eq. (9.58) and rearranging

$$\frac{\partial \ddot{Z}_i}{\partial X_j} + 2\omega_i \zeta_i \frac{\partial \dot{Z}_i}{\partial X_j} + \omega_i^2 \frac{\partial Z_i}{\partial X_j} = \frac{\partial T_i}{\partial X_j} - 2 \frac{\partial \omega_i}{\partial X_j} \zeta_i \dot{Z}_i - \frac{\partial \omega_i^2}{\partial X_j} Z_i . \tag{9.62}$$

Denoting

$$q_i = \partial Z_i / \partial X_j \qquad \dot{q}_i = \partial \dot{Z}_i / \partial X_j \qquad \ddot{q}_i = \partial \ddot{Z}_i / \partial X_j , \tag{9.63}$$

and substituting Eqs. (9.59), (9.63) into Eq. (9.62) yields

$$\ddot{q}_i + 2\omega_i \zeta_i \dot{q}_i + \omega_i^2 q_i = \mathbf{\Phi}_i^T \frac{\partial \mathbf{R}}{\partial X_j} + \frac{\partial \mathbf{\Phi}_i^T}{\partial X_j} \mathbf{R} - 2 \frac{\partial \omega_i}{\partial X_j} \zeta_i \dot{Z}_i - \frac{\partial \omega_i^2}{\partial X_j} Z_i . \tag{9.64}$$

Note that, whereas the right hand sides are different, the left hand sides and the initial conditions of Eqs. (9.58) and (9.64) are similar (e.g. $q_i = \dot{q}_i = 0$ for $t = 0$). This similarity will be used in Sect. 9.4.3 to reduce the number differential equations that must be solved during the solution process.

Given the eigenpairs and the response for a certain design and time, evaluation of the displacement derivatives involves the following steps.

- Evaluate the eigenpairs derivatives $\partial \mathbf{\Phi}_i / \partial X_j$, $\partial \lambda_i / \partial X_j$.
- Compute the right side of Eq. (9.64).
- Compute the derivatives $q_i = \partial Z_i / \partial X_j$ by solving Eq. (9.64).
- Evaluate the displacement derivatives $\partial \mathbf{r} / \partial X_j$ by Eq. (9.61).

Assuming a problem with p mode shapes and m design variables, the main computational effort is involved in evaluation of pm eigenpair derivatives and solution of the pm differential equations (9.64). Efficient evaluation of the eigenpair derivatives, using finite-difference and the CA approach, has been presented in Sect. 9.3. A procedure intended to reduce the number of differential equations to be solved during the solution process [15] is introduced in the following section.

9.4.3 Reducing the Number of Differential Equations

Due to the linearity of Eq. (9.64), we can use superposition and divide it into the following 3 equations with identical initial conditions

$$\ddot{q}_i^{(k)} + 2\omega_i\zeta_i\dot{q}_i^{(k)} + \omega_i^2 q_i^{(k)} = F_i^{(k)} \qquad k = 1, 2, 3, \tag{9.65}$$

where

$$F_i^{(1)} = \mathbf{\Phi}_i^T \frac{\partial \mathbf{R}}{\partial X_j} + \frac{\partial \mathbf{\Phi}_i^T}{\partial X_j}\mathbf{R} \qquad F_i^{(2)} = -2\frac{\partial \omega_i}{\partial X_j}\zeta_i \dot{Z}_i \qquad F_i^{(3)} = -\frac{\partial \omega_i^2}{\partial X_j} Z_i, \tag{9.66}$$

$$q_i = \sum_{k=1}^{3} q_i^{(k)} \qquad \dot{q}_i = \sum_{k=1}^{3} \dot{q}_i^{(k)} \qquad \ddot{q}_i = \sum_{k=1}^{3} \ddot{q}_i^{(k)}. \tag{9.67}$$

Noting that the right hand sides of Eq. (9.58) and Eq. (9.65) for $k = 1$ are

$$T_i(t) = \mathbf{\Phi}_i^T \mathbf{R}(t), \tag{9.68}$$

$$F_i^{(1)} = \mathbf{\Phi}_i^T \frac{\partial \mathbf{R}}{\partial X_j} + \frac{\partial \mathbf{\Phi}_i^T}{\partial X_j}\mathbf{R}, \tag{9.69}$$

and assuming that the load vector can be expressed as $\mathbf{R}(X_j, t) = \mathbf{R}(X_j)\, g(t)$, then Eqs. (9.68), (9.69) describe similar functions in time with different amplitudes. For zero initial conditions (or, if we neglect the influence of the homogeneous solution), the ratio between the two displacement functions of Eqs. (9.58) and (9.65) is equal to the ratio between the right-hand side terms. Thus, given the solutions Z_i of Eq. (9.58) for all p modes, the solutions $q_i^{(1)}$ of Eq. (9.65) for $k = 1$ can be determined directly by

$$q_i^{(1)} = Z_i \frac{\mathbf{\Phi}_i^T \dfrac{\partial \mathbf{R}}{\partial X_j} + \dfrac{\partial \mathbf{\Phi}_i^T}{\partial X_j}\mathbf{R}}{\mathbf{\Phi}_i^T \mathbf{R}}. \tag{9.70}$$

To find $q_i^{(2)}, q_i^{(3)}$, Eq. (9.65) must be solved for $k = 2$ and $k = 3$. For X_1 we have to solve the two equations

$$\ddot{q}_i^{(2)} + 2\omega_i\zeta_i\dot{q}_i^{(2)} + \omega_i^2 q_i^{(2)} = -2\frac{\partial \omega_i}{\partial X_1}\zeta_i \dot{Z}_i, \tag{9.71}$$

$$\ddot{q}_i^{(3)} + 2\omega_i\zeta_i\dot{q}_i^{(3)} + \omega_i^2 q_i^{(3)} = -\frac{\partial\omega_i^2}{\partial X_1} Z_i . \qquad (9.72)$$

Given the solutions of Eqs. (9.71), (9.72) with respect to X_1, it is observed that the solutions for any other variable X_j can be determined directly by

$$q_i^{(2)}(X_j) = \frac{\partial\omega_i / \partial X_j}{\partial\omega_i / \partial X_1} q_i^{(2)}(X_1) \qquad q_i^{(3)}(X_j) = \frac{\partial\omega_i^2 / \partial X_j}{\partial\omega_i^2 / \partial X_1} q_i^{(3)}(X_1) . \qquad (9.73)$$

If $\mathbf{\Phi}_i$ and $\mathbf{R}$ are orthogonal, we obtain $T_i = \mathbf{\Phi}_i^T \mathbf{R} = 0$. From Eq. (9.58) we have $Z_i = \dot{Z}_i = \ddot{Z}_i = 0$, and from Eqs. (9.70) – (9.72), $q_i^{(1)} = q_i^{(2)} = q_i^{(3)} = 0$.

In summary, considering a problem with p dominant mode shapes and m design variables, the number of times that the differential Eqs. (9.64) must be solved in order to perform sensitivity analysis is usually pm. Considering the procedure presented in this section and assuming that the solution of the analysis problem [Eq. (9.58)] is known, the number of times that the differential equations must be solved is reduced to $2p$ [Eqs. (9.71), (9.72)]. Thus, the ratio between the two numbers is $pm / 2p = m / 2$, which means a significant reduction in the computational cost. For example, for a problem with 10 design variables, the procedure presented requires about 20% of the effort involved in complete sensitivity analysis.

Example 9.6

Consider again the column of example 9.5, shown in Fig. 9.5, consisting of n equally sized elements. The column length is L =100, the bending stiffness is $EI = 10^9$, the distributed mass is 10^3 and the length of the elements is the design variable $X = L/n$. The structure is subjected to the ground acceleration of the El Centro earthquake shown in Fig 9.6, scaled to have a 10% probability of occurrence in 50 years [16]. The relative perturbation of the design variable is $\eta = \delta X/X$, where δX is the perturbation.

The object is to evaluate the derivatives of the horizontal displacement at the top of the column with respect to X for the following 2 cases:

- $n = 100$, $\eta = 0.0001$.
- $n = 300$, $\eta = 0.0001$.

Considering 2 and 3 basis vectors (CA2 and CA3, respectively) and the usual $\delta\mathbf{K}$ [not the improved $\delta\overline{\mathbf{K}}$, see Eqs. (9.52), (9.53)], the results shown in Fig. 9.7 demonstrate the high accuracy achieved by the CA procedure with very small numbers of basis vectors.

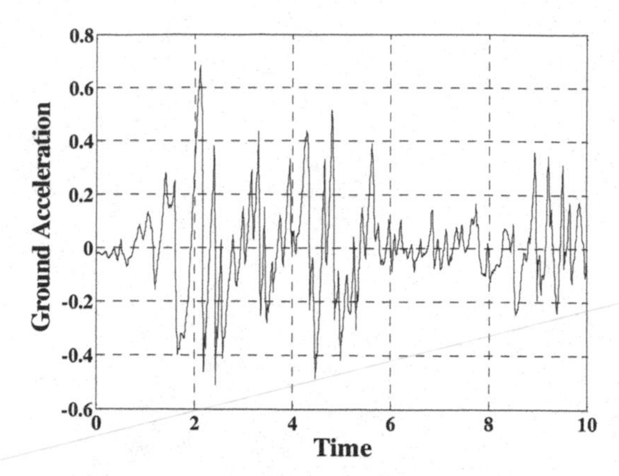

Fig. 9.6. El Centro Earthquake

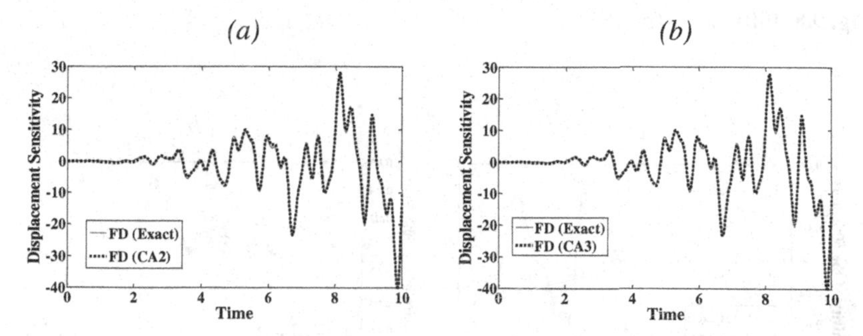

Fig. 9.7. Displacement derivatives ***a.*** $n = 100$, $\eta = 0.0001$ ***b.*** $n = 300$, $\eta = 0.0001$

Example 9.7

To demonstrate results for a larger structure, consider the fifty-story frame shown in Fig. 9.8. The number of degrees of freedom is 600, and the damping ratios are 0.05. The masses are assumed to be concentrated at the joints, and only horizontal inertia forces are considered. The inertia force is due to the frame self-weight and an additional concentrated mass of 50*ton* at an internal joint and 25*ton* at an external joint. The width of all elements is 0.5*m*, the depth of all columns is 1.0*m* and the depth of all beams is 0.8*m*. The modulus of elasticity is $3\ 10^7 kNm^2$ and the time-step is Δt = 0.02*sec*. The structure is subjected to the ground acceleration of the El Centro earthquake shown in Fig 9.6, scaled to have a 10% probability of occurrence in 50 years [16].

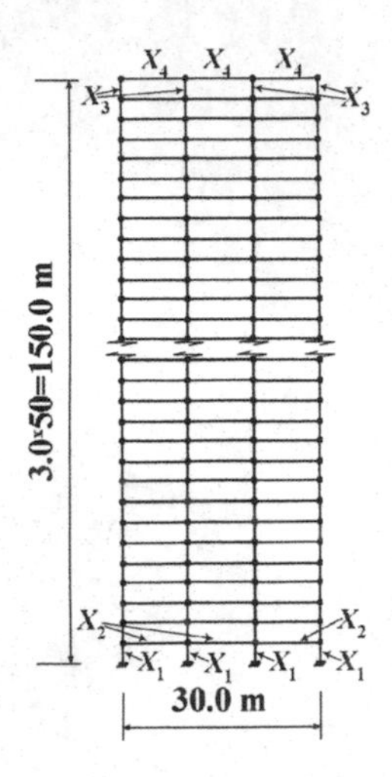

Fig. 9.8. Fifty-story frame

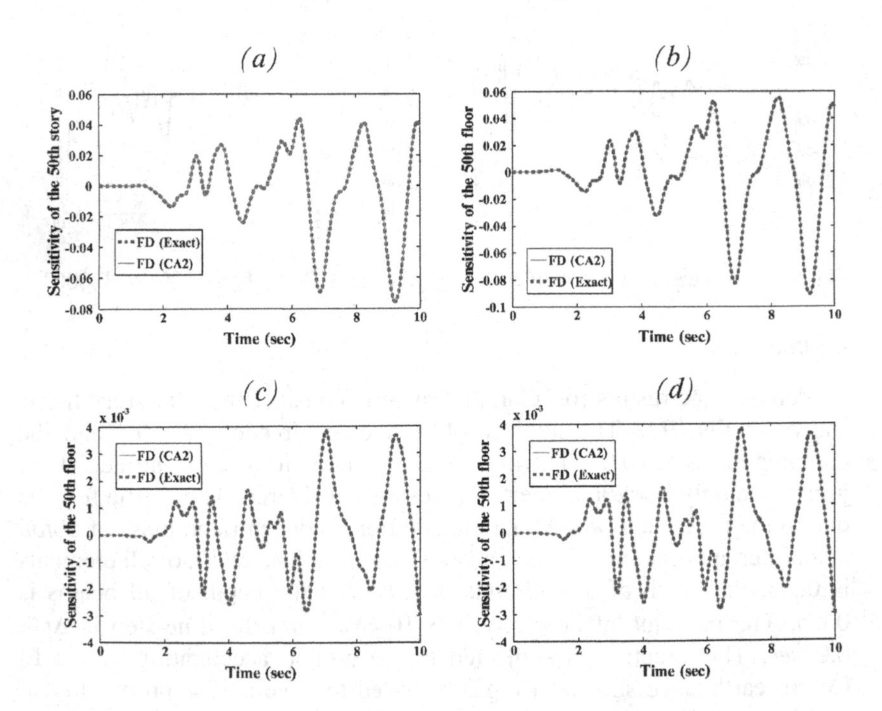

Fig. 9.9. Displacement derivatives with respect to: ***a***. X_1 ***b***. X_2 ***c***. X_3 ***d***. X_4

Considering the first 8 mode shapes, the object is to evaluate the derivatives of the horizontal displacements at the top of the frame with respect to the following four design variables

- X_1 – depth of the columns in the 1st story.
- X_2 – depth of the beams in the 1st story.
- X_3 – depth of the columns in the 50th story.
- X_4 – depth of the beams in the 50th story.

The results achieved by forward-difference derivatives using the CA approach with 2 basis vectors [FD(CA2)] are compared with those obtained by forward-difference derivatives using exact analysis formulation [FD(Exact)]. From the results shown in Fig. 9.9 it is observed that the procedure presented provides high accuracy with only 2 basis vectors.

9.5 Nonlinear Dynamic Problems

9.5.1 Modal Analysis Equations

Calculation of nonlinear dynamic response by modal analysis, presented in Sect. 3.3.2, involves the following steps for each time interval:

- Assemble the updated stiffness matrix ${}^t\mathbf{K}$ and solve the corresponding eigenproblem [Eq. (3.104)] to find the p eigenpairs ${}^t\lambda_i$, ${}^t\mathbf{\Phi}_i$ $(i = 1, ..., p)$

$$ {}^t\mathbf{K}\,{}^t\mathbf{\Phi}_i = {}^t\lambda_i\,\mathbf{M}\,{}^t\mathbf{\Phi}_i\,. \tag{9.74}$$

- Find the diagonal damping matrix ${}^t\mathbf{C}_d = {}^t\mathbf{\Phi}^T\,\mathbf{C}\,{}^t\mathbf{\Phi}$ and stiffness matrix ${}^t\mathbf{\Omega}^2 = {}^t\mathbf{\Phi}^T\,{}^t\mathbf{K}\,{}^t\mathbf{\Phi}$ [Eqs. (3.106), (3.107)].
- Evaluate ${}^t\Delta\mathbf{Z}$, ${}^t\Delta\dot{\mathbf{Z}}$, ${}^t\Delta\ddot{\mathbf{Z}}$ by solving the p uncoupled equations in the generalized coordinates [Eq. (3.105)]

$$\mathbf{I}\,{}^t\Delta\ddot{\mathbf{Z}} + {}^t\mathbf{C}_d\,{}^t\Delta\dot{\mathbf{Z}} + {}^t\mathbf{\Omega}^2\,{}^t\Delta\mathbf{Z} = {}^t\mathbf{\Phi}^T\,{}^t\Delta\mathbf{R}\,. \tag{9.75}$$

- Calculate ${}^t\Delta\mathbf{r}$, ${}^t\Delta\dot{\mathbf{r}}$, ${}^t\Delta\ddot{\mathbf{r}}$, by [Eq. (3.103)]

$$ {}^t\Delta\mathbf{r} = {}^t\mathbf{\Phi}\,{}^t\Delta\mathbf{Z} \qquad {}^t\Delta\dot{\mathbf{r}} = {}^t\mathbf{\Phi}\,{}^t\Delta\dot{\mathbf{Z}} \qquad {}^t\Delta\ddot{\mathbf{r}} = {}^t\mathbf{\Phi}\,{}^t\Delta\ddot{\mathbf{Z}}\,. \tag{9.76}$$

- Evaluate ${}^{t+\Delta t}\mathbf{r}$, ${}^{t+\Delta t}\dot{\mathbf{r}}$, ${}^{t+\Delta t}\ddot{\mathbf{r}}$ by [Eq. (3.97)]

$$ {}^{t+\Delta t}\mathbf{r} = {}^t\mathbf{r} + {}^t\Delta\mathbf{r} \qquad {}^{t+\Delta t}\dot{\mathbf{r}} = {}^t\dot{\mathbf{r}} + {}^t\Delta\dot{\mathbf{r}} \qquad {}^{t+\Delta t}\ddot{\mathbf{r}} = {}^t\ddot{\mathbf{r}} + {}^t\Delta\ddot{\mathbf{r}}\,. \tag{9.77}$$

- Evaluate the forces and check the properties of the members. If all stiffness coefficients do not change, start the calculations for the next time interval by solving Eq. (9.75). If the stiffness coefficients of any member change, reduce Δt and repeat the solution from the first step.

9.5.2 Nonlinear-Dynamic Sensitivity Analysis

We solve the nonlinear-dynamic sensitivity analysis problem incrementally, starting with the initial conditions $\partial\, {}^0\mathbf{r}/\partial X_j = \mathbf{0}$, $\partial\, {}^0\dot{\mathbf{r}}/\partial X_j = \mathbf{0}$, where X_j is a design variable. Differentiation of Eqs. (9.77) gives

$$\frac{\partial\, {}^{t+\Delta t}\mathbf{r}}{\partial X_j} = \frac{\partial\, {}^t\mathbf{r}}{\partial X_j} + \frac{\partial\, {}^t\Delta\mathbf{r}}{\partial X_j}, \tag{9.78}$$

$$\frac{\partial\, {}^{t+\Delta t}\dot{\mathbf{r}}}{\partial X_j} = \frac{\partial\, {}^t\dot{\mathbf{r}}}{\partial X_j} + \frac{\partial\, {}^t\Delta\dot{\mathbf{r}}}{\partial X_j}, \tag{9.79}$$

$$\frac{\partial\, {}^{t+\Delta t}\ddot{\mathbf{r}}}{\partial X_j} = \frac{\partial\, {}^t\ddot{\mathbf{r}}}{\partial X_j} + \frac{\partial\, {}^t\Delta\ddot{\mathbf{r}}}{\partial X_j}. \tag{9.80}$$

Considering the transformation of coordinates

$${}^t\Delta\mathbf{r} = \sum_{i=1}^{p} {}^t\mathbf{\Phi}_i\, {}^t\Delta Z_i \qquad \Delta\dot{\mathbf{r}} = \sum_{i=1}^{p} {}^t\mathbf{\Phi}_i\, {}^t\Delta \dot{Z}_i \qquad \Delta\ddot{\mathbf{r}} = \sum_{i=1}^{p} {}^t\mathbf{\Phi}_i\, {}^t\Delta \ddot{Z}_i, \tag{9.81}$$

and differentiating Eqs. (9.81) with respect to X_j, we obtain

$$\frac{\partial\, {}^t\Delta\mathbf{r}}{\partial X_j} = \sum_{i=1}^{p}\left(\frac{\partial\, {}^t\mathbf{\Phi}_i}{\partial X_j}\Delta Z_i + {}^t\mathbf{\Phi}_i \frac{\partial \Delta Z_i}{\partial X_j}\right), \tag{9.82}$$

$$\frac{\partial\, {}^t\Delta\dot{\mathbf{r}}}{\partial X_j} = \sum_{i=1}^{p}\left(\frac{\partial\, {}^t\mathbf{\Phi}_i}{\partial X_j}\Delta \dot{Z}_k + {}^t\mathbf{\Phi}_i \frac{\partial \Delta \dot{Z}_i}{\partial X_j}\right), \tag{9.83}$$

$$\frac{\partial\, {}^t\Delta\ddot{\mathbf{r}}}{\partial X_j} = \sum_{i=1}^{p}\left(\frac{\partial\, {}^t\mathbf{\Phi}_i}{\partial X_j}\Delta \ddot{Z}_i + {}^t\mathbf{\Phi}_i \frac{\partial \Delta \ddot{Z}_i}{\partial X_j}\right). \tag{9.84}$$

The values of ${}^t\mathbf{\Phi}_i$, ${}^t\Delta\mathbf{Z}$, ${}^t\Delta\dot{\mathbf{Z}}$, ${}^t\Delta\ddot{\mathbf{Z}}$ are known from solution of the analysis equations (9.74), (9.75). The derivatives at the right-hand side terms are calculated as follows.

The term $\partial\, {}^t\mathbf{\Phi}_i / \partial X_j$ is calculated by finite-differences and the CA approach (Sect. 9.3). To compute $\partial\, {}^t\Delta Z_i / \partial X_j, \partial\, {}^t\Delta\dot{Z}_i / \partial X_j, \partial\, {}^t\Delta\ddot{Z}_i / \partial X_j$ we differentiate the individual equations (9.75) to obtain

$$\frac{\partial \Delta^t\ddot{Z}_i}{\partial X_j} + 2\, {}^t\omega_i \zeta_i \frac{\partial \Delta^t\dot{Z}_i}{\partial X_j} + {}^t\omega_i^2 \frac{\partial\, {}^t\Delta Z_i}{\partial X_j} = \tag{9.85}$$

$$= \frac{\partial\, {}^tT_i}{\partial X_j} - 2\frac{\partial\, {}^t\omega_i}{\partial X_j}\zeta_i\, \Delta^t\dot{Z}_i - \frac{\partial\, {}^t\omega_i^2}{\partial X_j}\Delta^t Z_i.$$

Equation (9.85) is solved by Newmark's method, which involves evaluation of $\partial\, {}^tZ_i / \partial X_j, \partial\, {}^t\dot{Z}_i / \partial X_j, \partial\, {}^t\ddot{Z}_i / \partial X_j$. Using the transformation

$${}^t\mathbf{r} = \sum_{i=1}^{p} {}^t\mathbf{\Phi}_i\, {}^tZ_i \tag{9.86}$$

and differentiating Eq. (9.86) we obtain

$$\frac{\partial\, {}^t\mathbf{r}}{\partial X_j} = \sum_{i=1}^{p}\left(\frac{\partial\, {}^t\mathbf{\Phi}_i}{\partial X_j} Z_i + {}^t\mathbf{\Phi}_i \frac{\partial Z_i}{\partial X_j}\right). \tag{9.87}$$

Premultiplying Eq. (9.87) by ${}^t\mathbf{\Phi}_i^T\, \mathbf{M}$, noting the orthogonality property ${}^t\mathbf{\Phi}_i^T \mathbf{M}\, {}^t\mathbf{\Phi}_i = \mathbf{I}$ and rearranging, gives

$$\frac{\partial\, {}^tZ_i}{\partial X_j} = \frac{{}^t\mathbf{\Phi}_i^T \mathbf{M}\left(\dfrac{\partial\, {}^t\mathbf{r}}{\partial X_j} - \sum\limits_{i=1}^{p} \dfrac{\partial\, {}^t\mathbf{\Phi}_i}{\partial X_j}\, {}^tZ_i\right)}{{}^t\mathbf{\Phi}_i^T\, \mathbf{M}\, {}^t\mathbf{\Phi}_i}. \tag{9.88}$$

Similarly, we obtain for $\partial\, {}^t\dot{Z}_i / \partial X_j, \partial\, {}^t\ddot{Z}_i / \partial X_j$

$$\frac{\partial\, {}^t\dot{Z}_i}{\partial X_j} = \frac{{}^t\mathbf{\Phi}_i^T \mathbf{M}\left(\dfrac{\partial\, {}^t\dot{\mathbf{r}}}{\partial X_j} - \sum\limits_{k=1}^{p} \dfrac{\partial\, {}^t\mathbf{\Phi}_i}{\partial X_j}\, {}^t\dot{Z}_i\right)}{{}^t\mathbf{\Phi}_i^T\, \mathbf{M}\, {}^t\mathbf{\Phi}_i}, \tag{9.89}$$

$$\frac{\partial\, {}^t\ddot{Z}_i}{\partial X_j} = \frac{{}^t\mathbf{\Phi}_i^T \mathbf{M}\left(\dfrac{\partial\, {}^t\ddot{\mathbf{r}}}{\partial X_j} - \sum_{i=1}^{p} \dfrac{\partial\, {}^t\mathbf{\Phi}_i}{\partial X_j}\, {}^t\ddot{Z}_i\right)}{{}^t\mathbf{\Phi}_i^T \mathbf{M}\, {}^t\mathbf{\Phi}_i}. \tag{9.90}$$

In summary, starting with the initial values $\partial\, {}^0\mathbf{r}/\partial X_j = \partial\, {}^0\dot{\mathbf{r}}/\partial X_j = \mathbf{0}$, we calculate the following derivatives at each time step.

- $\partial\, {}^t\mathbf{\Phi}_i/\partial X_j$, $\partial\, {}^t\lambda_i/\partial X_j$ $(i = 1, ..., p)$.
- $\partial\, {}^t Z_i/\partial X_j$, $\partial\, {}^t\dot{Z}_i/\partial X_j$, $\partial\, {}^t\ddot{Z}_i/\partial X_j$ [Eqs. (9.88) – (9.90)].
- $\partial\, {}^t\Delta Z_i/\partial X_j$, $\partial\, {}^t\Delta\dot{Z}_i/\partial X_j$, $\partial\, {}^t\Delta\ddot{Z}_i/\partial X_j$ [Eq. (9.85)].
- $\partial\, {}^t\Delta\mathbf{r}/\partial X_j$, $\partial\, {}^t\Delta\dot{\mathbf{r}}/\partial X_j$, $\partial\, {}^t\Delta\ddot{\mathbf{r}}/\partial X_j$ [Eqs. (9.82) – (9.84)].
- $\partial\, {}^{t+\Delta t}\mathbf{r}/\partial X_j$, $\partial\, {}^{t+\Delta t}\dot{\mathbf{r}}/\partial X_j$, $\partial\, {}^{t+\Delta t}\ddot{\mathbf{r}}/\partial X_j$ [Eqs. (9.78) – (9.80].

9.5.3 Efficient Solution Procedures

It has been noted that solution of nonlinear dynamic analysis problems by the mode superposition approach is usually not efficient. In this section several solution procedures, based on the CA approach and intended to improve the efficiency of nonlinear-dynamic sensitivity analysis using mode superposition, are presented [17]. The accuracy of the approximations is demonstrated by a numerical example. It is shown later in Sect. 10.1.3 that these procedures reduce significantly the computational effort. The following solution procedures are considered and compared:

- DI = *Direct Integration.* Analytical derivatives are obtained by Eqs. (9.78) – (9.80). Since the analysis and the sensitivity analysis equations have the same coefficient matrices in the left-hand side, we can solve efficiently the latter equations.
- MS(EX) = *Mode Superposition using exact analysis formulation.* The eigenpair derivatives $\partial\, {}^t\mathbf{\Phi}_i/\partial X_j$, $\partial\, {}^t\lambda_i/\partial X_j$ are calculated by forward finite-differences, using exact analysis for both the original and the perturbed eigenproblems

$$\frac{\partial\, {}^t\mathbf{\Phi}_i}{\partial X_j} = \frac{{}^t\mathbf{\Phi}_{iEx}(X_0 + \delta X) - {}^t\mathbf{\Phi}_{iEx}(X_0)}{\delta X}, \tag{9.91}$$

$$\frac{\partial\, {}^t\lambda_i}{\partial X_j} = \frac{{}^t\lambda_{iEx}(X_0 + \delta X) - {}^t\lambda_{iEx}(X_0)}{\delta X}. \tag{9.92}$$

- MS(CA) = *Mode Superposition using CA analysis.* The eigenpair derivatives are calculated by forward finite-differences, using CA analysis for both the original and the perturbed eigenproblems

$$\frac{\partial\, {}^t\mathbf{\Phi}_i}{\partial X_j} = \frac{{}^t\mathbf{\Phi}_{iCA}(X_0 + \delta X) - {}^t\mathbf{\Phi}_{iCA}(X_0)}{\delta X}, \tag{9.93}$$

$$\frac{\partial\, {}^t\lambda_i}{\partial X_j} = \frac{{}^t\lambda_{iCA}(X_0 + \delta X) - {}^t\lambda_{iCA}(X_0)}{\delta X}. \tag{9.94}$$

- MS(EX)+MS(CA) = *Mode Superposition using exact analysis for the original eigenproblem and CA analysis for the perturbed eigenproblem*

$$\frac{\partial\, {}^t\mathbf{\Phi}_i}{\partial X_j} = \frac{{}^t\mathbf{\Phi}_{iCA}(X_0 + \delta X) - {}^t\mathbf{\Phi}_{iEx}(X_0)}{\delta X}, \tag{9.95}$$

$$\frac{\partial\, {}^t\lambda_i}{\partial X_j} = \frac{{}^t\lambda_{iCA}(X_0 + \delta X) - {}^t\lambda_{iEx}(X_0)}{\delta X}. \tag{9.96}$$

- MS(EX)+APP = *Mode Superposition using exact analysis and approximate eigenvector sensitivities.* The eigenpair derivatives are calculated using the exact eigenpairs and the initial eigenpair derivatives $\partial\, {}^0\mathbf{\Phi} / \partial X$, $\partial\, {}^0\lambda / \partial X$. We first calculate the exact $\partial\, {}^t\lambda / \partial X$ by

$$\frac{\partial\, {}^t\lambda_i}{\partial X_j} = \frac{{}^t\mathbf{\Phi}_{iEx}^T \left(\dfrac{\partial\, {}^t\mathbf{K}}{\partial X_j} - {}^t\lambda_{iEx} \dfrac{\partial \mathbf{M}}{\partial X_j} \right) {}^t\mathbf{\Phi}_{iEx}}{{}^t\mathbf{\Phi}_{iEx}^T \mathbf{M}\, {}^t\mathbf{\Phi}_{iEx}} = \frac{{}^t\mathbf{\Phi}_{iEx}^T \left(\dfrac{\partial\, {}^t\mathbf{K}}{\partial X_j} - {}^t\lambda_{iEx} \dfrac{\partial \mathbf{M}}{\partial X_j} \right) {}^t\mathbf{\Phi}_{iEx}}{{}^0\mathbf{\Phi}_{iEx}^T \left(\dfrac{\partial\, {}^0\mathbf{K}}{\partial X_j} - {}^0\lambda_{iEx} \dfrac{\partial \mathbf{M}}{\partial X_j} \right) {}^0\mathbf{\Phi}_{iEx}} \left(\frac{\partial\, {}^0\lambda_i}{\partial X_j} \right). \tag{9.97}$$

The eigenvector derivatives are assumed to be constant, using the approximation

$$\partial\, {}^t\boldsymbol{\Phi}_i / \partial X_j \cong \partial\, {}^0\boldsymbol{\Phi}_i / \partial X_j . \tag{9.98}$$

- MS(CA)+APP = *Mode Superposition using CA analysis and approximations*. The eigenpair derivatives are calculated by

$$\frac{\partial\, {}^t\lambda_i}{\partial X_j} = \frac{{}^t\boldsymbol{\Phi}^T_{iCA}\left(\dfrac{\partial\, {}^t\mathbf{K}}{\partial X_j} - {}^t\lambda_{iCA}\dfrac{\partial \mathbf{M}}{\partial X_j}\right){}^t\boldsymbol{\Phi}_{iCA}}{{}^0\boldsymbol{\Phi}^T_{iEx}\left(\dfrac{\partial\, {}^0\mathbf{K}}{\partial X_j} - {}^0\lambda_{iEx}\dfrac{\partial \mathbf{M}}{\partial X_j}\right){}^0\boldsymbol{\Phi}_{iEx}}\left(\frac{\partial\, {}^0\lambda_i}{\partial X_j}\right), \tag{9.99}$$

$$\partial\, {}^t\boldsymbol{\Phi}_i / \partial X_j \cong \partial\, {}^0\boldsymbol{\Phi}_i / \partial X_j . \tag{9.100}$$

The efficiency of the various procedures is demonstrated in Sect. 10.1.3.

Example 9.8

Consider again the fifty-story frame shown in Fig. 9.8. The modulus of elasticity is $3\ 10^7 kN/m^2$, the elastic limit stress is $\sigma_Y = 20000 kN/m^2$ and the assumed moment-curvature relation is bi-linear as shown in Fig. 9.10. The damping ratios are 0.05 and the assumed time step is 0.02*sec*. The width and depth of all elements are 0.5*m* and 1.0*m*, respectively.

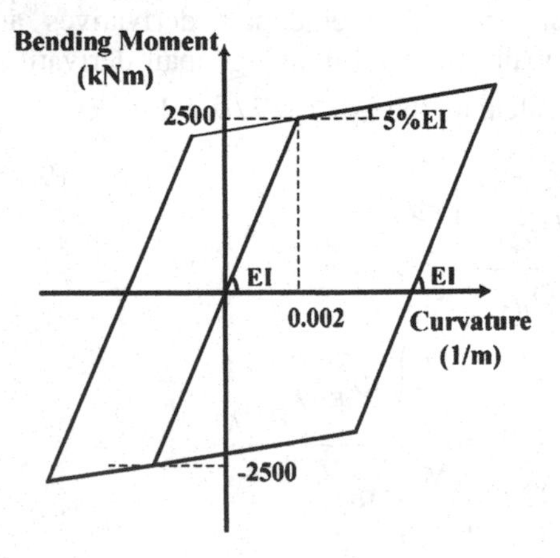

Fig. 9.10. Moment-Curvature relation

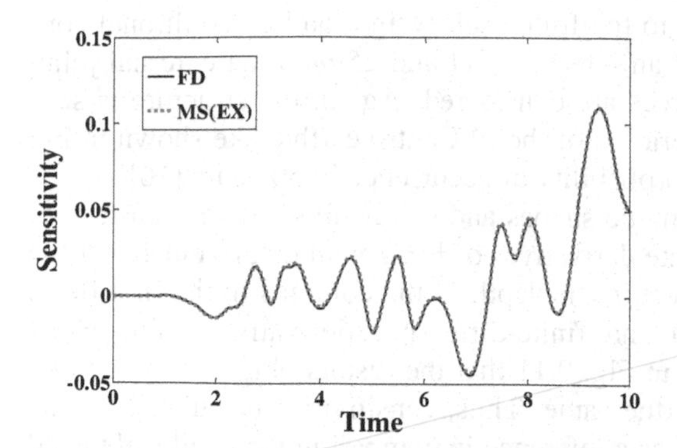

Fig. 9.11. Displacement derivatives obtained by FD and MS(EX)

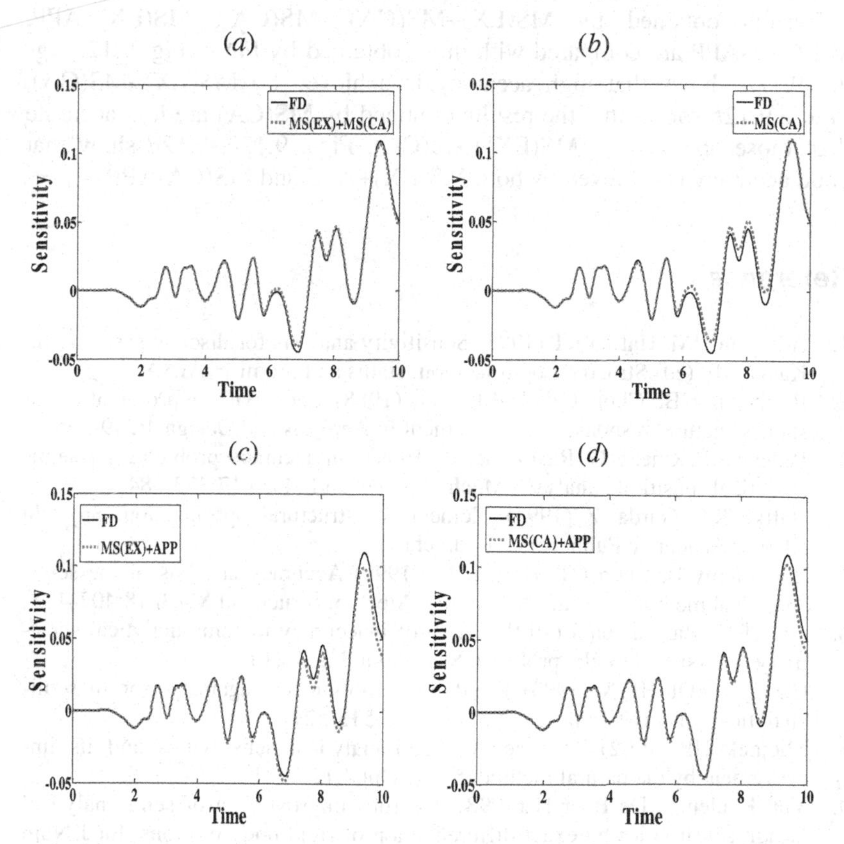

Fig. 9.12. Displacement derivatives obtained by FD and: **a.** MS(EX)+MS(CA) ***b.*** MS(CA) ***c.*** MS(EX)+APP ***d.*** MS(CA)+APP

The inertia force is due to the frame self-weight and an additional concentrated mass of 50*ton* at an internal joint and 25*ton* at an external joint. Only horizontal inertia forces are considered. Again, the structure is subjected to the ground acceleration of the El Centro earthquake shown in Fig 9.6, scaled to have a 10% probability of occurrence in 50 years [16].

Considering the first 8 mode shapes and CA analysis with 9 basis vectors, the object is to evaluate derivatives of horizontal displacements at the top of the frame with respect to the depth of the columns in the first floor. Solving first by MS(EX) and finite-difference derivatives using exact analysis (FD), it is shown in Fig. 9.11 that the results obtained by the two procedures are practically the same. Thus, sensitivities calculated by the FD procedure will be used as a reference in comparing the results obtained by the other procedures.

Results obtained by MS(EX)+MS(CA), MS(CA), MS(EX)+APP, MS(CA)+APP are compared with those obtained by FD in Fig. 9.12. Figure 9.12*a* shows that high accuracy is achieved by MS(EX)+MS(CA). Figure 9.12*b* shows that the results obtained by MS(CA) are less accurate than those obtained by MS(EX)+MS(CA). Figs. 9.12*c*, 9.12*d* show that good accuracy is achieved by both MS(EX)+APP and MS(CA)APP.

References

1. Adelman HM, Haftka RT (1993) Sensitivity analysis for discrete systems. In: Kamat MP (ed) Structural optimization: status and promise. AIAA
2. Barthelemy B, Chon CT, Haftka RT (1988) Sensitivity approximation of static structural response. Finite Element in Analysis and Design 4:249-265.
3. Pedersen P, Cheng G, Rasmussen J (1989) On accuracy problems for semi-analytical sensitivity analysis. Mech of Struct and Mach 17:373-384
4. Haftka RT, Gurdal Z (1993) Elements of structural optimization. 3rd edn Kluwer Academic Publishers, Dordrecht
5. Barthelemy B, Chon CT, Haftka RT (1990) Accuracy analysis of the semi-analytical method for shape sensitivity. Mech of Struct and Mach 18:407-432
6. Olhoff N, Rasmussen J (1991) Study of inaccuracy in semi-analytical sensitivity analysis – a model problem. Struct Opt 3:203-213
7. Cheng Y, Olhoff N (1993) Rigid body motion test against error in semi-analytical sensitivity analysis. Struct Opt 6:515-527
8. Mlejnek HP (1992) Accuracy of semi-analytical sensitivities and its improvement by the natural method. Struct Opt 4:128-131
9. Van Keulen F, De Boer H (1998) Rigorous improvement of semi-analytical design sensitivities by exact differentiation of rigid body motions. Int J Num Meth Engrg 42:71-91

10. Kirsch U (1994) Effective sensitivity analysis for structural optimization. Comp Meth Appl Mech Engrg 117:143-156
11. Kirsch U, Papalambros PY (2001) Accurate displacement derivatives for structural optimization using approximate reanalysis. Comp Meth Appl Mech Engrg 190:3945-3956
12. Kirsch U, Bogomolni M, Sheinman I (2007) Efficient procedures for repeated calculations of structural sensitivities. Engrg Opt 39:307–325
13. Kirsch U, Bogomolni M, van Keulen F (2005) Efficient finite-difference design-sensitivities. AIAA J 43:399-405
14. van Keulen F, Vervenne K, Cerulli C (2004) Improved combined approach for shape modifications. In proc 10th AIAA/ISSMO Multidisciplinary Analysis and Optimization Conference, Albany, New York, AIAA 2004-4378
15. Bogomolni M, Kirsch U, Sheinman I (2006) Efficient design-sensitivities of structures subjected to dynamic loading. Int J Sol & Str 43:5485-5500
16. Somerville P et al (1997) Development of ground motion time histories for phase 2 of the FEMA/SAC steel project. Report No SAC/BD-97/04
17. Bogomolni M, Kirsch U, Sheinman I (2006) Nonlinear-dynamic sensitivities of structures using combined approximations. AIAA J 44: 2675-2772

10 Computational Considerations

The advantages of the CA approach can be studied from several different points of view, such as efficiency, accuracy flexibility and ease-of-implementation. In this chapter, some computational considerations, related to the efficiency of the calculations and the accuracy of the results, are discussed. It has been noted that in general the efficiency and the accuracy are conflicting considerations, that is, better accuracy is often achieved at the expense of additional computational effort. Using the CA approach, the efficiency and the accuracy can be controlled in various ways. Sophisticated or simplified versions of the approach may be considered, depending on the problem to be solved.

In Sect. 10.1 the efficiency of the calculations is demonstrated for various problems of repeated analysis and sensitivity reanalysis, including linear, nonlinear, static and dynamic systems. Since the calculations are based on results of a single exact analysis, significant reductions in the computational effort are achieved, compared with complete analysis of modified designs. Further efficiency can be achieved in cases where reanalysis is required for numerous modified design points.

Some topics related to the accuracy of the results are discussed in Sect. 10.2. It has been shown in previous chapters via numerical examples that accurate approximations can be achieved for significant changes in the properties of large structures. In addition, various means may be considered to improve the accuracy of the results, and exact solutions can be achieved in certain cases. Some typical cases where accurate results can be expected are discussed in Sect. 10.2. These include cases where the high-order basis vectors come close to being linearly dependent on previous vectors, and nearly scaled designs. In Sect. 10.3 it is shown that the CA procedure and a Preconditioned Conjugate Gradient (PCG) method are equivalent. The particular pre-conditioner used by the CA procedure explains the improved convergence properties and the high accuracy of the results. Several topics related to error evaluation for static and vibration reanalysis problems are presented in Sect. 10.4. The accuracy of forces and stresses, calculated from displacements evaluated by the CA approach, is demonstrated in Sect. 10.5.

10.1 Efficiency of the Calculations

The following symbols are used in this section:

- m = number of design variables.
- m_k = half band-width of the stiffness matrix ($m_k = n^{1/2}$ for 2 dimensional problems, $m_k = n^{3/2}$ for 3 dimensional problems).
- N = number of time steps in dynamic reanalysis.
- n = number of degrees of freedom (DOF).
- p = number of considered mode shapes (usually $p << n$).
- q = number of reanalyses.
- s = number of basis vectors considered in the CA approach.
- α = the fraction of time in which the stiffness matrix is updated in nonlinear dynamic analysis (α = 0.2 means that the stiffness matrix is updated during 20% of the loading time history).

10.1.1 Static Reanalysis

The number of algebraic operations and the total CPU effort involved in solution by the CA approach are usually much smaller than those needed to carry out complete analysis of modified designs. It was found that most of the computational effort invested in reanalysis by the approach is involved in calculation of the basis vectors. Solution of various problems indicates that calculation of each basis vector requires about 2% of the CPU time needed for complete analysis. In addition, small numbers of vectors are often sufficient to achieve accurate results. In some common cases, such as reanalysis along a line in the design space, repeated calculations of the basis vectors involve almost no computational effort. In more general cases where calculation of the basis vectors must be repeated for numerous modified designs, it is possible to use various procedures where these vectors are calculated only for a limited number of design points, and then used to evaluate the response for any modified design [1].

Considering *linear static reanalysis*, the numbers of algebraic operations required by the following procedures are compared in this section:

- EX = exact reanalysis. The numbers of operations are as follows:
  Matrix factorizations = $q(1/2\, n\, m_k^2) = 1/2\, q\, n^2$.
  Forward and backward substitutions = $q(2\, n\, m_k) = 2\, q\, n^{3/2}$.

- CA = the common CA procedure. The numbers of operations for s basis vectors are as follows:
  Matrix factorizations = $q\, s^3$.
  Forward and backward substitutions = $q(s-1)(2\, n\, m_k) = 2\, q(s-1)\, n^{3/2}$.
- CARS = a CA procedure for response surface calculations. We calculate constant basis vectors for a limited number of design points. These vectors are then used to introduce explicit approximations of the basis vectors for any modified design. The numbers of operations are as follows:
  Matrix factorizations = $q\, [m(s-1)+1]^3$.
  Forward and backward substitutions = $m(s-1)(2\, n\, m_k) = 2\, m(s-1)\, n^{3/2}$.
  (For rank-one changes $s = 2$.)

Typical relations between the numbers of operations EX/CARS, EX/CA for various numbers of DOF, rank-one changes in the stiffness matrix, 10000 reanalyses and 10 design variables are shown in Fig. 10.1. It is observed that the CARS procedure is significantly more efficient than the usual EX. In addition, the efficiency of the CA procedures is higher for large numbers of DOF. Similar results have been obtained for vibration reanalysis and for repeated sensitivity analysis.

Considering *nonlinear static reanalysis*, the numbers of algebraic operations required by the CA procedure, the Full Newton-Raphson (FNR) and the Modified Newton-Raphson (MNR) procedures have been compared. The following assumptions are based on various numerical examples.

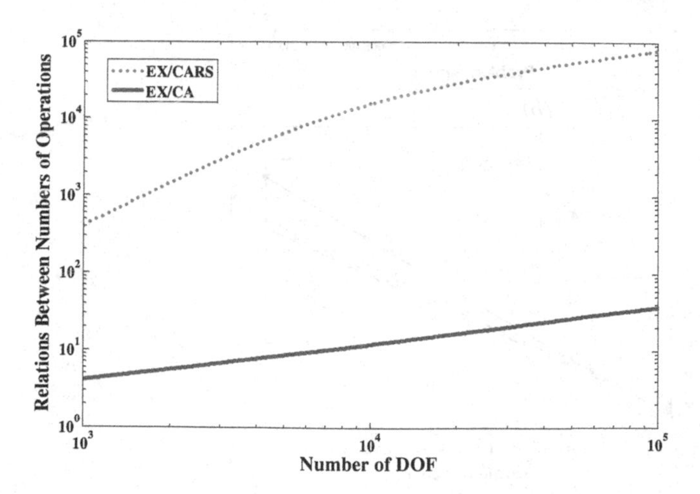

Fig. 10.1. Relations between numbers of algebraic operations, EX/CA, EX/CARS

- For material nonlinear reanalysis the CA procedure is compared with the MNR procedure. It is assumed that the load is divided into 20 increments and the CA procedure requires 3 times the number of iterations per increment compared with MNR.
- For geometric nonlinear reanalysis the CA procedure is compared with the FNR procedure. The load is divided into 30 increments and the CA procedure requires the same number of increments and twice the number of iterations per increment compared with FNR.

The required numbers of operations versus the number of DOF for CA4, CA6 and CA10 (CA with 4, 6 and 10 basis vectors, respectively) are shown in Fig. 10.2. Again, significant reductions in the computational effort can be achieved by the CA approach,

(a)

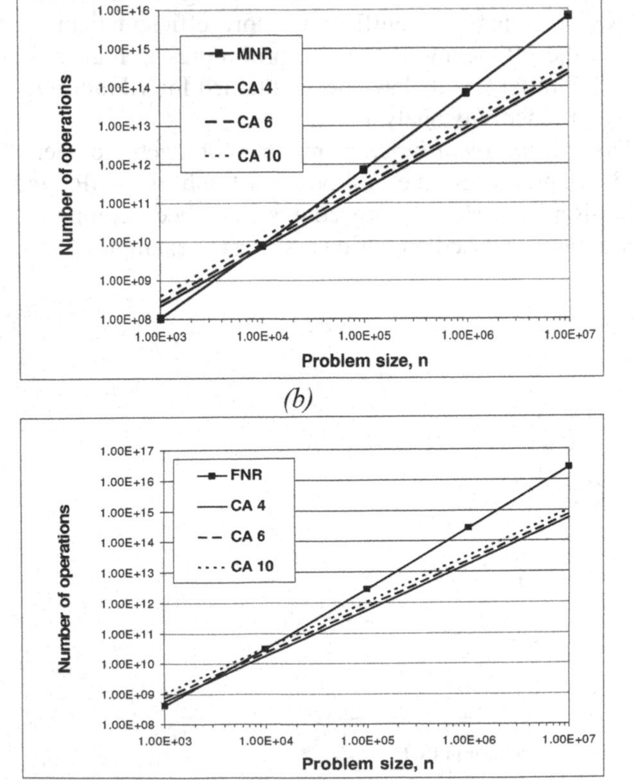

Fig. 10.2 Numbers of algebraic operations versus number of DOF: ***a***. Material nonlinear reanalysis ***b***. Geometric nonlinear reanalysis

10.1.2 Dynamic Reanalysis

We assume $m_k = n^{1/2}$ and the following numbers of algebraic operations:

- Numerical integration at each step = $3n\, m_k$.
- Matrix factorization = $n\, m_k^2/2$.
- Forward and backward substitutions = $2n\, m_k$.

The numbers of operations involved in dynamic reanalysis by the following procedures are summarized in Table 10.1:

- DI = complete direct integration.
- MS(EX) = mode superposition using exact formulation of reanalysis.
- MS(CA) = mode superposition using the CA approach.

Note that the expressions shown in Table 10.1 can be used either for linear ($\alpha = 0$) or nonlinear ($0 < \alpha \leq 1$) dynamic reanalysis. For illustrative purposes assume $N = 5000$, $p = 5$, $s = 5$. The numbers of operations required by the various methods versus the number of degrees of freedom for linear and nonlinear ($\alpha = 0.5$) dynamic reanalysis are shown in Fig. 10.3. It is observed that significant reductions in the computational effort (1–2 orders of magnitude) can be achieved by the CA approach for large scale structures.

Table 10.1. Numbers of operations, dynamic reanalysis

Procedure	Number of Operations
DI	$(3nm_k)N + [(1/2)nm_k^2 + (3/2)nm_k](1+\alpha N)$
MS(EX)	$(3p)N + [(nm_k^2) + (15pnm_k)](1+\alpha N)$
MS(CA)	$(3p)N + [2(nm_k)(s-1)p + ps^3](1+\alpha N)$

Table 10.2. Numbers of operations, repeated-dynamic sensitivity analysis

Procedure	Sensitivity Analysis
DI	$(3nm_k)Nm + 3nm_k m(1+\alpha N)$
MS(EX)	$(pm)N + (nm_k^2 + 15pnm_k)m(1+\alpha N)$
MS(CA)	$(pm)N + [2(nm_k)(s-1)p + ps^3]m(1+\alpha N)$

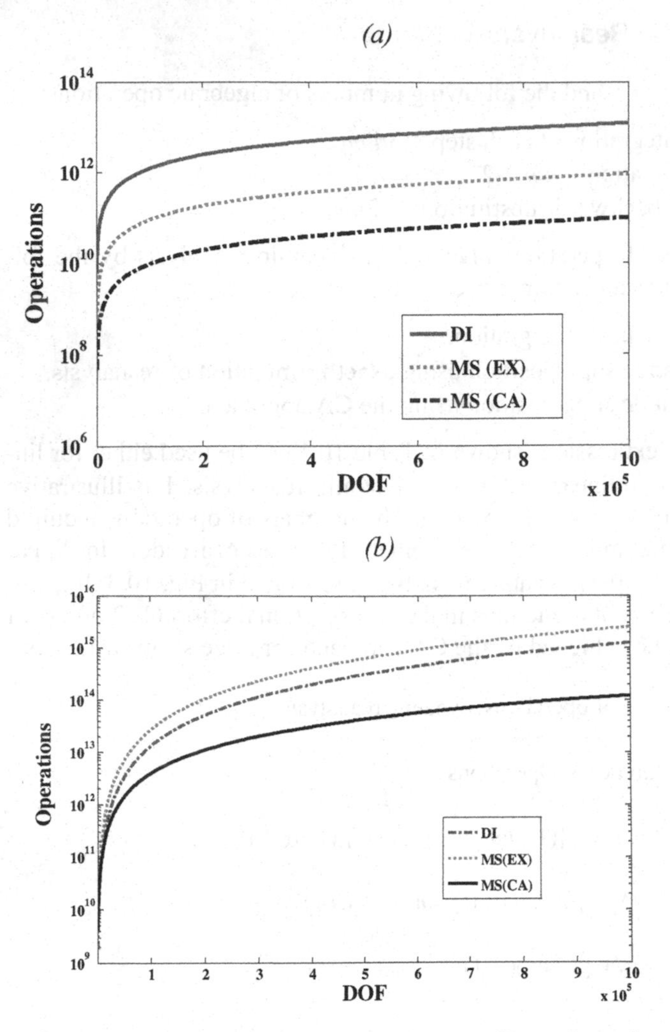

Fig. 10.3. Numbers of algebraic operations: ***a***. linear dynamic reanalysis ($\alpha = 0$) ***b***. nonlinear dynamic reanalysis ($\alpha = 0.5$)

10.1.3 Dynamic Sensitivity Analysis

The numbers of algebraic operations involved in (linear and nonlinear) repeated dynamic sensitivity analysis by the procedures DI, MS(EX) and MS(CA) are shown in Table 10.2.

Consider the following procedures, previously described in Sect. 9.5.3:

Table 10.3. Numbers of algebraic operations, repeated dynamic analysis and sensitivity analysis

Type	Procedure	Number of operations
Analysis	DI	$(3nm_k)N+(1/2nm_k^2+3/2nm_k)(1+\alpha N)$
	MS(EX)	$(3p)N+(nm_k^2+15pnm_k)(1+\alpha N)$
	MS(CA)	$(3p)N+[2(nm_k)(s\text{-}1)p+ps^3](1+\alpha N)$
	MS(EX)+MS(CA)	$(3p)N+(nm_k^2+15pnm_k)(1+\alpha N)$
	MS(EX)+APP	$(3p)N+(nm_k^2+15pnm_k)(1+\alpha N)$
	MS(CA)+APP	$(3p)N+[2(nm_k)(s\text{-}1)p+ps^3](1+\alpha N)$
Sensitivity analysis	DI	$(3nm_k)Nm+3nm_km(1+\alpha N)$
	MS(EX)	$(pm)N+(nm_k^2+15pnm_k)m(1+\alpha N)$
	MS(CA)	$(pm)N+[2(nm_k)(s-1)p+ps^3]m(1+\alpha N)$
	MS(EX)+MS(CA)	$(2p)N+[2(nm_k)(s-1)p+ps^3]m(1+\alpha N)$
	MS(EX)+APP	$(2p)N+[2(nm_k)sp+ps^3]m$
	MS(CA)+APP	$(2p)N+[2(nm_k)sp+ps^3]m$

- DI = Direct Integration.
- MS(EX) = Mode Superposition using exact analysis.
- MS(CA) = Mode Superposition using CA analysis.
- MS(EX)+MS(CA) = Mode Superposition using exact analysis for the original eigenproblem and CA analysis for the perturbed eigenproblem.
- MS(EX)+APP= Mode Superposition using exact analysis and approximations.
- MS(CA)+APP = Mode Superposition using CA and approximations.

The numbers of operations required for dynamic analysis and sensitivity analysis are summarized in Table 10.3 [1–3]. Assume for illustrative purposes $N = 5000$, $p = 5$, $m = 20, 100$, $m_k = n^{1/2}$, $s = 5$, $\alpha = 0.5$. The numbers of algebraic operations versus the number of DOF for nonlinear dynamic sensitivity analysis are shown in Fig. 10.4. Comparing solutions by direct integration DI and mode superposition MS(EX) it is observed that the number of operations in the latter case is significantly larger. This number is reduced by the MS(EX)+MS(CA) procedure and even more by the MS(CA) procedure. The best efficiency is obtained by the MS(CA)+APP procedure. The relations between the numbers of algebraic operations involved in DI and MS(CA)+APP (Fig. 10.5) show that sensitivity analysis by the latter procedure significantly reduces the computational effort.

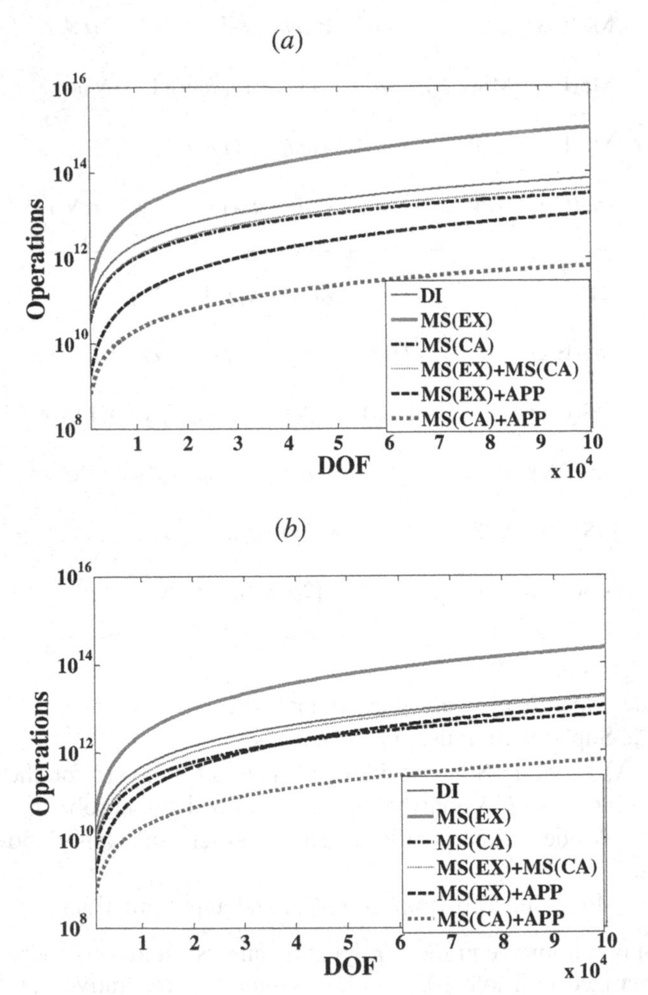

Fig. 10.4. Numbers of operations, nonlinear dynamic sensitivity analysis: ***a.*** m =100 ***b.*** $m = 20$

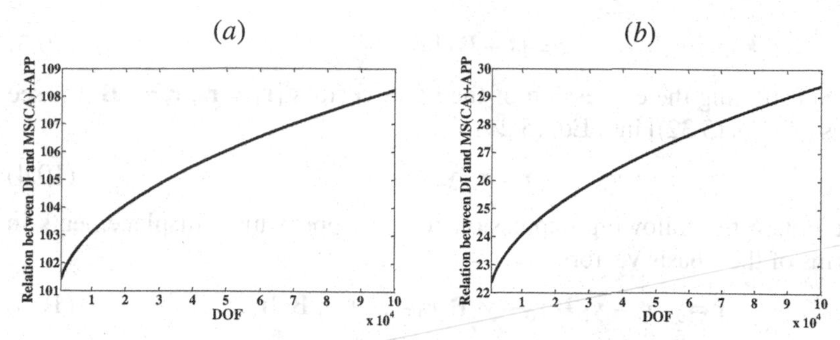

Fig. 10.5. Relations between numbers of operations in DI and MS(CA)+APP: *a*. m =100 *b*. m = 20

10.2 Accuracy Considerations

10.2.1 Linearly Dependent Basis Vectors

It has been noted that linear independence of the basis vectors is necessary for convergence of reduced-basis approximations. The vector of approximate displacements evaluated by the CA approach is a linear combination of the basis vectors $\mathbf{r}_1, \mathbf{r}_2, ..., \mathbf{r}_s$ [Eq. (5.24)]. The latter vectors are said to be linearly independent if the relation,

$$y_1\mathbf{r}_1 + y_2\mathbf{r}_2 + ... + y_s\mathbf{r}_s = \mathbf{0}, \tag{10.1}$$

can be satisfied only for the trivial case, that is, only for the case where all the coefficients $y_1, y_2, ..., y_s$ are identically zero. If the relation is satisfied and part of the coefficients are different from zero, then the basis vectors $\mathbf{r}_1, \mathbf{r}_2, ... , \mathbf{r}_s$ are said to be linearly dependent, with the implication that one vector is a linear combination of the remaining vectors.

It is shown in this section that an exact solution is obtained by the CA procedure in cases where a newly created vector becomes a linear combination of the previous vectors [4]. For simplicity of presentation assume that $\mathbf{R} = \mathbf{R}_0$ and therefore $\mathbf{r}_1 = \mathbf{r}_0$. Pre-multiplying the modified equations $(\mathbf{K}_0+\Delta\mathbf{K})\ \mathbf{r} = \mathbf{R}_0$ by $\mathbf{K}_0^{-1}$ and substituting $\mathbf{B} = \mathbf{K}_0^{-1}\Delta\mathbf{K}$, we obtain the following exact expression for the modified design

$$(\mathbf{I} + \mathbf{B})\ \mathbf{r} = \mathbf{r}_0, \tag{10.2}$$

where $\mathbf{I}$ is an identity matrix. Premultipling Eq. (10.2) by $(\mathbf{I}+\mathbf{B})^{-1}$ gives the exact modified displacements

$$\mathbf{r} = (\mathbf{I} + \mathbf{B})^{-1}\, \mathbf{r}_0. \tag{10.3}$$

Substituting the expression of the basis vectors [$\mathbf{r}_1 = \mathbf{r}_0$, $\mathbf{r}_i = -\mathbf{B}\, \mathbf{r}_{i-1}$, see Eqs. (5.31), (5.32)] into Eq. (5.24)

$$\mathbf{r} = \mathbf{r}_B\, \mathbf{y}, \tag{10.4}$$

we obtain the following expression for the approximate displacements in terms of the s basis vectors

$$\mathbf{r} = y_1\, \mathbf{r}_0 - y_2\, \mathbf{B}\, \mathbf{r}_0 + y_3\, \mathbf{B}^2\mathbf{r}_0 - \ldots + y_s\, \mathbf{B}^{s-1}\mathbf{r}_0. \tag{10.5}$$

Assuming that the approximate solution involving s terms [Eq. (10.5)] is equal to the exact solution [Eq. (10.3)], pre-multiplying both equations by ($\mathbf{I}$+$\mathbf{B}$) and rearranging we obtain the expression for an additional term $\mathbf{r}_{s+1}$

$$\mathbf{r}_{s+1} = \sum_{i=1}^{s} a_i \mathbf{r}_i\,, \tag{10.6}$$

where a_i are scalar multipliers given by

$$a_1 = (y_1 - 1)/\, y_s, \tag{10.7}$$

$$a_i = (y_i - y_{i-1})/\, y_s \qquad i = 2,3,\ldots,s.$$

Equation (10.6) shows that when the reduced basis expression with s terms [Eq. (10.5)] is equal to the exact solution [Eq. (10.3)], the (s+1)th basis vector is a linear combination of the previous s vectors. That is, the (s+1) basis vectors are linearly dependent.

In general the CA procedure provides approximate solutions, but accurate solutions are often achieved with a small number of basis vectors. It is expected that accurate (nearly exact) solutions will be achieved when the high-order basis vectors come close to being linearly dependent on previous vectors. Two basis vectors $\mathbf{r}_i$ and $\mathbf{r}_{i+1} = -\mathbf{B}\, \mathbf{r}_i$ are close to being linearly dependent if

$$cos\, \beta_{i,\, i+1} = (\mathbf{r}_i^T\, \mathbf{B}\, \mathbf{r}_i) \,/\, (|\mathbf{r}_i|\, |\mathbf{B}\, \mathbf{r}_i|) \approx 1, \tag{10.8}$$

where $\beta_{i,\, i+1}$ is the angle between the two vectors. Various numerical examples show that the basis vectors determined by the CA approach satisfy the condition of Eq. (10.8), as the basis vectors index i is increased, even for very large changes in the design.

Example 10.1

Consider the ten-bar truss shown in Fig. 10.6 with 2 geometrical variables: D = the truss depth, W = the panel width. Assuming the initial design $W = D = 360$, the following cases of modified geometries have been solved:

Case a. $W = 360$, $D = 540$ (50% increase in the depth).
Case b. $W = 360$, $D = 720$ (100% increase in the depth).
Case c $W = 180$, $D = 720$ (50% decrease in the width, 100% increase in the depth).

The results obtained for 2 and 3 basis vectors (CA2 and CA3, respectively) summarized in Table 10.4 show that the changes in the geometry lead to significant changes in the displacements. It was found that the basis vectors are close to being linearly dependent. In Case *b*, for example, we obtain $\cos \beta_{1,2} = 0.9912$, $\cos \beta_{2,3} = 0.9999$ for the first 3 basis vectors, which explains the high accuracy achieved for these large design changes.

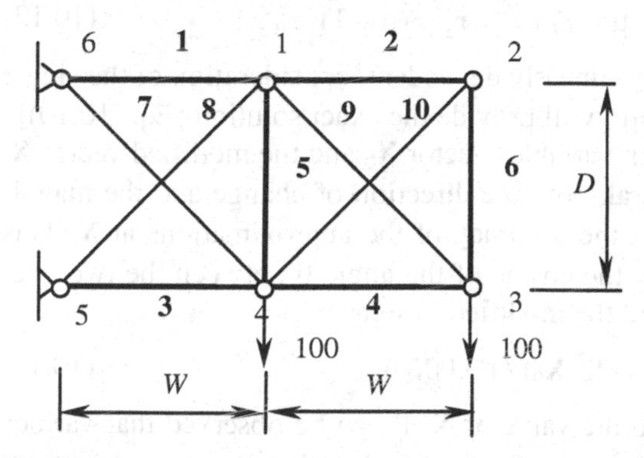

Fig. 10.6. Ten-bar truss

Table 10.4. Geometrical changes, ten-bar truss

Case	W	D	Method	1	2	3	4	5	6	7	8
Initial	360	360	Exact	2.34	5.58	2.82	12.65	-3.17	13.13	-2.46	6.01
a	360	540	CA2	1.49	4.02	1.71	7.86	-2.06	8.40	-1.60	4.48
			CA3	1.53	3.93	1.81	7.83	-2.17	8.46	-1.64	4.45
			Exact	1.55	3.94	1.82	7.84	-2.18	8.47	-1.66	4.44
b	360	720	CA2	1.17	3.78	1.26	6.72	-1.61	7.29	-1.28	4.27
			CA3	1.14	3.67	1.34	6.62	-1.68	7.35	-1.24	4.25
			Exact	1.15	3.67	1.34	6.60	-1.66	7.36	-1.25	4.24
c	180	720	CA2	0.43	2.59	0.36	3.83	-0.59	4.27	-0.50	2.97
			CA3	0.31	2.52	0.43	3.86	-0.55	4.44	-0.35	2.98
			Exact	0.29	2.47	0.33	3.85	-0.42	4.53	-0.31	2.94

10.2.2 Scaled and Nearly Scaled Designs

In this section we first show that if the modified design is a scaled initial design, then the basis vectors are linearly dependent. A scaled design is obtained by multiplying the initial stiffness matrix $\mathbf{K}_0$ by a positive scaling multiplier μ to obtain the modified matrix

$$\mathbf{K} = \mu\, \mathbf{K}_0. \tag{10.9}$$

For $\mathbf{R} = \mathbf{R}_0$ the exact displacements after scaling are given by

$$\mathbf{r} = \mu^{-1}\, \mathbf{r}_0. \tag{10.10}$$

It is observed that in this case matrix $\mathbf{B}$ becomes

$$\mathbf{B} = \mathbf{K}_0^{-1}\Delta\mathbf{K} = (\mu - 1)\mathbf{I}, \tag{10.11}$$

where $\mathbf{I}$ is the identity matrix, and the resulting basis vectors are

$$\mathbf{r}_1 = \mathbf{r}_0 \qquad \mathbf{r}_2 = -\,(\mu - 1)\,\mathbf{r}_0 \qquad \mathbf{r}_3 = (\mu - 1)^2\,\mathbf{r}_0\ \ldots \tag{10.12}$$

Since these basis vectors are linearly dependent, consideration of the single basis vector $\mathbf{r}_1$=$\mathbf{r}_0$ with y_1=μ^{-1} will provide the exact solution [Eq. (10.10)].

Assume the initial design variables vector $\mathbf{X}_0$ and the modified vector $\mathbf{X}$, due a change $\Delta\mathbf{X}$. In general, both the direction of change and the magnitude of change may affect the accuracy of the approximations at $\mathbf{X}$. This effect can be quantified by the cosine of the angle θ between the two vectors of the initial design and the modified design

$$cos\,\theta = (\mathbf{X}^T\,\mathbf{X}_0)\,/\,(|\mathbf{X}|\,|\mathbf{X}_0|), \tag{10.13}$$

where $|\mathbf{X}|$ denotes the absolute value of $\mathbf{X}$. It can be observed that various designs, obtained by scaling a certain modified design, provide identical θ angles. It was found [4] that high accuracy is achieved with a small number of basis vectors for designs correspond to small θ values. More basis vectors are needed for designs that correspond to larger θ. These observations are limited to the space formed by the vectors $\mathbf{X}_0$ and $\mathbf{X}$. For the complete design space, smaller θ values do not guarantee better approximations.

Example 10.2

Consider the initial geometry of the fifty-bar truss shown in Fig. 10.7. The member cross section areas equal unity, the modulus of elasticity is 10000 and the 40 unknowns are the horizontal (X direction) and the vertical (Y direction) displacements of joints 2 through 21.

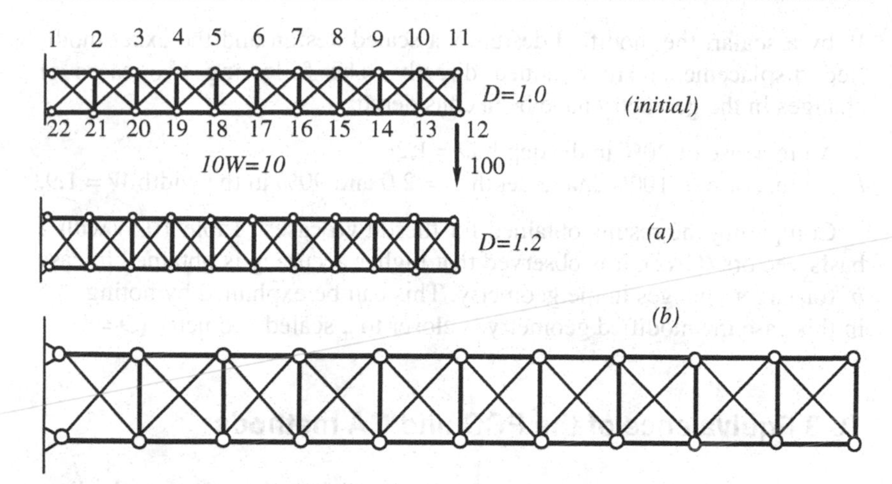

Fig. 10.7. Initial and modified geometries, 50-bar truss

Table 10.5. Approximations of displacements, 50-bar truss.

Joint	Direction	Case *a*		Case *b*	
		CA2	Exact	CA2	Exact
2	*X*	0.09	0.08	0.20	0.20
	Y	0.11	0.08	0.25	0.24
3	*X*	0.16	0.15	0.38	0.38
	Y	0.35	0.28	0.90	0.88
4	*X*	0.22	0.21	0.54	0.54
	Y	0.69	0.60	1.90	1.87
5	*X*	0.26	0.27	0.68	0.67
	Y	1.12	1.01	3.21	3.18
6	*X*	0.29	0.31	0.79	0.79
	Y	1.60	1.51	4.79	4.74
7	*X*	0.32	0.35	0.88	0.88
	Y	2.13	2.07	6.58	6.54
8	*X*	0.34	0.38	0.95	0.96
	Y	2.69	2.69	8.54	8.50
9	*X*	0.35	0.40	1.01	1.01
	Y	3.27	3.36	10.63	10.60
10	*X*	0.35	0.41	1.04	1.04
	Y	3.86	4.05	12.81	12.79
11	*X*	0.35	0.42	1.05	1.05
	Y	4.45	4.75	15.02	15.02

The truss depth D and the panel width W are the two geometric variables. Assuming the initial design $D = W = 1.0$ and multiplying both D and

W by a scalar, the modified design is a scaled design and the exact modified displacements are obtained directly. The following two cases of changes in the geometry have been considered:

a. An increase of 20% in the depth $D = 1.2$.
b. An increase of 100% in the depth $D = 2.0$ and 90% in the width $W = 1.9$.

Comparing the results obtained for these two cases (Table 10.5) with 2 basis vectors (CA2), it is observed that higher accuracy is obtained in case *b*, for larger changes in the geometry. This can be explained by noting that in this case the modified geometry is closer to a scaled geometry ($D = W$).

10.3 Equivalence of the PCG and CA methods

It is shown in this section that the Preconditioned Conjugate Gradient (PCG) method, described in Sect. 1.3.2, and the CA approach are equivalent and provide identical results [5] if matrix **C** is chosen as [Eq. (1.53)]

$$\mathbf{C} = \mathbf{U}_0^{-1}, \tag{10.14}$$

where $\mathbf{U}_0$ is an upper triangular matrix, given by factorization of the initial stiffness matrix $\mathbf{K}_0 = \mathbf{U}_0^T\,\mathbf{U}_0$.

Applying k iterations of the Conjugate Gradient (CG) method to Eq. (1.31) is equivalent to the minimization of the quadratic function Q [Eq. (1.32)] on the Krylov subspace of degree k, defined as

$$\mathcal{D}_k = \mathbf{r}_0 + \text{span}\,\{\boldsymbol{\delta}_0, \mathbf{K}\,\boldsymbol{\delta}_0, \mathbf{K}^2\boldsymbol{\delta}_0, \ldots, \mathbf{K}^{k-1}\,\boldsymbol{\delta}_0\}, \tag{10.15}$$

where the residual vector $\boldsymbol{\delta}_0$ is given by Eq. (1.35). Consider the case where the CG method is applied to the preconditioned system [Eq. (1.46)]

$$\tilde{\mathbf{K}}\,\tilde{\mathbf{r}} = \tilde{\mathbf{R}}, \tag{10.16}$$

where [Eq. (1.45)]

$$\tilde{\mathbf{K}} = \mathbf{C}^T\mathbf{K}\mathbf{C} \qquad \tilde{\mathbf{R}} = \mathbf{C}^T\,\mathbf{R}, \tag{10.17}$$

with $\mathbf{C} = \mathbf{U}_0^{-1}$ [Eq. (10.14)]. The quadratic function,

$$\tilde{Q} = 1/2\,\tilde{\mathbf{r}}^T\,\tilde{\mathbf{K}}\,\tilde{\mathbf{r}} - \tilde{\mathbf{r}}^T\,\tilde{\mathbf{R}}, \tag{10.18}$$

is minimized in the kth iteration on the Krylov subspace

$$\tilde{\mathcal{D}}_k = \tilde{\mathbf{r}}_0 + \text{span}\left\{\tilde{\boldsymbol{\delta}}_0,\ \tilde{\mathbf{K}}\,\tilde{\boldsymbol{\delta}}_0,\ \tilde{\mathbf{K}}^2\tilde{\boldsymbol{\delta}}_0,\ \ldots,\ \tilde{\mathbf{K}}^{k-1}\,\tilde{\boldsymbol{\delta}}_0\right\}. \tag{10.19}$$

Assuming the initial point $\tilde{\mathbf{r}}_0 = 0$ and denoting $\mathbf{P} = \boldsymbol{\Delta}\mathbf{K}\,\mathbf{K}_0^{-1}$, then [Eqs. (10.16), (10.17)] $\tilde{\boldsymbol{\delta}}_0 = \tilde{\mathbf{R}} - \tilde{\mathbf{K}}\,\tilde{\mathbf{r}}_0 = \mathbf{C}^T\mathbf{R}$. Since $\mathbf{C}\mathbf{C}^{\mathrm{T}} = \mathbf{K}_0^{-1}$ we obtain

$$\tilde{\mathbf{K}}\,\tilde{\boldsymbol{\delta}}_0 = \mathbf{C}^T(\mathbf{K}_0 + \boldsymbol{\Delta}\mathbf{K})\mathbf{C}\mathbf{C}^T\mathbf{R} = \mathbf{C}^T\mathbf{R} + \mathbf{C}^T\mathbf{P}\mathbf{R}, \tag{10.20}$$

$$\tilde{\mathbf{K}}^2\,\tilde{\boldsymbol{\delta}}_0 = \mathbf{C}^T\mathbf{R} + 2\mathbf{C}^T\mathbf{P}\mathbf{R} + \mathbf{C}^T\mathbf{P}^2\mathbf{R},$$

and so on, so that

$$\tilde{\mathcal{D}}_k = \mathbf{C}^{\mathrm{T}}\text{span}\left\{\mathbf{R},\ \mathbf{P}\mathbf{R},\ \mathbf{P}^2\mathbf{R},\ \ldots,\ \mathbf{P}^{k-1}\mathbf{R}\right\}. \tag{10.21}$$

Hence, the minimizer of $\tilde{Q}$ on $\tilde{\mathcal{D}}_k$ [Eqs. (10.18), (10.21)] is of the form

$$\tilde{\mathbf{r}}_k = \mathbf{C}^{\mathrm{T}}\left(\alpha_0\,\mathbf{R} + \alpha_1\,\mathbf{P}\mathbf{R} + \alpha_2\mathbf{P}^2\,\mathbf{R} + \ldots + \alpha_{k-1}\mathbf{P}^{k-1}\,\mathbf{R}\right). \tag{10.22}$$

To return to the original variables, pre-multiply Eq. (10.22) [and thus Eq. (10.21)] by $\mathbf{C}$ to obtain

$$\mathcal{D}_k = \text{span}\left(\mathbf{K}_0^{-1}\,\mathbf{R},\ \mathbf{K}_0^{-1}\,\mathbf{P}\mathbf{R},\ \mathbf{K}_0^{-1}\mathbf{P}^2\mathbf{R},\ \ldots,\ \mathbf{K}_0^{-1}\mathbf{P}^{k-1}\mathbf{R}\right), \tag{10.23}$$

$$\mathbf{r}_k = \mathbf{K}_0^{-1}\left(\alpha_0\,\mathbf{R} + \alpha_1\,\mathbf{P}\mathbf{R} + \alpha_2\mathbf{P}^2\mathbf{R} + \ldots + \alpha_{k-1}\mathbf{P}^{k-1}\mathbf{R}\right). \tag{10.24}$$

In summary, in the kth iteration of the PCG method applied to $\mathbf{K}\,\mathbf{r} = \mathbf{R}$, with the preconditioned matrix $\mathbf{C} = \mathbf{U}_0^{-1}$ [Eq. (10.14)] and the initial value $\tilde{\mathbf{r}}_0 = \mathbf{0}$, we minimize the quadratic function Q [Eq. (1.32)] on the Krylov subspace of Eq. (10.23). In the CA approach we introduce an $n \times k$ matrix $\mathbf{r}_B$ by means of vectors $\mathbf{r}_i$ defined recurrently. The approximate solution of $\mathbf{K}\,\mathbf{r} = \mathbf{R}$ is obtained by solving the reduced system $\mathbf{K}_R\,\mathbf{y} = \mathbf{R}_R$, and interpolating by $\mathbf{r} = \mathbf{r}_B\,\mathbf{y}$. But this is the same as minimizing the quadratic functional Q [Eq. (1.32)] on a subspace defined by Eq. (10.23). The conclusion is that the solution found by the CA approach with an $n \times k$ matrix $\mathbf{r}_B$ is fully equivalent to k iterations of the PCG method with the preconditioned matrix $\mathbf{C} = \mathbf{U}_0^{-1}$ and the initial value $\tilde{\mathbf{r}}_0 = \mathbf{0}$.

Example 10.3

Consider a minimum compliance (external work) topology design problem for a truss with a ground structure of 51×11 nodes [5]. The potential bars connect all neighboring nodes as shown in Figure 10.8*a*. All the nodes on the left-hand-side are fixed and the bottom node at the right-hand-side is subject to a vertical load. Altogether there are 2050 bars and 1100 degrees of freedom (nodal displacements). The optimal topology is shown in Fig. 10.8*b* and the compliance of the optimal design is 6.2165 10^{-6}.

Results achieved by the CA approach are demonstrated for different systems defined by various values of the initial and the modified stiffness matrices. These matrices are given (through the design variables **X**) by a parameter α according to $\mathbf{X} = \mathbf{X}_0 + \alpha(\mathbf{X}_{opt} - \mathbf{X}_0)$ where $\mathbf{X}_0$ is a vector of initial constant numbers 1/2050, and $\mathbf{X}_{opt}$ is the optimal design vector from Fig. 10.8*b*. All design vectors satisfy the feasibility constraint $\sum X_i = 1$. Table 10.6 shows minimal and maximal values of the design variables for various α values, and indicates the significant changes in **X**.

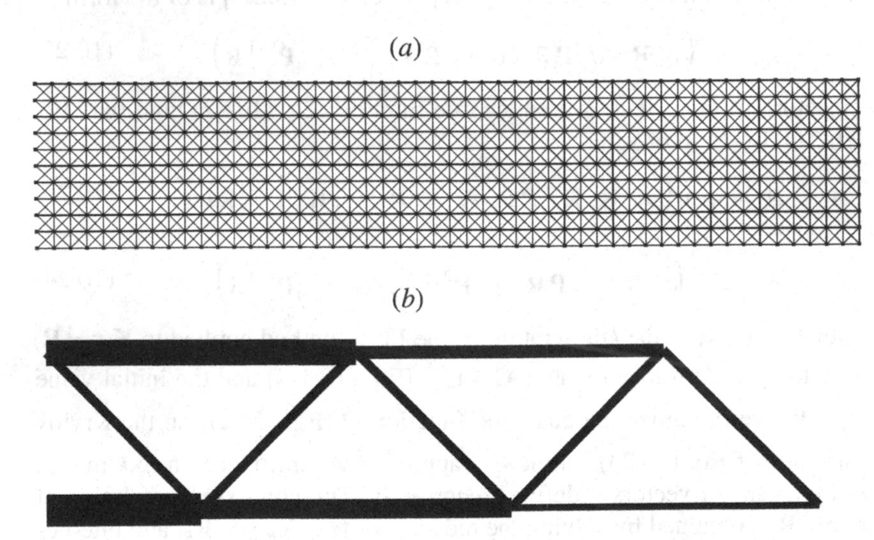

Fig. 10.8. 2050-bar truss: ***a.*** Initial ground structure ***b.*** Optimal topology

Table 10.6. Minimal and maximal **X** for various α values

	α = 0	α = 0.1	α = 0.5	α = 0.9	α = 0.99	α = 1.0
X_{min}	0.0004878	0.000439	0.000243	4.9 10^{-5}	4.9 10^{-6}	2.7 10^{-12}
X_{max}	0.0004878	0.014725	0.071672	0.12862	0.14143	0.14286

Table 10.7. Condition numbers of **K** and $\tilde{\mathbf{K}}$ for various modified designs

Initial α	Modified α	$\psi(\mathbf{K})$	$\psi(\tilde{\mathbf{K}})$
0.10	0.50	$1.67\ 10^6$	8.8
0.10	0.90	$2.29\ 10^6$	78.5
0.50	0.90	$2.29\ 10^6$	9.0
0.50	0.99	$1.08\ 10^7$	98.7

Table 10.8. Compliance (times 10^6)

Initial α	Modified α	CA3 solution	Exact solution
0.10	0.50	11.44	11.44
0.10	0.90	6.84	6.84
0.50	0.90	6.84	6.84
0.50	0.99	6.27	6.27

The condition numbers (defined by the ratio of the maximum and minimum eigenvalues) of the original matrix, $\psi(\mathbf{K})$, and the preconditioned matrix, $\psi(\tilde{\mathbf{K}})$, for various modified designs are shown in Table 10.7. It is observed that the condition numbers of matrices $\tilde{\mathbf{K}}$ are much smaller than those of **K**. The compliance for various initial designs and modified designs, considering 3 basis vectors (CA3), are shown in Table 10.8. For the initial designs $\alpha = 0.10 - 0.50$ and modified designs $\alpha = 0.50 - 0.99$, only 3 basis vectors are required to achieve accurate results.

10.4 Error Evaluation

10.4.1 Static Reanalysis

The Conjugate Gradient (CG) Method

In exact arithmetic the CG method, described in Sect. 1.3.2, will terminate at the solution in at most n iterations. What is more remarkable is that when the distribution of the eigenvalues of **K** has certain favorable features, the method will identify the solution in much less than n iteration cycles. It has been shown [e.g. 6, 7] that if **K** has only p distinct eigenvalues, then the CG iteration will terminate at the solution in at most p iterations. In addition, if the eigenvalues of **K** occur in p distinct clusters, the CG method will approximately solve the problem after p steps.

Define the usual energy norm

$$\| \mathbf{r} \|_{\mathbf{K}} = \left(\mathbf{r}^T \mathbf{K}\, \mathbf{r}\right)^{1/2}. \tag{10.25}$$

If **K** has eigenvalues $\lambda_1 \leq \lambda_2 \leq \ldots \leq \lambda_n$, a useful estimate of the convergence behavior for k steps of the CG method is given by [8]

$$\| \mathbf{r}_{k+1} - \mathbf{r}_{ex} \|_{\mathbf{K}}^2 \leq \left(\frac{\lambda_{n-k} - \lambda_1}{\lambda_{n-k} + \lambda_1} \right)^2 \left\| \mathbf{r}_0 - \mathbf{r}_{ex} \right\|_{\mathbf{K}}^2. \tag{10.26}$$

Another, more approximate, convergence expression for the CG method is

$$\| \mathbf{r}_k - \mathbf{r}_{ex} \|_{\mathbf{K}} \leq 2 \left(\frac{\sqrt{\psi(\mathbf{K})} - 1}{\sqrt{\psi(\mathbf{K})} + 1} \right)^k \left\| \mathbf{r}_0 - \mathbf{r}_{ex} \right\|_{\mathbf{K}}, \tag{10.27}$$

where $\psi(\mathbf{K})$ is the condition number of **K**, defined by the ratio of the maximum and minimum eigenvalues

$$\psi(\mathbf{K}) = \lambda_{max} / \lambda_{min}. \tag{10.28}$$

This bound gives an overestimate of the error, but it can be useful when the only information about **K** is estimates of the extreme eigenvalues.

The CA Procedure

It has been noted in Sect. 5.1.3 that the series of basis vectors [Eq. (5.3)] converges if and only if *lim* $\mathbf{B}^k = \mathbf{0}$ as $k \to \infty$, which in turn holds if and only if $\rho(\mathbf{B}) < 1$, where $\rho(\mathbf{B})$ is the spectral radius (the largest eigenvalue) of matrix **B** [9]. To evaluate the errors involved in the binomial series approximations, it has been shown that the sum $\Delta\mathbf{r}$ of the additional terms in the series of Eq. (5.3), beyond the first s terms, is bounded from above by Eq. (5.23) [10]

$$\Delta\mathbf{r} \leq \| \mathbf{B} \|^s \frac{1}{1 - \| \mathbf{B} \|} \| \mathbf{r}_0 \|. \tag{10.29}$$

Evidently, for large changes $\Delta\mathbf{K}$ (and corresponding large elements of **B**) this bound may become very large and the series diverges.

In the CA procedure the binomial series terms can be normalized, and very fast convergence is obtained even in cases where the series of basis vectors [Eq. (5.3)] diverges. The reduced problem is [Eqs. (5.28), (5.29)]

$$\mathbf{K}_R\, \mathbf{y} = \mathbf{R}_R, \tag{10.30}$$

where

$$\mathbf{K}_R = \mathbf{r}_B^T\,\mathbf{K}\,\mathbf{r}_B \qquad \mathbf{R}_R = \mathbf{r}_B^T\,\mathbf{R}. \tag{10.31}$$

It has been noted in Sect. 10.3 that the CA solution with k basis vectors is equivalent to k iterations of the PCG method. Applying results from the CG method, the error bound for k basis vectors in the CA procedure can be evaluated by an expression similar to Eq. (10.27). Considering $\mathbf{C} = \mathbf{U}_0^{-1}$ [Eq. (10.14)], we obtain the preconditioned system of Eqs. (10.16), (10.17). Thus, the corresponding expression giving the error bound for k basis vectors in the CA approach is

$$\|\mathbf{r}_k - \mathbf{r}_{ex}\|_{\tilde{\mathbf{K}}} \le 2\left(\frac{\sqrt{\psi(\tilde{\mathbf{K}})}-1}{\sqrt{\psi(\tilde{\mathbf{K}})}+1}\right)^k \|\mathbf{r}_0 - \mathbf{r}_{ex}\|_{\tilde{\mathbf{K}}}, \tag{10.32}$$

where

$$\|\mathbf{r}\|_{\tilde{\mathbf{K}}} = \left(\mathbf{r}^T\tilde{\mathbf{K}}\,\mathbf{r}\right)^{1/2}. \tag{10.33}$$

Consider the approximate displacements expressed in the uncoupled form of Eq. (5.51). The errors in the results for a specific number of basis vectors s can be evaluated by assessing the size of the elements of the sth term of Eq. (5.51)

$$\mathbf{r}^{(s)} = \mathbf{V}_s(\mathbf{V}_s^T\mathbf{R}). \tag{10.34}$$

If the solution process converges, the relative size of the elements of the vector $\mathbf{r}^{(s)}$ can be used as a measure for evaluating the error, $\varepsilon\mathbf{r}^{(s)}$, namely

$$\varepsilon\mathbf{r}^{(s)} \equiv \frac{\left\|\mathbf{r}^{(s)}\right\|}{\left\|\sum_{i=1}^{s}\mathbf{r}^{(i)}\right\|}. \tag{10.35}$$

The true percentage error $\varepsilon(\mathbf{r})$ in the approximate displacements $\mathbf{r}(appr)$ relative to the exact displacements $\mathbf{r}(exact)$ is defined as

$$\varepsilon(\mathbf{r}) = 100\frac{\|\mathbf{r}(appr) - \mathbf{r}(exact)\|}{\|\mathbf{r}(exact)\|}. \tag{10.36}$$

To evaluate the errors involved in the approximations, we define the error vector of the equilibrium equations

$$\varepsilon\mathbf{R} = \mathbf{K}\,\mathbf{r}(appr) - \mathbf{R}. \tag{10.37}$$

The norm of $\varepsilon\mathbf{R}$, defined as

$$\|\varepsilon\mathbf{R}\| = (\varepsilon\mathbf{R}^T\varepsilon\mathbf{R})^{1/2}, \tag{10.38}$$

can be used as a measure for evaluating the error. It should be noted that although $\|\varepsilon\mathbf{R}\|$ may be very small, the error in the solution might still be large. On the other hand, for an accurate solution $\|\varepsilon\mathbf{R}\|$ must be small. Therefore, a small residual $\varepsilon\mathbf{R}$ is a necessary but not a sufficient condition for an accurate solution. To obtain more information on the solution errors the residual displacements vector $\varepsilon\mathbf{r}$ is expressed as $\varepsilon\mathbf{r}= \mathbf{K}^{-1}\varepsilon\mathbf{R}$. An analysis can be performed that uses the condition number of $\mathbf{K}$, $\psi(\mathbf{K})$, to evaluate the solution errors. It has been noted [2] that a large $\psi(\mathbf{K})$ means that solution errors are more likely.

10.4.2 Vibration Reanalysis

The true percentage error, $\varepsilon(\lambda)$, in the approximate eigenvalues, $\lambda(appr)$, relative to the exact eigenvalues, $\lambda(exact)$, is defined as [see Eq. (10.36)]

$$\varepsilon(\lambda) = 100\frac{|\lambda(appr) - \lambda(exact)|}{|\lambda(exact)|}, \tag{10.39}$$

and the error vector $\varepsilon\mathbf{R}$ of the eigenproblem equations is [see Eq. (10.37)]

$$\varepsilon\mathbf{R} = \mathbf{K}\,\mathbf{r}(appr) - \lambda(appr)\,\mathbf{M}\,\mathbf{r}(appr), \tag{10.40}$$

where $\mathbf{r}(appr)$ is the approximate mode shape. Using the definition of the norm [Eq. (10.38)], we may define the following relative norm as a measure for evaluating the error

$$\varepsilon(\mathbf{R}) = \frac{\|\varepsilon\mathbf{R}\|}{\|\mathbf{Kr}\|}. \tag{10.41}$$

To evaluate the quality of the results, it is possible to calculate lower and upper bounds on the eigenvalues [11, 12]. Denote the $(i–1)$th, (i)th, $(i+1)$th approximate eigenvectors by $\mathbf{r}(i-1)$, $\mathbf{r}(i)$, $\mathbf{r}(i+1)$, respectively, and the corresponding eigenvalues by $\lambda(i-1)$, $\lambda(i)$, $\lambda(i+1)$. Defining

$$A \equiv \frac{\mathbf{r}(i)^T\mathbf{M}\mathbf{K}^{-1}\mathbf{M}\mathbf{r}(i)}{\mathbf{r}(i)^T\,\mathbf{K}\mathbf{r}(i)}, \tag{10.42}$$

the following lower and upper bounds on $\lambda(i)$ can be established [11]

$$\lambda(i)^L = \frac{\lambda(i)}{1+\lambda(i+1)[\lambda(i)^2 A - 1]/[\lambda(i+1)-\lambda(i)]}, \tag{10.43}$$

$$\lambda(i)^U = \frac{\lambda(i)}{1-\lambda(i-1)[\lambda(i)^2 A - 1]/[\lambda(i)-\lambda(i-1)]}. \tag{10.44}$$

For the first eigenvalue we assume $\lambda(0) = 0$.

Example 10.4

Consider the eight-story frame shown in Fig. 10.9. The mass of the frames is lumped in the girders, with initial values $M_1 = 1.0$, $M_2 = 1.5$, $M_3 = 2.0$. The girders are assumed to be non-deformable and the initial lateral stiffness of each of the stories is $EI/L^3 = 5.0$.

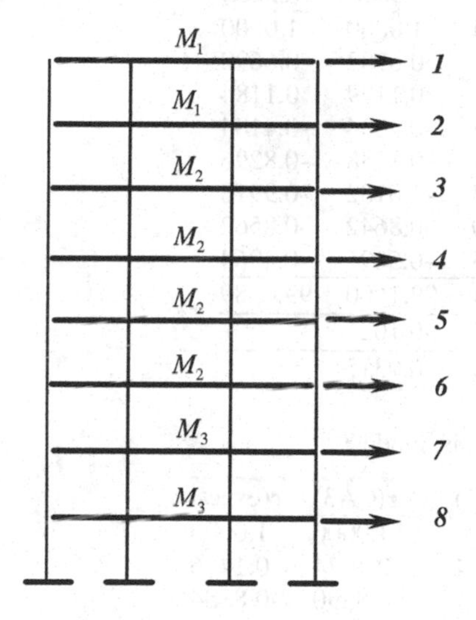

Fig. 10.9. Eight-story frame

Table 10.9. Eigenproblem reanalysis, norms of the first 4 binomial series terms

Mode	$\|\mathbf{r}_1\|$	$\|\mathbf{r}_2\|$	$\|\mathbf{r}_3\|$	$\|\mathbf{r}_4\|$
1	2	22	226	2374
2	0.4	4	42	402
3	0.2	2	21	213

Table 10.10. Results, eigenproblem reanalysis, mode 1

Shape	Initial	**r**(CA2)	**r**(CA3)	**r**(exact)
Displacements	1.0000	1.0000	1.0000	1.0000
	0.9738	0.9622	0.9614	0.9612
	0.9221	0.8876	0.8853	0.8852
	0.8342	0.7885	0.7866	0.7849
	0.7134	0.6608	0.6595	0.6570
	0.5647	0.5087	0.5081	0.5059
	0.3937	0.3373	0.3375	0.3370
	0.2022	0.1718	0.1718	0.1713
λ	1.5718	11.6277	11.6274	11.6269
True error $\varepsilon(\lambda)$		0.0068	0.0043	
Relative norm $\varepsilon(R)$		0.0169	0.0155	

Table 10.11. Results, eigenproblem reanalysis, mode 2

Shape	Initial	**r**(CA2)	**r**(CA3)	**r**(exact)
Displacements	1.0000	1.0000	1.0000	1.0000
	0.7935	0.6784	0.6782	0.6698
	0.4232	0.1204	0.1199	0.1184
	-0.0781	-0.4035	-0.4034	-0.4184
	-0.5553	-0.8344	-0.8338	-0.8296
	-0.8605	-1.0187	-1.0182	-0.9918
	-0.8992	-0.8640	-0.8642	-0.8562
	-0.5666	-0.5093	-0.5096	-0.4964
λ	12.3880	99.1601	99.1600	99.0589
True error $\varepsilon(\lambda)$		0.1022	0.1021	
Relative norm $\varepsilon(R)$		0.0314	0.0315	

Table 10.12. Results, eigenproblem reanalysis, mode 3

Shape	Initial	**r**(CA2)	**r**(CA3)	**r**(exact)
Displacements	1.0000	1.0000	1.0000	1.0000
	0.4800	0.1754	0.1574	0.1416
	-0.2895	-0.8431	-0.8590	-0.8384
	-0.8333	-1.0958	-1.0538	-1.0750
	-0.7271	-0.5436	-0.5050	-0.4727
	-0.0539	0.4948	0.4806	0.4985
	0.6614	1.1817	1.1172	1.0807
	0.6889	0.8550	0.8668	0.8412
λ	31.1972	258.4644	257.9363	257.5310
True error $\varepsilon(\lambda)$		0.3624	0.1574	
Relative norm $\varepsilon(R)$		0.0292	0.0207	

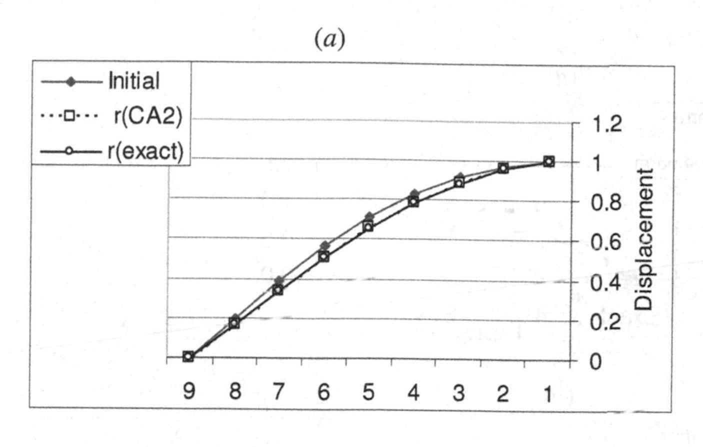

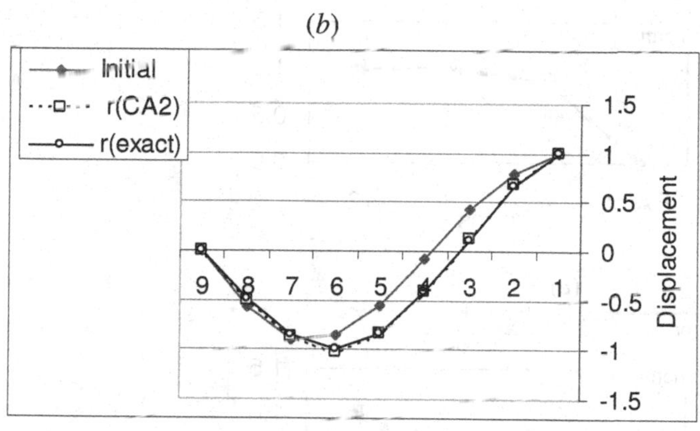

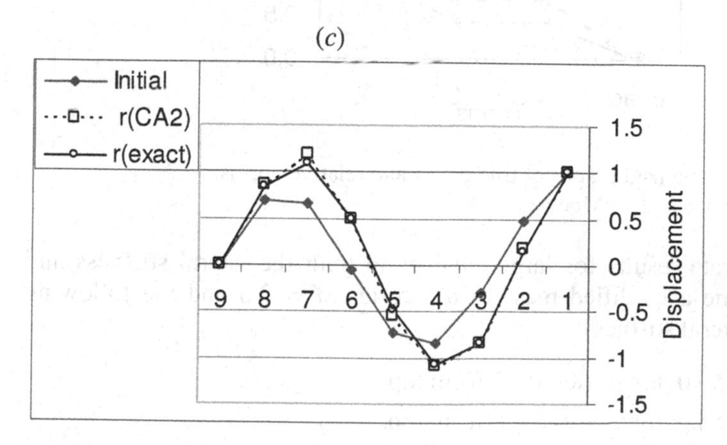

Fig. 10.10. Eigenproblem reanalysis: ***a***. Mode 1 ***b***. Mode 2 ***c***. Mode 3

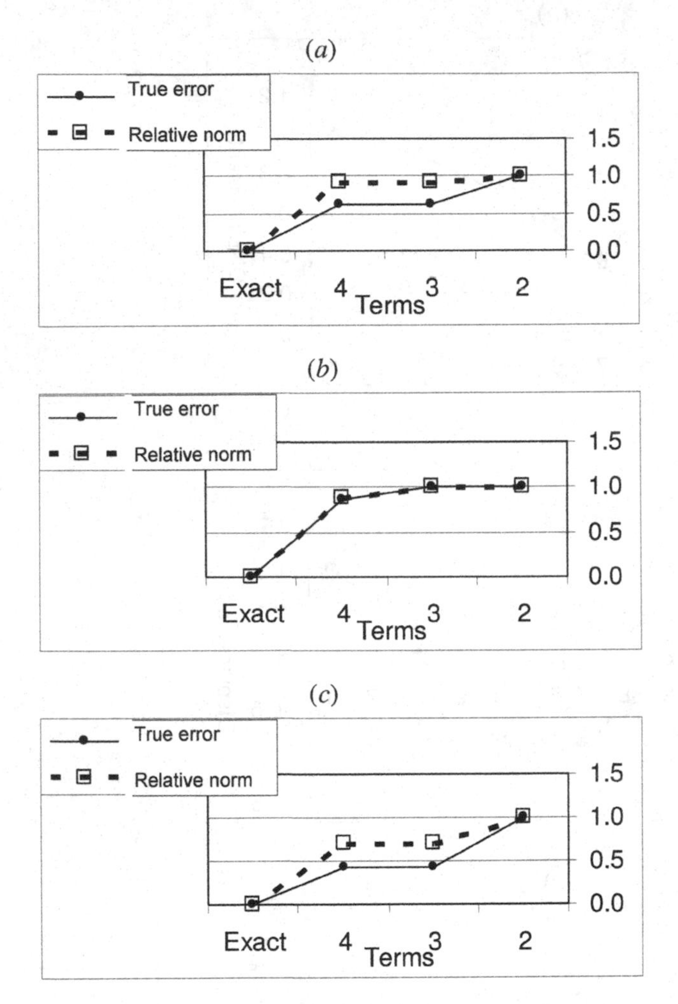

Fig. 10.11. Normalized graphs of true errors and relative norms:
a. Mode 1 ***b***. Mode 2 ***c***. Mode 3

To illustrate results for large changes in both the lateral stiffness and mass, assume a modified mass in all stories M = 2.0 and the following modified lateral stiffness:

- $EI/L^3 = 50.0$ for stories 1 – 2 from top.
- $EI/L^3 = 55.0$ for stories 3 – 6 from top.
- $EI/L^3 = 60.0$ for stories 7 – 8 from top.

Calculating the norms $\|\mathbf{r}_i\|$ of the first 4 basis vectors for the first 3 mode shapes (Table 10.9) we observe that the norm of a basis vector is larger by an order of magnitude from the norm of the previous one. It was found that even in this case, where the series of basis vectors diverges, the CA solution converges with a small number of basis vectors.

Solving the reduced eigenproblem by the CA approach with 2 and 3 basis vectors (CA2 and CA3, respectively), the results for the first 3 mode shapes are shown in Tables 10.10, 10.11, 10.12 and in Fig. 10.10. It was found that the basis vectors for each of the 3 mode shapes are close to being linearly dependent, which explains the relatively good accuracy achieved with only 2 basis vectors.

The normalized graphs of the true percentage error [Eq. (10.39)] and the relative norm [Eq. (10.41)] for the eigenvalues of the first 3 mode shapes and various numbers basis vectors are demonstrated in Fig. 10.11. It is observed that the two graphs are similar.

Finally, the lower and upper bounds on $\lambda(i)$ are computed by Eqs. (10.43), (10.44). The results in Table 10.13 show that the bounds obtained by the CA approach provide good estimation of the eigenvalues.

Table 10.13. Bounds on eigenvalues

Mode	Terms	λ^L	λ^U	λ(CA)	λ(exact)
1	3	11.6269	11.6274	11.6274	11.6269
2	3	99.0188	99.1716	99.1600	99.0589
3	4	257.301	258.097	257.920	257.531

10.5 Accuracy of Forces and Stresses

The CA approach is mainly intended to reduce the computational effort involved in repeated calculations of the displacements. In many applications it is important to evaluate also the forces and the stresses. Once the displacements are available, calculation of the forces and the stresses is often straightforward, using force-displacement and stress-displacement relations. However, it should be noted that in some applications evaluation of forces and stresses might involve significant computational effort or accuracy problems. Moreover, it is possible that small errors in displacements can magnify to large errors in stresses.

Since the accuracy of forces and stresses depends on the accuracy of displacements, it is important to obtain sufficiently accurate approximations of displacements. Various means introduced in this text can be used for this purpose. These include improved basis vectors and consideration of additional vectors. The numerical example presented in this subsection illustrates the accuracy of forces and stresses, calculated from displacements evaluated by the CA approach.

Example 10.5

Consider again the ten-bar truss shown in Fig. 10.12. The modulus of elasticity is 30000 and the eight unknown displacements are the horizontal and vertical displacements at joints 1, 2, 3 and 4, respectively. The design variables **X** are the member cross-sectional areas. The initial areas are all unity and the modified areas are as follows

$$\mathbf{X}^T = \{3.8, 0.6, 3.8, 2.2, 0.6, 0.6, 2.867, 2.867, 2.867, 0.6\} \qquad (a)$$

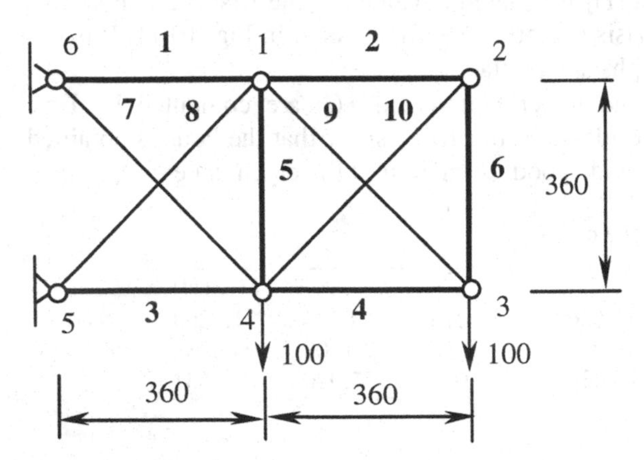

Fig. 10.12. Ten-bar truss

Results obtained by the CA approach with 2, 3, and 4 basis vectors (CA2, CA3 and CA4, respectively) and the corresponding percentage errors [E(CA2), E(CA3), E(CA4)], relative to exact solutions (Exact), are summarized in Table 10.14. The forces and the stresses are computed by the force-displacement and stress-displacement relations. It is observed that solving by CA2, the force and stress errors in some members (members 2, 4, 5, 6, 10) are very large. Solving by CA3, the displacement errors are small but the force and stress errors in member 5 are still large. A practically exact solution is achieved by CA4.

Table 10.14. Percentage errors in modified displacements, forces and stresses

Method	CA2	CA3	CA4	Exact	E(CA2)	E(CA3)	E(CA4)
Displacements	0.59	0.61	0.61	0.61	3.4	0.0	0
	1.78	1.71	1.73	1.73	2.8	1.2	0
	0.84	0.88	0.89	0.89	6.0	1.1	0
	4.14	4.20	4.21	4.21	1.7	0.2	0
	-1.04	-1.12	-1.12	-1.12	7.7	0.0	0
	4.38	4.47	4.49	4.49	2.5	0.4	0
	-0.64	-0.65	-0.65	-0.65	1.6	0.0	0
	1.98	1.91	1.90	1.90	4.0	0.5	0
Forces	188.0	193.5	194.5	194.5	3.5	0.5	0
	12.1	13.4	13.7	13.7	13.9	3.0	0
	-201.3	-206.6	-205.5	-205.5	2.1	0.5	0
	-73.7	-85.8	-86.3	-86.3	17.1	0.6	0
	9.9	10.0	8.2	8.2	17.5	21.9	0
	12.1	13.4	13.7	13.7	13.9	3.0	0
	160.5	150.6	149.2	149.2	7.0	0.9	0
	-141.8	-131.6	-133.6	-133.6	5.8	1.5	0
	115.5	122.6	122.0	122.0	5.6	0.5	0
	-17.1	-19.0	-19.4	-19.4	13.3	2.5	0
Stresses	49.4	50.9	51.2	51.2	3.5	0.5	0
	20.2	22.3	23.0	23.0	13.9	3.0	0
	-53.0	-54.4	-54.1	-54.1	2.1	0.5	0
	-33.5	-39.0	-39.2	-39.2	17.1	0.6	0
	16.6	16.7	13.7	13.7	17.5	21.9	0
	20.2	22.3	23.0	23.0	13.9	3.0	0
	56.0	52.5	52.0	52.0	7.0	0.9	0
	-49.5	-45.9	-46.6	-46.6	5.8	1.5	0
	40.3	42.8	42.5	42.5	5.6	0.5	0
	-28.6	-31.6	-32.4	-32.4	13.3	2.5	0

References

1. Kirsch U, Bogomolni M, Sheinman I (2006) Efficient procedures for repeated calculations of the structural response. Struct and Multidis Opt 32:435-446
2. Bathe KJ (1996) Finite element procedures. Prentice Hall, NJ
3. Kirsch U, Bogomolni M, Sheinman I (2007) Efficient procedures for repeated calculations of structural sensitivities. Engrg Opt 39 No 3:307–325
4. Kirsch U, Papalambros PY (2001) Exact and accurate Solutions in the approximate reanalysis of structures. AIAA J 39:2198-2205
5. Kirsch U, Kocvara M, Zowe J (2002) Accurate reanalysis of structures by a preconditioned conjugate gradient method. Int J Num Meth Engrg 55:233-251

6. Golub GH, Van Loan CF (1996) Matrix computations. 3rd ed the Johns Hopkins University Press, Baltimore
7. Nocedal J, Wright SJ (1999) Numerical optimization. Springer Verlag, NY
8. Leunberger DG (1984) Introduction to linear and nonlinear programming. Addison-Wesley, Reading, Mass
9. Wilkinson W (1965) The algebraic eigenvalue problem. Oxford Univ Press, Oxford
10. Ortega JM, Rheinboldt WC (1970) Iterative solutions of nonlinear equations in several variables. Academic Press, New York
11. Geradin M (1971) Error bounds for eigenvalue analysis by elimination of variables. J Sound Vib19:111-132
12. Kirsch U, Bogomolni M (2004) Error evaluation in approximate reanalysis of structures. Struct and Multidis Opt 28:77-86

Index

N

O

P

Q

R

T

U

V

Mechanics

SOLID MECHANICS AND ITS APPLICATIONS

Series Editor: G.M.L. Gladwell

Aims and Scope of the Series

The fundamental questions arising in mechanics are: *Why?*, *How?*, and *How much?* The aim of this series is to provide lucid accounts written by authoritative researchers giving vision and insight in answering these questions on the subject of mechanics as it relates to solids. The scope of the series covers the entire spectrum of solid mechanics. Thus it includes the foundation of mechanics; variational formulations; computational mechanics; statics, kinematics and dynamics of rigid and elastic bodies; vibrations of solids and structures; dynamical systems and chaos; the theories of elasticity, plasticity and viscoelasticity; composite materials; rods, beams, shells and membranes; structural control and stability; soils, rocks and geomechanics; fracture; tribology; experimental mechanics; biomechanics and machine design.

1. R.T. Haftka, Z. Gürdal and M.P. Kamat: *Elements of Structural Optimization*. 2nd rev.ed., 1990 ISBN 0-7923-0608-2
2. J.J. Kalker: *Three-Dimensional Elastic Bodies in Rolling Contact*. 1990 ISBN 0-7923-0712-7
3. P. Karasudhi: *Foundations of Solid Mechanics*. 1991 ISBN 0-7923-0772-0
4. *Not published*
5. *Not published.*
6. J.F. Doyle: *Static and Dynamic Analysis of Structures*. With an Emphasis on Mechanics and Computer Matrix Methods. 1991 ISBN 0-7923-1124-8; Pb 0-7923-1208-2
7. O.O. Ochoa and J.N. Reddy: *Finite Element Analysis of Composite Laminates*. ISBN 0-7923-1125-6
8. M.H. Aliabadi and D.P. Rooke: *Numerical Fracture Mechanics*. ISBN 0-7923-1175-2
9. J. Angeles and C.S. López-Cajún: *Optimization of Cam Mechanisms*. 1991 ISBN 0-7923-1355-0
10. D.E. Grierson, A. Franchi and P. Riva (eds.): *Progress in Structural Engineering*. 1991 ISBN 0-7923-1396-8
11. R.T. Haftka and Z. Gürdal: *Elements of Structural Optimization*. 3rd rev. and exp. ed. 1992 ISBN 0-7923-1504-9; Pb 0-7923-1505-7
12. J.R. Barber: *Elasticity*. 1992 ISBN 0-7923-1609-6; Pb 0-7923-1610-X
13. H.S. Tzou and G.L. Anderson (eds.): *Intelligent Structural Systems*. 1992 ISBN 0-7923-1920-6
14. E.E. Gdoutos: *Fracture Mechanics*. An Introduction. 1993 ISBN 0-7923-1932-X
15. J.P. Ward: *Solid Mechanics*. An Introduction. 1992 ISBN 0-7923-1949-4
16. M. Farshad: *Design and Analysis of Shell Structures*. 1992 ISBN 0-7923-1950-8
17. H.S. Tzou and T. Fukuda (eds.): *Precision Sensors, Actuators and Systems*. 1992 ISBN 0-7923-2015-8
18. J.R. Vinson: *The Behavior of Shells Composed of Isotropic and Composite Materials*. 1993 ISBN 0-7923-2113-8
19. H.S. Tzou: *Piezoelectric Shells*. Distributed Sensing and Control of Continua. 1993 ISBN 0-7923-2186-3
20. W. Schiehlen (ed.): *Advanced Multibody System Dynamics*. Simulation and Software Tools. 1993 ISBN 0-7923-2192-8
21. C.-W. Lee: *Vibration Analysis of Rotors*. 1993 ISBN 0-7923-2300-9
22. D.R. Smith: *An Introduction to Continuum Mechanics*. 1993 ISBN 0-7923-2454-4
23. G.M.L. Gladwell: *Inverse Problems in Scattering*. An Introduction. 1993 ISBN 0-7923-2478-1

Mechanics

SOLID MECHANICS AND ITS APPLICATIONS

Series Editor: G.M.L. Gladwell

24. G. Prathap: *The Finite Element Method in Structural Mechanics*. 1993 ISBN 0-7923-2492-7
25. J. Herskovits (ed.): *Advances in Structural Optimization*. 1995 ISBN 0-7923-2510-9
26. M.A. González-Palacios and J. Angeles: *Cam Synthesis*. 1993 ISBN 0-7923-2536-2
27. W.S. Hall: *The Boundary Element Method*. 1993 ISBN 0-7923-2580-X
28. J. Angeles, G. Hommel and P. Kovács (eds.): *Computational Kinematics*. 1993 ISBN 0-7923-2585-0
29. A. Curnier: *Computational Methods in Solid Mechanics*. 1994 ISBN 0-7923-2761-6
30. D.A. Hills and D. Nowell: *Mechanics of Fretting Fatigue*. 1994 ISBN 0-7923-2866-3
31. B. Tabarrok and F.P.J. Rimrott: *Variational Methods and Complementary Formulations in Dynamics*. 1994 ISBN 0-7923-2923-6
32. E.H. Dowell (ed.), E.F. Crawley, H.C. Curtiss Jr., D.A. Peters, R. H. Scanlan and F. Sisto: *A Modern Course in Aeroelasticity*. Third Revised and Enlarged Edition. 1995 ISBN 0-7923-2788-8; Pb: 0-7923-2789-6
33. A. Preumont: *Random Vibration and Spectral Analysis*. 1994 ISBN 0-7923-3036-6
34. J.N. Reddy (ed.): *Mechanics of Composite Materials*. Selected works of Nicholas J. Pagano. 1994 ISBN 0-7923-3041-2
35. A.P.S. Selvadurai (ed.): *Mechanics of Poroelastic Media*. 1996 ISBN 0-7923-3329-2
36. Z. Mróz, D. Weichert, S. Dorosz (eds.): *Inelastic Behaviour of Structures under Variable Loads*. 1995 ISBN 0-7923-3397-7
37. R. Pyrz (ed.): *IUTAM Symposium on Microstructure-Property Interactions in Composite Materials*. Proceedings of the IUTAM Symposium held in Aalborg, Denmark. 1995 ISBN 0-7923-3427-2
38. M.I. Friswell and J.E. Mottershead: *Finite Element Model Updating in Structural Dynamics*. 1995 ISBN 0-7923-3431-0
39. D.F. Parker and A.H. England (eds.): *IUTAM Symposium on Anisotropy, Inhomogeneity and Nonlinearity in Solid Mechanics*. Proceedings of the IUTAM Symposium held in Nottingham, U.K. 1995 ISBN 0-7923-3594-5
40. J.-P. Merlet and B. Ravani (eds.): *Computational Kinematics '95*. 1995 ISBN 0-7923-3673-9
41. L.P. Lebedev, I.I. Vorovich and G.M.L. Gladwell: *Functional Analysis*. Applications in Mechanics and Inverse Problems. 1996 ISBN 0-7923-3849-9
42. J. Menčik: *Mechanics of Components with Treated or Coated Surfaces*. 1996 ISBN 0-7923-3700-X
43. D. Bestle and W. Schiehlen (eds.): *IUTAM Symposium on Optimization of Mechanical Systems*. Proceedings of the IUTAM Symposium held in Stuttgart, Germany. 1996 ISBN 0-7923-3830-8
44. D.A. Hills, P.A. Kelly, D.N. Dai and A.M. Korsunsky: *Solution of Crack Problems*. The Distributed Dislocation Technique. 1996 ISBN 0-7923-3848-0
45. V.A. Squire, R.J. Hosking, A.D. Kerr and P.J. Langhorne: *Moving Loads on Ice Plates*. 1996 ISBN 0-7923-3953-3
46. A. Pineau and A. Zaoui (eds.): *IUTAM Symposium on Micromechanics of Plasticity and Damage of Multiphase Materials*. Proceedings of the IUTAM Symposium held in Sèvres, Paris, France. 1996 ISBN 0-7923-4188-0
47. A. Naess and S. Krenk (eds.): *IUTAM Symposium on Advances in Nonlinear Stochastic Mechanics*. Proceedings of the IUTAM Symposium held in Trondheim, Norway. 1996 ISBN 0-7923-4193-7
48. D. Ieşan and A. Scalia: *Thermoelastic Deformations*. 1996 ISBN 0-7923-4230-5

Mechanics

SOLID MECHANICS AND ITS APPLICATIONS

Series Editor: G.M.L. Gladwell

49. J.R. Willis (ed.): *IUTAM Symposium on Nonlinear Analysis of Fracture*. Proceedings of the IUTAM Symposium held in Cambridge, U.K. 1997 ISBN 0-7923-4378-6
50. A. Preumont: *Vibration Control of Active Structures*. An Introduction. 1997 ISBN 0-7923-4392-1
51. G.P. Cherepanov: *Methods of Fracture Mechanics: Solid Matter Physics*. 1997 ISBN 0-7923-4408-1
52. D.H. van Campen (ed.): *IUTAM Symposium on Interaction between Dynamics and Control in Advanced Mechanical Systems*. Proceedings of the IUTAM Symposium held in Eindhoven, The Netherlands. 1997 ISBN 0-7923-4429-4
53. N.A. Fleck and A.C.F. Cocks (eds.): *IUTAM Symposium on Mechanics of Granular and Porous Materials*. Proceedings of the IUTAM Symposium held in Cambridge, U.K. 1997 ISBN 0-7923-4553-3
54. J. Roorda and N.K. Srivastava (eds.): *Trends in Structural Mechanics*. Theory, Practice, Education. 1997 ISBN 0-7923-4603-3
55. Yu.A. Mitropolskii and N. Van Dao: *Applied Asymptotic Methods in Nonlinear Oscillations*. 1997 ISBN 0-7923-4605-X
56. C. Guedes Soares (ed.): *Probabilistic Methods for Structural Design*. 1997 ISBN 0-7923-4670-X
57. D. François, A. Pineau and A. Zaoui: *Mechanical Behaviour of Materials*. Volume I: Elasticity and Plasticity. 1998 ISBN 0-7923-4894-X
58. D. François, A. Pineau and A. Zaoui: *Mechanical Behaviour of Materials*. Volume II: Viscoplasticity, Damage, Fracture and Contact Mechanics. 1998 ISBN 0-7923-4895-8
59. L.T. Tenek and J. Argyris: *Finite Element Analysis for Composite Structures*. 1998 ISBN 0-7923-4899-0
60. Y.A. Bahei-El-Din and G.J. Dvorak (eds.): *IUTAM Symposium on Transformation Problems in Composite and Active Materials*. Proceedings of the IUTAM Symposium held in Cairo, Egypt. 1998 ISBN 0-7923-5122-3
61. I.G. Goryacheva: *Contact Mechanics in Tribology*. 1998 ISBN 0-7923-5257-2
62. O.T. Bruhns and E. Stein (eds.): *IUTAM Symposium on Micro- and Macrostructural Aspects of Thermoplasticity*. Proceedings of the IUTAM Symposium held in Bochum, Germany. 1999 ISBN 0-7923-5265-3
63. F.C. Moon: *IUTAM Symposium on New Applications of Nonlinear and Chaotic Dynamics in Mechanics*. Proceedings of the IUTAM Symposium held in Ithaca, NY, USA. 1998 ISBN 0-7923-5276-9
64. R. Wang: *IUTAM Symposium on Rheology of Bodies with Defects*. Proceedings of the IUTAM Symposium held in Beijing, China. 1999 ISBN 0-7923-5297-1
65. Yu.I. Dimitrienko: *Thermomechanics of Composites under High Temperatures*. 1999 ISBN 0-7923-4899-0
66. P. Argoul, M. Frémond and Q.S. Nguyen (eds.): *IUTAM Symposium on Variations of Domains and Free-Boundary Problems in Solid Mechanics*. Proceedings of the IUTAM Symposium held in Paris, France. 1999 ISBN 0-7923-5450-8
67. F.J. Fahy and W.G. Price (eds.): *IUTAM Symposium on Statistical Energy Analysis*. Proceedings of the IUTAM Symposium held in Southampton, U.K. 1999 ISBN 0-7923-5457-5
68. H.A. Mang and F.G. Rammerstorfer (eds.): *IUTAM Symposium on Discretization Methods in Structural Mechanics*. Proceedings of the IUTAM Symposium held in Vienna, Austria. 1999 ISBN 0-7923-5591-1

Mechanics

SOLID MECHANICS AND ITS APPLICATIONS

Series Editor: G.M.L. Gladwell

69. P. Pedersen and M.P. Bendsøe (eds.): *IUTAM Symposium on Synthesis in Bio Solid Mechanics.* Proceedings of the IUTAM Symposium held in Copenhagen, Denmark. 1999 ISBN 0-7923-5615-2
70. S.K. Agrawal and B.C. Fabien: *Optimization of Dynamic Systems.* 1999 ISBN 0-7923-5681-0
71. A. Carpinteri: *Nonlinear Crack Models for Nonmetallic Materials.* 1999 ISBN 0-7923-5750-7
72. F. Pfeifer (ed.): *IUTAM Symposium on Unilateral Multibody Contacts.* Proceedings of the IUTAM Symposium held in Munich, Germany. 1999 ISBN 0-7923-6030-3
73. E. Lavendelis and M. Zakrzhevsky (eds.): *IUTAM/IFToMM Symposium on Synthesis of Nonlinear Dynamical Systems.* Proceedings of the IUTAM/IFToMM Symposium held in Riga, Latvia. 2000 ISBN 0-7923-6106-7
74. J.-P. Merlet: *Parallel Robots.* 2000 ISBN 0-7923-6308-6
75. J.T. Pindera: *Techniques of Tomographic Isodyne Stress Analysis.* 2000 ISBN 0-7923-6388-4
76. G.A. Maugin, R. Drouot and F. Sidoroff (eds.): *Continuum Thermomechanics.* The Art and Science of Modelling Material Behaviour. 2000 ISBN 0-7923-6407-4
77. N. Van Dao and E.J. Kreuzer (eds.): *IUTAM Symposium on Recent Developments in Non-linear Oscillations of Mechanical Systems.* 2000 ISBN 0-7923-6470-8
78. S.D. Akbarov and A.N. Guz: *Mechanics of Curved Composites.* 2000 ISBN 0-7923-6477-5
79. M.B. Rubin: *Cosserat Theories: Shells, Rods and Points.* 2000 ISBN 0-7923-6489-9
80. S. Pellegrino and S.D. Guest (eds.): *IUTAM-IASS Symposium on Deployable Structures: Theory and Applications.* Proceedings of the IUTAM-IASS Symposium held in Cambridge, U.K., 6–9 September 1998. 2000 ISBN 0-7923-6516-X
81. A.D. Rosato and D.L. Blackmore (eds.): *IUTAM Symposium on Segregation in Granular Flows.* Proceedings of the IUTAM Symposium held in Cape May, NJ, U.S.A., June 5–10, 1999. 2000 ISBN 0-7923-6547-X
82. A. Lagarde (ed.): *IUTAM Symposium on Advanced Optical Methods and Applications in Solid Mechanics.* Proceedings of the IUTAM Symposium held in Futuroscope, Poitiers, France, August 31–September 4, 1998. 2000 ISBN 0-7923-6604-2
83. D. Weichert and G. Maier (eds.): *Inelastic Analysis of Structures under Variable Loads.* Theory and Engineering Applications. 2000 ISBN 0-7923-6645-X
84. T.-J. Chuang and J.W. Rudnicki (eds.): *Multiscale Deformation and Fracture in Materials and Structures.* The James R. Rice 60th Anniversary Volume. 2001 ISBN 0-7923-6718-9
85. S. Narayanan and R.N. Iyengar (eds.): *IUTAM Symposium on Nonlinearity and Stochastic Structural Dynamics.* Proceedings of the IUTAM Symposium held in Madras, Chennai, India, 4–8 January 1999 ISBN 0-7923-6733-2
86. S. Murakami and N. Ohno (eds.): *IUTAM Symposium on Creep in Structures.* Proceedings of the IUTAM Symposium held in Nagoya, Japan, 3-7 April 2000. 2001 ISBN 0-7923-6737-5
87. W. Ehlers (ed.): *IUTAM Symposium on Theoretical and Numerical Methods in Continuum Mechanics of Porous Materials.* Proceedings of the IUTAM Symposium held at the University of Stuttgart, Germany, September 5-10, 1999. 2001 ISBN 0-7923-6766-9
88. D. Durban, D. Givoli and J.G. Simmonds (eds.): *Advances in the Mechanis of Plates and Shells The Avinoam Libai Anniversary Volume.* 2001 ISBN 0-7923-6785-5
89. U. Gabbert and H.-S. Tzou (eds.): *IUTAM Symposium on Smart Structures and Structonic Systems.* Proceedings of the IUTAM Symposium held in Magdeburg, Germany, 26–29 September 2000. 2001 ISBN 0-7923-6968-8

Mechanics

SOLID MECHANICS AND ITS APPLICATIONS

Series Editor: G.M.L. Gladwell

90. Y. Ivanov, V. Cheshkov and M. Natova: *Polymer Composite Materials – Interface Phenomena & Processes.* 2001 ISBN 0-7923-7008-2
91. R.C. McPhedran, L.C. Botten and N.A. Nicorovici (eds.): *IUTAM Symposium on Mechanical and Electromagnetic Waves in Structured Media.* Proceedings of the IUTAM Symposium held in Sydney, NSW, Australia, 18-22 Januari 1999. 2001 ISBN 0-7923-7038-4
92. D.A. Sotiropoulos (ed.): *IUTAM Symposium on Mechanical Waves for Composite Structures Characterization.* Proceedings of the IUTAM Symposium held in Chania, Crete, Greece, June 14-17, 2000. 2001 ISBN 0-7923-7164-X
93. V.M. Alexandrov and D.A. Pozharskii: *Three-Dimensional Contact Problems.* 2001 ISBN 0-7923-7165-8
94. J.P. Dempsey and H.H. Shen (eds.): *IUTAM Symposium on Scaling Laws in Ice Mechanics and Ice Dynamics.* Proceedings of the IUTAM Symposium held in Fairbanks, Alaska, U.S.A., 13-16 June 2000. 2001 ISBN 1-4020-0171-1
95. U. Kirsch: *Design-Oriented Analysis of Structures.* A Unified Approach. 2002 ISBN 1-4020-0443-5
96. A. Preumont: *Vibration Control of Active Structures.* An Introduction (2nd Edition). 2002 ISBN 1-4020-0496-6
97. B.L. Karihaloo (ed.): *IUTAM Symposium on Analytical and Computational Fracture Mechanics of Non-Homogeneous Materials.* Proceedings of the IUTAM Symposium held in Cardiff, U.K., 18-22 June 2001. 2002 ISBN 1-4020-0510-5
98. S.M. Han and H. Benaroya: *Nonlinear and Stochastic Dynamics of Compliant Offshore Structures.* 2002 ISBN 1-4020-0573-3
99. A.M. Linkov: *Boundary Integral Equations in Elasticity Theory.* 2002 ISBN 1-4020-0574-1
100. L.P. Lebedev, I.I. Vorovich and G.M.L. Gladwell: *Functional Analysis.* Applications in Mechanics and Inverse Problems (2nd Edition). 2002 ISBN 1-4020 0667-5; Pb: 1-4020-0756-6
101. Q.P. Sun (ed.): *IUTAM Symposium on Mechanics of Martensitic Phase Transformation in Solids.* Proceedings of the IUTAM Symposium held in Hong Kong, China, 11-15 June 2001. 2002 ISBN 1-4020-0741-8
102. M.L. Munjal (ed.): *IUTAM Symposium on Designing for Quietness.* Proceedings of the IUTAM Symposium held in Bangkok, India, 12-14 December 2000. 2002 ISBN 1-4020-0765-5
103. J.A.C. Martins and M.D.P. Monteiro Marques (eds.): *Contact Mechanics.* Proceedings of the 3rd Contact Mechanics International Symposium, Praia da Consolação, Peniche, Portugal, 17-21 June 2001. 2002 ISBN 1-4020-0811-2
104. H.R. Drew and S. Pellegrino (eds.): *New Approaches to Structural Mechanics, Shells and Biological Structures.* 2002 ISBN 1-4020-0862-7
105. J.R. Vinson and R.L. Sierakowski: *The Behavior of Structures Composed of Composite Materials.* Second Edition. 2002 ISBN 1-4020-0904-6
106. Not yet published.
107. J.R. Barber: *Elasticity.* Second Edition. 2002 ISBN Hb 1-4020-0964-X; Pb 1-4020-0966-6
108. C. Miehe (ed.): *IUTAM Symposium on Computational Mechanics of Solid Materials at Large Strains.* Proceedings of the IUTAM Symposium held in Stuttgart, Germany, 20-24 August 2001. 2003 ISBN 1-4020-1170-9

Mechanics

SOLID MECHANICS AND ITS APPLICATIONS

Series Editor: G.M.L. Gladwell

109. P. Ståhle and K.G. Sundin (eds.): *IUTAM Symposium on Field Analyses for Determination of Material Parameters – Experimental and Numerical Aspects.* Proceedings of the IUTAM Symposium held in Abisko National Park, Kiruna, Sweden, July 31 – August 4, 2000. 2003 ISBN 1-4020-1283-7
110. N. Sri Namachchivaya and Y.K. Lin (eds.): *IUTAM Symposium on Nonlinear Stochastic Dynamics.* Proceedings of the IUTAM Symposium held in Monticello, IL, USA, 26 – 30 August, 2000. 2003 ISBN 1-4020-1471-6
111. H. Sobieckzky (ed.): *IUTAM Symposium Transsonicum IV.* Proceedings of the IUTAM Symposium held in Göttingen, Germany, 2–6 September 2002, 2003 ISBN 1-4020-1608-5
112. J.-C. Samin and P. Fisette: *Symbolic Modeling of Multibody Systems.* 2003 ISBN 1-4020-1629-8
113. A.B. Movchan (ed.): *IUTAM Symposium on Asymptotics, Singularities and Homogenisation in Problems of Mechanics.* Proceedings of the IUTAM Symposium held in Liverpool, United Kingdom, 8-11 July 2002. 2003 ISBN 1-4020-1780-4
114. S. Ahzi, M. Cherkaoui, M.A. Khaleel, H.M. Zbib, M.A. Zikry and B. LaMatina (eds.): *IUTAM Symposium on Multiscale Modeling and Characterization of Elastic-Inelastic Behavior of Engineering Materials.* Proceedings of the IUTAM Symposium held in Marrakech, Morocco, 20-25 October 2002. 2004 ISBN 1-4020-1861-4
115. H. Kitagawa and Y. Shibutani (eds.): *IUTAM Symposium on Mesoscopic Dynamics of Fracture Process and Materials Strength.* Proceedings of the IUTAM Symposium held in Osaka, Japan, 6-11 July 2003. Volume in celebration of Professor Kitagawa's retirement. 2004 ISBN 1-4020-2037-6
116. E.H. Dowell, R.L. Clark, D. Cox, H.C. Curtiss, Jr., K.C. Hall, D.A. Peters, R.H. Scanlan, E. Simiu, F. Sisto and D. Tang: *A Modern Course in Aeroelasticity.* 4th Edition, 2004 ISBN 1-4020-2039-2
117. T. Burczyński and A. Osyczka (eds.): *IUTAM Symposium on Evolutionary Methods in Mechanics.* Proceedings of the IUTAM Symposium held in Cracow, Poland, 24-27 September 2002. 2004 ISBN 1-4020-2266-2
118. D. Ieşan: *Thermoelastic Models of Continua.* 2004 ISBN 1-4020-2309-X
119. G.M.L. Gladwell: *Inverse Problems in Vibration.* Second Edition. 2004 ISBN 1-4020-2670-6
120. J.R. Vinson: *Plate and Panel Structures of Isotropic, Composite and Piezoelectric Materials, Including Sandwich Construction.* 2005 ISBN 1-4020-3110-6
121. *Forthcoming*
122. G. Rega and F. Vestroni (eds.): *IUTAM Symposium on Chaotic Dynamics and Control of Systems and Processes in Mechanics.* Proceedings of the IUTAM Symposium held in Rome, Italy, 8–13 June 2003. 2005 ISBN 1-4020-3267-6
123. E.E. Gdoutos: *Fracture Mechanics. An Introduction. 2nd edition.* 2005 ISBN 1-4020-3267-6
124. M.D. Gilchrist (ed.): *IUTAM Symposium on Impact Biomechanics from Fundamental Insights to Applications.* 2005 ISBN 1-4020-3795-3
125. J.M. Huyghe, P.A.C. Raats and S. C. Cowin (eds.): *IUTAM Symposium on Physicochemical and Electromechanical Interactions in Porous Media.* 2005 ISBN 1-4020-3864-X
126. H. Ding, W. Chen and L. Zhang: *Elasticity of Transversely Isotropic Materials.* 2005 ISBN 1-4020-4033-4
127. W. Yang (ed): *IUTAM Symposium on Mechanics and Reliability of Actuating Materials.* Proceedings of the IUTAM Symposium held in Beijing, China, 1–3 September 2004. 2005 ISBN 1-4020-4131-6

Mechanics

SOLID MECHANICS AND ITS APPLICATIONS

Series Editor: G.M.L. Gladwell

128. J.-P. Merlet: *Parallel Robots*. 2006 ISBN 1-4020-4132-2
129. G.E.A. Meier and K.R. Sreenivasan (eds.): *IUTAM Symposium on One Hundred Years of Boundary Layer Research*. Proceedings of the IUTAM Symposium held at DLR-Göttingen, Germany, August 12–14, 2004. 2006 ISBN 1-4020-4149-7
130. H. Ulbrich and W. Günthner (eds.): *IUTAM Symposium on Vibration Control of Nonlinear Mechanisms and Structures*. 2006 ISBN 1-4020-4160-8
131. L. Librescu and O. Song: *Thin-Walled Composite Beams*. Theory and Application. 2006 ISBN 1-4020-3457-1
132. G. Ben-Dor, A. Dubinsky and T. Elperin: *Applied High-Speed Plate Penetration Dynamics*. 2006 ISBN 1-4020-3452-0
133. X. Markenscoff and A. Gupta (eds.): *Collected Works of J. D. Eshelby*. Mechanics of Defects and Inhomogeneities. 2006 ISBN 1-4020-4416-X
134. R.W. Snidle and H.P. Evans (eds.): *IUTAM Symposium on Elastohydrodynamics and Microelastohydrodynamics*. Proceedings of the IUTAM Symposium held in Cardiff, UK, 1–3 September, 2004. 2006 ISBN 1-4020-4532-8
135. T. Sadowski (ed.): *IUTAM Symposium on Multiscale Modelling of Damage and Fracture Processes in Composite Materials*. Proceedings of the IUTAM Symposium held in Kazimierz Dolny, Poland, 23–27 May 2005. 2006 ISBN 1-4020-4565-4
136. A. Preumont: *Mechatronics*. Dynamics of Electromechanical and Piezoelectric Systems. 2006 ISBN 1-4020-4695-2
137. M.P. Bendsøe, N. Olhoff and O. Sigmund (eds.): *IUTAM Symposium on Topological Design Optimization of Structures, Machines and Materials*. Status and Perspectives. 2006 ISBN 1-4020-4729-0
138. A. Klarbring: *Models of Mechanics*. 2006 ISBN 1-4020-4834-3
139. H.D. Bui: *Fracture Mechanics*. Inverse Problems and Solutions. 2006 ISBN 1-4020-4836-X
140. M. Pandey, W.-C. Xie and L. Xu (eds.): *Advances in Engineering Structures, Mechanics and Construction*. Proceedings of an International Conference on Advances in Engineering Structures, Mechanics & Construction, held in Waterloo, Ontario, Canada, May 14–17, 2006. 2006 ISBN 1-4020-4890-4
141. G.Q. Zhang, W.D. van Driel and X. J. Fan: *Mechanics of Microelectronics*. 2006 ISBN 1-4020-4934-X
142. Q.P. Sun and P. Tong (eds.): *IUTAM Symposium on Size Effects on Material and Structural Behavior at Micron- and Nano-Scales*. Proceedings of the IUTAM Symposium held in Hong Kong, China, 31 May–4 June, 2004. 2006 ISBN 1-4020-4945-5
143. A.P. Mouritz and A.G. Gibson: *Fire Properties of Polymer Composite Materials*. 2006 ISBN 1-4020-5355-X
144. Y.L. Bai, Q.S. Zheng and Y.G. Wei (eds.): *IUTAM Symposium on Mechanical Behavior and Micro-Mechanics of Nanostructured Materials*. Proceedings of the IUTAM Symposium held in Beijing, China, 27–30 June 2005. 2007 ISBN 1-4020-5623-0
145. L.P. Pook: *Metal Fatigue*. What It Is, Why It Matters. 2007 ISBN 1-4020-5596-6
146. H.I. Ling, L. Callisto, D. Leshchinsky, J. Koseki (eds.): *Geomechanics: Laboratory Testing, Modeling and Applications*. A Collection of Papers of the Geotechnical Symposium in Rome, March 16-17, 2006 ISBN 978-1-4020-6145-5
147. A.S. Kravchuk and P.J. Neittaanmäki: *Variational and Quasi-Variational Inequalities in Mechanics*. 2007. ISBN 978-1-4020-6376-3

Mechanics

***SOLID* MECHANICS AND ITS APPLICATIONS**

Series Editor: G.M.L. Gladwell

148. S.K. Kanaun and V. Levin: *Self-Consistent Methods for Composites.* Vol. 1: Static Problems. 2008 ISBN 978-1-4020-6663-4
149. G. Gogu: *Structural Synthesis of Parallel Robots.* Part 1: Methodology. 2008 ISBN 978-1-4020-5102-9
150. S.K. Kanaun and V. Levin: *Self-Consistent Methods for Composites.* Vol. 2: Wave Propagation in Heterogeneous Materials. 2008 ISBN 978-1-4020-6967-3
151. U. Kirsch: *Reanalysis of Structures.* A Unified Approach for Linear, Nonlinear, Static and Dynamic Systems. 2008 ISBN 978-1-4020-8197-2